# Suzuki GSX/GS1000, 1100 & 1150 4-valve Fours Owners Workshop Manual

by Pete Shoemark
with an additional Chapter on the 1984-on 1135cc-engined GSX1100 and GS1150 models
by Mark Coombs

**Models covered**
GSX1000 S Katana. 998cc. UK September 1981 to March 1985
GSX1100 S Katana. 1075cc. UK February 1981 to June 1985
GSX1100 E. 1075cc. UK August 1979 to March 1984
GSX1100 ES. 1075cc. UK February 1983 to March 1985
GSX1100 E. 1135cc. UK July 1985 to October 1988
GSX1100 EF. 1135cc. UK March 1984 to October 1988
GS1000 S Katana. 998cc. US September 1981 to 1982
GS1100 S Katana. 1075cc. US October 1982 to 1983
GS1100 E. 1075cc. US January 1980 to 1983
GS1100 ES. 1075cc. US 1983 only
GS1100 L. 1075cc. US 1980 only
GS1150 E. 1135cc. US July 1984 to 1986
GS1150 ES. 1135cc. US February 1984 to 1985

**ISBN 978 1 85010 574 9**

*(737-6AR2)*

**Haynes Group Limited**
**Haynes North America, Inc**

**www.haynes.com**

| British Library Cataloguing in Publication Data |
| --- |
| Shoemark, Pete 1952–<br>  Suzuki GSX/GS1000, 1100 & 1150 4-valve fours owners<br>  workshop manual.–2nd. ed.<br>  1. Motorcycles. Maintenance & repair<br>  I. Title II. Coombs. Mark 1969–<br>  629.28775<br>  ISBN 1-85010-574-X |
| Library of Congress Catalog Card Number |
| 90-83670 |

# Acknowledgements

Our thanks are due to P R Taylor and Sons of Chippenham who supplied the GSX1100 EZ model featured throughout this manual, Bridge Garage (Exeter) Ltd who supplied the GSX1100 EG shown on the front cover, and to Heron Suzuki (GB) Ltd who provided the service literature and who gave permission for the reproduction of many of the original line drawings.

The Avon Rubber Company supplied information on tyre care and fitting, and NGK Spark Plugs (UK) Ltd provided information on plug maintenance and electrode conditions.

# About this manual

The purpose of this manual is to present the owner with a concise and graphic guide which will enable him to tackle any operation from basic routine maintenance to a major overhaul. It has been assumed that any work would be undertaken without the luxury of a well-equipped workshop and a range of manufacturer's service tools.

To this end, the machine featured in the manual was stripped and rebuilt in our own workshop, by a team comprising a mechanic, a photographer and the author. The resulting photographic sequence depicts events as they took place, the hands shown being those of the author and the mechanic.

The use of specialised, and expensive, service tools was avoided unless their use was considered to be essential due to risk of breakage or injury. There is usually some way of improvising a method of removing a stubborn component, providing that a suitable degree of care is exercised.

The author learnt his motorcycle mechanics over a number of years, faced with the same difficulties and using similar facilities to those encountered by most owners. It is hoped that this practical experience can be passed on through the pages of this manual.

Where possible, a well-used example of the machine is chosen for the workshop project, as this highlights any areas which might be particularly prone to giving rise to problems. In this way, any such difficulties are encountered and resolved before the text is written, and the techniques used to deal with them can be incorporated in the relevant section. Armed with a working knowledge of the machine, the author undertakes a considerable amount of research in order that the maximum amount of data can be included in the manual.

A comprehensive section, preceding the main part of the manual, describes procedures for carrying out the routine maintenance of the machine at intervals of time and mileage. This section is included particularly for those owners who wish to ensure the efficient day-to-day running of their motorcycle, but who choose not to undertake overhaul or renovation work.

Each Chapter is divided into numbered sections. Within these sections are numbered paragraphs. Cross reference throughout the manual is quite straightforward and logical. When reference is made 'See Section 6.10' it means Section 6, paragraph 10 in the same Chapter. If another Chapter were intended, the reference would read, for example, 'See Chapter 2, Section 6.10'. All the photographs are captioned with a section/paragraph number to which they refer and are relevant to the Chapter text adjacent.

Figures (usually line illustrations) appear in a logical but numerical order, within a given Chapter. Fig. 1.1 therefore refers to the first figure in Chapter 1.

Left-hand and right-hand descriptions of the machines and their components refer to the left and right of a given machine when the rider is seated normally.

Motorcycle manufacturers continually make changes to specifications and recommendations, and these, when notified, are incorporated into our manuals at the earliest opportunity.

**We take great pride in the accuracy of information given in this manual, but motorcycle manufacturers make alterations and design changes during the production run of a particular motorcycle of which they do not inform us. No liability can be accepted by the authors or publishers for loss, damage or injury caused by any errors in, or omissions from, the information given.**

# Contents

**Left-hand view of the GSX/GS 1100 EZ model**

**Right-hand view of the GSX/GS 1000 SZ model**

Engine and gearbox unit of the GSX/GS 1100 EZ model

# Introduction to the Suzuki GSX/GS 1000 & 1100 models

The 1000cc and 1100cc four-valve Suzuki fours, known as GS models in the US and as GSX models in the UK, are developments of the earlier and successful GS1000 range. A number of modifications have been made, most significant of which is the introduction of a four-valve head design, called TSCC by Suzuki.

In the patented TSCC (Twin Swirl Combustion Chamber) arrangement Suzuki have taken advantage of the benefits of better cylinder filling conferred by a four-valve design. In addition, the combustion chamber has been carefully shaped to provoke turbulence. This, in conjunction with the central placing of the spark plug and the shallow valve angle ensures rapid and even burning of the fuel/air mixture, enhancing the efficiency of the engine.

The more conventional models in the range feature varying levels of sophistication, encompassing electronic check panels and LCD gear indicators. The Katana models are a step in the opposite direction, and in some respects are strictly functional by comparison. Underneath the innovative European styling, there is little in the way of gadgetry, the aim being a fast and relatively light sports machine. This ambition has been pursued at a detailed level; even components like the alternator rotor have been lightened to improve throttle response.

The chassis components are of a high standard throughout the range, with various permutations of spring and damping adjustment front and rear. On the later machines, an anti-dive arrangement is incorporated in the front suspension. This allows compliant suspension movement in normal use, whilst under heavy braking the damping rate is greatly increased to resist fork compression.

Throughout this manual all models are identified by their suffix letters (eg ET, EX, EZ) rather than the date of import. To assist the owner in identiíying his or her machine exactly, given below are the initial frame numbers with which each model's production run commenced, together with the approximate dates of import for UK models or the model year for US models. Note that the latter may not necessarily coincide with the machine's date of registration or sale.

| Model | Initial frame no. | Dates of import |
|---|---|---|
| UK GSX1000 SZ | GS10X-500001 | Sep 81 to Mar 84 |
| UK GSX1000 SD | Not available | Mar 84 to Mar 85 |
| UK GSX1100 SZ | GS110X-522093 | Feb 81 to Jan 83 |
| UK GSX1100 SD | GS110X-535683 | Jan 83 to Jun 85 |
| UK GSX1100 T | GS110X-100001 | Aug 79 to Oct 80 |
| UK GSX1100 ET | GS110X-500001 | Aug 79 to Oct 80 |
| UK GSX1100 EX | GS110X-513756 | Oct 80 to Jul 82 |
| UK GSX1100 EZ | GU71B-100006 | Jan 83 to Mar 84 |
| UK GSX1100 ESD | GU71B-535728 | Feb 83 to Mar 85 |
| US GS1000 SZ | JS1GT72A B2100001 | 1982 |
| US GS1100 SD | Not available | 1983 |
| US GS1100 ET | GS110X-500001 | 1980 |
| US GS1100 EX | GU71A B2100001 | 1981 |
| US GS1100 EZ | JS1GU73A C2100001 | 1982 |
| US GS1100 ED/ESD | JS1GU73A D2100001 | 1983 |
| US GS1100 LT | Not available | 1980 |

For all later models, namely the 1984-on 1135 cc-engined GSX1100 and GS1150, refer to Chapter 7.

# Dimensions and weights

| | 1100 T, ET, EX | 1100 LT |
|---|---|---|
| Overall length | 2245 mm (88.4 in) | 2255 mm (88.8 in) |
| Overall width | 870 mm (34.3 in) | 905 mm (35.6 in) |
| Overall height | 1190 mm (46.9 in) | 1230 mm (48.4 in) |
| Wheelbase | 1520 mm (59.8 in) | 1530 mm (60.2 in) |
| Ground clearance | 155 mm (6.1 in) | 155 mm (6.1 in) |
| Seat height | 830 mm (32.7 in) | 810 mm (31.9 in) |
| Dry weight | 243 kg (536 lbs) | 237 kg (522 lbs) |
| Gross weight | 467 kg (1030 lbs) | 516 kg (1137 lbs) |

| | GSX 1100 EZ | All Katana models |
|---|---|---|
| Overall length | 2225 mm (87.6 in) | 2260 mm (89.0 in) |
| Overall width – UK | 770 mm (30.3 in) | 715 mm (28.1 in) |
|          – US | 850 mm (33.5 in) | 715 mm (28.1 in) |
| Overall height | 1165 mm (45.9 in) | 1205 mm (47.4 in) |
| Wheelbase | 1510 mm (59.5 in) | 1520 mm (59.8 in) |
| Ground clearance | 155 mm (6.1 in) | 175 mm (6.9 in) |
| Seat height | 790 mm (31.1 in) | 775 mm (30.5 in) |
| Dry weight | 237 kg (522 lbs) | 232 kg (511 lbs) |
| Gross weight | not available | not available |

**Note:** *No information on the remaining models was available at the time of publication.*

# Ordering spare parts

When ordering spare parts for any Suzuki model it is advisable to deal direct with an official Suzuki dealer who will be able to supply most items ex-stock. Where parts have to be ordered, an authorized dealer will be able to obtain them as quickly as possible. The engine and frame numbers must always be quoted in full. This avoids the risk of incorrect parts being supplied and is particularly important where detail modifications have been made in the middle of production runs. In some instances it will be necessary for the dealer to check compatibility of later parts designs with earlier models. The frame number is stamped into the steering head and the engine number on a raised boss on the crankcase.

It is recommended that genuine Suzuki parts are used. Although pattern parts are often cheaper, there is no guarantee that they are of the same specification as the original, and in some instances may be positively dangerous. Note also that the use of non-standard parts may invalidate the warranty in the event of subsequent failure.

Some of the more expendable parts such as oils, greases, spark plugs, tyres and bulbs, can be obtained from auto accessory shops. These are often more conveniently located and may open during weekends. It is also possible to obtain parts on a mail order basis from specialists who advertise in the motorcycle magazines.

Engine number is stamped on raised boss on the crankcase

Frame number is stamped on steering head

# Safety first!

Professional motor mechanics are trained in safe working procedures. However enthusiastic you may be about getting on with the job in hand, do take the time to ensure that your safety is not put at risk. A moment's lack of attention can result in an accident, as can failure to observe certain elementary precautions.

There will always be new ways of having accidents, and the following points do not pretend to be a comprehensive list of all dangers; they are intended rather to make you aware of the risks and to encourage a safety-conscious approach to all work you carry out on your vehicle.

### Essential DOs and DON'Ts

**DON'T** start the engine without first ascertaining that the transmission is in neutral.

**DON'T** suddenly remove the filler cap from a hot cooling system – cover it with a cloth and release the pressure gradually first, or you may get scalded by escaping coolant.

**DON'T** attempt to drain oil until you are sure it has cooled sufficiently to avoid scalding you.

**DON'T** grasp any part of the engine, exhaust or silencer without first ascertaining that it is sufficiently cool to avoid burning you.

**DON'T** allow brake fluid or antifreeze to contact the machine's paintwork or plastic components.

**DON'T** syphon toxic liquids such as fuel, brake fluid or antifreeze by mouth, or allow them to remain on your skin.

**DON'T** inhale dust – it may be injurious to health (see *Asbestos* heading).

**DON'T** allow any spilt oil or grease to remain on the floor – wipe it up straight away, before someone slips on it.

**DON'T** use ill-fitting spanners or other tools which may slip and cause injury.

**DON'T** attempt to lift a heavy component which may be beyond your capability – get assistance.

**DON'T** rush to finish a job, or take unverified short cuts.

**DON'T** allow children or animals in or around an unattended vehicle.

**DON'T** inflate a tyre to a pressure above the recommended maximum. Apart from overstressing the carcase and wheel rim, in extreme cases the tyre may blow off forcibly.

**DO** ensure that the machine is supported securely at all times. This is especially important when the machine is blocked up to aid wheel or fork removal.

**DO** take care when attempting to slacken a stubborn nut or bolt. It is generally better to pull on a spanner, rather than push, so that if slippage occurs you fall away from the machine rather than on to it.

**DO** wear eye protection when using power tools such as drill, sander, bench grinder etc.

**DO** use a barrier cream on your hands prior to undertaking dirty jobs – it will protect your skin from infection as well as making the dirt easier to remove afterwards; but make sure your hands aren't left slippery. Note that long-term contact with used engine oil can be a health hazard.

**DO** keep loose clothing (cuffs, tie etc) and long hair well out of the way of moving mechanical parts.

**DO** remove rings, wristwatch etc, before working on the vehicle – especially the electrical system.

**DO** keep your work area tidy – it is only too easy to fall over articles left lying around.

**DO** exercise caution when compressing springs for removal or installation. Ensure that the tension is applied and released in a controlled manner, using suitable tools which preclude the possibility of the spring escaping violently.

**DO** ensure that any lifting tackle used has a safe working load rating adequate for the job.

**DO** get someone to check periodically that all is well, when working alone on the vehicle.

**DO** carry out work in a logical sequence and check that everything is correctly assembled and tightened afterwards.

**DO** remember that your vehicle's safety affects that of yourself and others. If in doubt on any point, get specialist advice.

**IF**, in spite of following these precautions, you are unfortunate enough to injure yourself, seek medical attention as soon as possible.

### Asbestos

Certain friction, insulating, sealing, and other products – such as brake linings, clutch linings, gaskets, etc – contain asbestos. *Extreme care must be taken to avoid inhalation of dust from such products since it is hazardous to health.* If in doubt, assume that they *do* contain asbestos.

### Fire

Remember at all times that petrol (gasoline) is highly flammable. Never smoke, or have any kind of naked flame around, when working on the vehicle. But the risk does not end there – a spark caused by an electrical short-circuit, by two metal surfaces contacting each other, by careless use of tools, or even by static electricity built up in your body under certain conditions, can ignite petrol vapour, which in a confined space is highly explosive.

Always disconnect the battery earth (ground) terminal before working on any part of the fuel or electrical system, and never risk spilling fuel on to a hot engine or exhaust.

It is recommended that a fire extinguisher of a type suitable for fuel and electrical fires is kept handy in the garage or workplace at all times. Never try to extinguish a fuel or electrical fire with water.

**Note:** *Any reference to a 'torch' appearing in this manual should always be taken to mean a hand-held battery-operated electric lamp or flashlight. It does **not** mean a welding/gas torch or blowlamp.*

### Fumes

Certain fumes are highly toxic and can quickly cause unconsciousness and even death if inhaled to any extent. Petrol (gasoline) vapour comes into this category, as do the vapours from certain solvents such as trichloroethylene. Any draining or pouring of such volatile fluids should be done in a well ventilated area.

When using cleaning fluids and solvents, read the instructions carefully. Never use materials from unmarked containers – they may give off poisonous vapours.

Never run the engine of a motor vehicle in an enclosed space such as a garage. Exhaust fumes contain carbon monoxide which is extremely poisonous; if you need to run the engine, always do so in the open air or at least have the rear of the vehicle outside the workplace.

### The battery

Never cause a spark, or allow a naked light, near the vehicle's battery. It will normally be giving off a certain amount of hydrogen gas, which is highly explosive.

Always disconnect the battery earth (ground) terminal before working on the fuel or electrical systems.

If possible, loosen the filler plugs or cover when charging the battery from an external source. Do not charge at an excessive rate or the battery may burst.

Take care when topping up and when carrying the battery. The acid electrolyte, even when diluted, is very corrosive and should not be allowed to contact the eyes or skin.

If you ever need to prepare electrolyte yourself, always add the acid slowly to the water, and never the other way round. Protect against splashes by wearing rubber gloves and goggles.

### Mains electricity and electrical equipment

When using an electric power tool, inspection light etc, always ensure that the appliance is correctly connected to its plug and that, where necessary, it is properly earthed (grounded). Do not use such appliances in damp conditions and, again, beware of creating a spark or applying excessive heat in the vicinity of fuel or fuel vapour. Also ensure that the appliances meet the relevant national safety standards.

### Ignition HT voltage

A severe electric shock can result from touching certain parts of the ignition system, such as the HT leads, when the engine is running or being cranked, particularly if components are damp or the insulation is defective. Where an electronic ignition system is fitted, the HT voltage is much higher and could prove fatal.

# Tools and working facilities

The first priority when undertaking maintenance or repair work of any sort on a motorcycle is to have a clean, dry, well-lit working area. Work carried out in peace and quiet in the well-ordered atmosphere of a good workshop will give more satisfaction and much better results than can usually be achieved in poor working conditions. A good workshop must have a clean flat workbench or a solidly constructed table of convenient working height. The workbench or table should be equipped with a vice which has a jaw opening of at least 4 in (100 mm). A set of jaw covers should be made from soft metal such as aluminium alloy or copper, or from wood. These covers will minimise the marking or damaging of soft or delicate components which may be clamped in the vice. Some clean, dry, storage space will be required for tools, lubricants and dismantled components. It will be necessary during a major overhaul to lay out engine/gearbox components for examination and to keep them where they will remain undisturbed for as long as is necessary. To this end it is recommended that a supply of metal or plastic containers of suitable size is collected. A supply of clean, lint-free, rags for cleaning purposes and some newspapers, other rags, or paper towels for mopping up spillages should also be kept. If working on a hard concrete floor note that both the floor and one's knees can be protected from oil spillages and wear by cutting open a large cardboard box and spreading it flat on the floor under the machine or workbench. This also helps to provide some warmth in winter and to prevent the loss of nuts, washers, and other tiny components which have a tendency to disappear when dropped on anything other than a perfectly clean, flat, surface.

Unfortunately, such working conditions are not always available to the home mechanic. When working in poor conditions it is essential to take extra time and care to ensure that the components being worked on are kept scrupulously clean and to ensure that no components or tools are lost or damaged.

A selection of good tools is a fundamental requirement for anyone contemplating the maintenance and repair of a motor vehicle. For the owner who does not possess any, their purchase will prove a considerable expense, offsetting some of the savings made by doing-it-yourself. However, provided that the tools purchased meet the relevant national safety standards and are of good quality, they will last for many years and prove an extremely worthwhile investment.

To help the average owner to decide which tools are needed to carry out the various tasks detailed in this manual, we have compiled three lists of tools under the following headings: *Maintenance and minor repair, Repair and overhaul,* and *Specialized.* The newcomer to practical mechanics should start off with the simpler jobs around the vehicle. Then, as his confidence and experience grow, he can undertake more difficult tasks, buying extra tools as and when they are needed. In this way, a *Maintenance and minor repair* tool kit can be built-up into a *Repair and overhaul* tool kit over a considerable period of time without any major cash outlays. The experienced home mechanic will have a tool kit good enough for most repair and overhaul procedures and will add tools from the specialized category when he feels the expense is justified by the amount of use these tools will be put to.

It is obviously not possible to cover the subject of tools fully here. For those who wish to learn more about tools and their use there is a book entitled *Motorcycle Workshop Practice Manual (Bk no 1454)* available from the publishers of this manual.

As a general rule, it is better to buy the more expensive, good quality tools. Given reasonable use, such tools will last for a very long time, whereas the cheaper, poor quality, item will wear out faster and need to be renewed more often, thus nullifying the original saving. There is also the risk of a poor quality tool breaking while in use, causing personal injury or expensive damage to the component being worked on.

For practically all tools, a tool factor is the best source since he will have a very comprehensive range compared with the average garage or accessory shop. Having said that, accessory shops often offer excellent quality tools at discount prices, so it pays to shop around. There are plenty of tools around at reasonable prices, but always aim to purchase items which meet the relevant national safety standards. If in doubt, seek the advice of the shop proprietor or manager before making a purchase.

The basis of any toolkit is a set of spanners. While open-ended spanners with their slim jaws, are useful for working on awkwardly-positioned nuts, ring spanners have advantages in that they grip the nut far more positively. There is less risk of the spanner slipping off the nut and damaging it, for this reason alone ring spanners are to be preferred. Ideally, the home mechanic should acquire a set of each, but if expense rules this out a set of combination spanners (open-ended at one end and with a ring of the same size at the other) will provide a good compromise. Another item which is so useful it should be considered an essential requirement for any home mechanic is a set of socket spanners. These are available in a variety of drive sizes. It is recommended that the $\frac{1}{2}$-inch drive type is purchased to begin with as although bulkier and more expensive than the $\frac{3}{8}$-inch type, the larger size is far more common and will accept a greater variety of torque wrenches, extension pieces and socket sizes. The socket set should comprise sockets of sizes between 8 and 24 mm, a reversible ratchet drive, an extension bar of about 10 inches in length, a spark plug socket with a rubber insert, and a universal joint. Other attachments can be added to the set at a later date.

### Maintenance and minor repair tool kit

Set of spanners 8 – 24 mm
Set of sockets and attachments
Spark plug spanner with rubber insert – 10, 12, or 14 mm as appropriate
Adjustable spanner
C-spanner/pin spanner
Torque wrench (same size drive as sockets)
Set of screwdrivers (flat blade)
Set of screwdrivers (cross-head)
Set of Allen keys 4 – 10 mm
Impact screwdriver and bits
Ball pein hammer – 2 lb
Hacksaw (junior)
Self-locking pliers – Mole grips or vice grips
Pliers – combination
Pliers – needle nose
Wire brush (small)
Soft-bristled brush
Tyre pump
Tyre pressure gauge
Tyre tread depth gauge
Oil can
Fine emery cloth
Funnel (medium size)
Drip tray
Grease gun
Set of feeler gauges
Brake bleeding kit
Strobe timing light
Continuity tester (dry battery and bulb)
Soldering iron and solder
Wire stripper or craft knife
PVC insulating tape
Assortment of split pins, nuts, bolts, and washers

### Repair and overhaul toolkit

The tools in this list are virtually essential for anyone undertaking major repairs to a motorcycle and are additional to the tools listed above. Concerning Torx driver bits, Torx screws are encountered on some of the more modern machines where their use is restricted to fastening certain components inside the engine/gearbox unit. It is therefore recommended that if Torx bits cannot be borrowed from a local dealer, they are purchased individually as the need arises. They are not in regular use in the motor trade and will therefore only be available in specialist tool shops.

Plastic or rubber soft-faced mallet
Torx driver bits
Pliers – electrician's side cutters
Circlip pliers – internal (straight or right-angled tips are available)
Circlip pliers – external
Cold chisel
Centre punch
Pin punch
Scriber
Scraper (made from soft metal such as aluminium or copper)
Soft metal drift
Steel rule/straight edge
Assortment of files

Electric drill and bits
Wire brush (large)
Soft wire brush (similar to those used for cleaning suede shoes)
Sheet of plate glass
Hacksaw (large)
Valve grinding tool
Valve grinding compound (coarse and fine)
Stud extractor set (E-Z out)

### Specialized tools

This is not a list of the tools made by the machine's manufacturer to carry out a specific task on a limited range of models. Occasional references are made to such tools in the text of this manual and, in general, an alternative method of carrying out the task without the manufacturer's tool is given where possible. The tools mentioned in this list are those which are not used regularly and are expensive to buy in view of their infrequent use. Where this is the case it may be possible to hire or borrow the tools against a deposit from a local dealer or tool hire shop. An alternative is for a group of friends or a motorcycle club to join in the purchase.

Valve spring compressor
Piston ring compressor
Universal bearing puller
Cylinder bore honing attachment (for electric drill)
Micrometer set
Vernier calipers
Dial gauge set
Cylinder compression gauge
Vacuum gauge set
Multimeter
Dwell meter/tachometer

### Care and maintenance of tools

Whatever the quality of the tools purchased, they will last much longer if cared for. This means in practice ensuring that a tool is used for its intended purpose; for example screwdrivers should not be used as a substitute for a centre punch, or as chisels. Always remove dirt or grease and any metal particles but remember that a light film of oil will prevent rusting if the tools are infrequently used. The common tools can be kept together in a large box or tray but the more delicate, and more expensive, items should be stored separately where they cannot be damaged. When a tool is damaged or worn out, be sure to renew it immediately. It is false economy to continue to use a worn spanner or screwdriver which may slip and cause expensive damage to the component being worked on.

### Fastening systems

Fasteners, basically, are nuts, bolts and screws used to hold two or more parts together. There are a few things to keep in mind when working with fasteners. Almost all of them use a locking device of some type; either a lock washer, lock nut, locking tab or thread adhesive. All threaded fasteners should be clean, straight, have undamaged threads and undamaged corners on the hexagon head where the spanner fits. Develop the habit of replacing all damaged nuts and bolts with new ones.

Rusted nuts and bolts should be treated with a rust penetrating fluid to ease removal and prevent breakage. After applying the rust penetrant, let it 'work' for a few minutes before trying to loosen the nut or bolt. Badly rusted fasteners may have to be chiseled off or removed with a special nut breaker, available at tool shops.

Flat washers and lock washers, when removed from an assembly should always be replaced exactly as removed. Replace any damaged washers with new ones. Always use a flat washer between a lock washer and any soft metal surface (such as aluminium), thin sheet metal or plastic. Special lock nuts can only be used once or twice before they lose their locking ability and must be renewed.

If a bolt or stud breaks off in an assembly, it can be drilled out and removed with a special tool called an E-Z out. Most dealer service departments and motorcycle repair shops can perform this task, as well as others (such as the repair of threaded holes that have been stripped out).

# Spanner size comparison

| Jaw gap (in) | Spanner size | Jaw gap (in) | Spanner size |
|---|---|---|---|
| 0.250 | $\frac{1}{4}$ in AF | 0.945 | 24 mm |
| 0.276 | 7 mm | 1.000 | 1 in AF |
| 0.313 | $\frac{5}{16}$ in AF | 1.010 | $\frac{9}{16}$ in Whitworth; $\frac{5}{8}$ in BSF |
| 0.315 | 8 mm | 1.024 | 26 mm |
| 0.344 | $\frac{11}{32}$ in AF; $\frac{1}{8}$ in Whitworth | 1.063 | $1\frac{1}{16}$ in AF; 27 mm |
| 0.354 | 9 mm | 1.100 | $\frac{5}{8}$ in Whitworth; $\frac{11}{16}$ in BSF |
| 0.375 | $\frac{3}{8}$ in AF | 1.125 | $1\frac{1}{8}$ in AF |
| 0.394 | 10 mm | 1.181 | 30 mm |
| 0.433 | 11 mm | 1.200 | $\frac{11}{16}$ in Whitworth; $\frac{3}{4}$ in BSF |
| 0.438 | $\frac{7}{16}$ in AF | 1.250 | $1\frac{1}{4}$ in AF |
| 0.445 | $\frac{3}{16}$ in Whitworth; $\frac{1}{4}$ in BSF | 1.260 | 32 mm |
| 0.472 | 12 mm | 1.300 | $\frac{3}{4}$ in Whitworth; $\frac{7}{8}$ in BSF |
| 0.500 | $\frac{1}{2}$ in AF | 1.313 | $1\frac{5}{16}$ in AF |
| 0.512 | 13 mm | 1.390 | $\frac{13}{16}$ in Whitworth; $\frac{15}{16}$ in BSF |
| 0.525 | $\frac{1}{4}$ in Whitworth; $\frac{5}{16}$ in BSF | 1.417 | 36 mm |
| 0.551 | 14 mm | 1.438 | $1\frac{7}{16}$ in AF |
| 0.563 | $\frac{9}{16}$ in AF | 1.480 | $\frac{7}{8}$ in Whitworth; 1 in BSF |
| 0.591 | 15 mm | 1.500 | $1\frac{1}{2}$ in AF |
| 0.600 | $\frac{5}{16}$ in Whitworth; $\frac{3}{8}$ in BSF | 1.575 | 40 mm; $\frac{15}{16}$ in Whitworth |
| 0.625 | $\frac{5}{8}$ in AF | 1.614 | 41 mm |
| 0.630 | 16 mm | 1.625 | $1\frac{5}{8}$ in AF |
| 0.669 | 17 mm | 1.670 | 1 in Whitworth; $1\frac{1}{8}$ in BSF |
| 0.686 | $\frac{11}{16}$ in AF | 1.688 | $1\frac{11}{16}$ in AF |
| 0.709 | 18 mm | 1.811 | 46 mm |
| 0.710 | $\frac{3}{8}$ in Whitworth; $\frac{7}{16}$ in BSF | 1.813 | $1\frac{13}{16}$ in AF |
| 0.748 | 19 mm | 1.860 | $1\frac{1}{8}$ in Whitworth; $1\frac{1}{4}$ in BSF |
| 0.750 | $\frac{3}{4}$ in AF | 1.875 | $1\frac{7}{8}$ in AF |
| 0.813 | $\frac{13}{16}$ in AF | 1.969 | 50 mm |
| 0.820 | $\frac{7}{16}$ in Whitworth; $\frac{1}{2}$ in BSF | 2.000 | 2 in AF |
| 0.866 | 22 mm | 2.050 | $1\frac{1}{4}$ in Whitworth; $1\frac{3}{8}$ in BSF |
| 0.875 | $\frac{7}{8}$ in AF | 2.165 | 55 mm |
| 0.920 | $\frac{1}{2}$ in Whitworth; $\frac{9}{16}$ in BSF | 2.362 | 60 mm |
| 0.938 | $\frac{15}{16}$ in AF | | |

# Standard torque settings

Specific torque settings will be found at the end of the specifications section of each chapter. Where no figure is given, bolts should be secured according to the table below.

| Fastener type (thread diameter) | kgf m | lbf ft |
|---|---|---|
| 5mm bolt or nut | 0.45 – 0.6 | 3.5 – 4.5 |
| 6 mm bolt or nut | 0.8 – 1.2 | 6 – 9 |
| 8 mm bolt or nut | 1.8 – 2.5 | 13 – 18 |
| 10 mm bolt or nut | 3.0 – 4.0 | 22 – 29 |
| 12 mm bolt or nut | 5.0 – 6.0 | 36 – 43 |
| 5 mm screw | 0.35 – 0.5 | 2.5 – 3.6 |
| 6 mm screw | 0.7 – 1.1 | 5 – 8 |
| 6 mm flange bolt | 1.0 – 1.4 | 7 – 10 |
| 8 mm flange bolt | 2.4 – 3.0 | 17 – 22 |
| 10 mm flange bolt | 3.0 – 4.0 | 22 – 29 |

# Choosing and fitting accessories

The range of accessories available to the modern motorcyclist is almost as varied and bewildering as the range of motorcycles. This Section is intended to help the owner in choosing the correct equipment for his needs and to avoid some of the mistakes made by many riders when adding accessories to their machines. It will be evident that the Section can only cover the subject in the most general terms and so it is recommended that the owner, having decided that he wants to fit, for example, a luggage rack or carrier, seeks the advice of several local dealers and the owners of similar machines. This will give a good idea of what makes of carrier are easily available, and at what price. Talking to other owners will give some insight into the drawbacks or good points of any one make. A walk round the motorcycles in car parks or outside a dealer will often reveal the same sort of information.

The first priority when choosing accessories is to assess exactly what one needs. It is, for example, pointless to buy a large heavy-duty carrier which is designed to take the weight of fully laden panniers and topbox when all you need is a place to strap on a set of waterproofs and a lunchbox when going to work. Many accessory manufacturers have ranges of equipment to cater for the individual needs of different riders and this point should be borne in mind when looking through a dealer's catalogues. Having decided exactly what is required and the use to which the accessories are going to be put, the owner will need a few hints on what to look for when making the final choice. To this end the Section is now sub-divided to cover the more popular accessories fitted. Note that it is in no way a customizing guide, but merely seeks to outline the practical considerations to be taken into account when adding aftermarket equipment to a motorcycle.

## Fairings and windscreens

A fairing is possibly the single, most expensive, aftermarket item to be fitted to any motorcycle and, therefore, requires the most thought before purchase. Fairings can be divided into two main groups: front fork mounted handlebar fairings and windscreens, and frame mounted fairings.

The first group, the front fork mounted fairings, are becoming far more popular than was once the case, as they offer several advantages over the second group. Front fork mounted fairings generally are much easier and quicker to fit, involve less modification to the motorcycle, do not as a rule restrict the steering lock, permit a wider selection of handlebar styles to be used, and offer adequate protection for much less money than the frame mounted type. They are also lighter, can be swapped easily between different motorcycles, and are available in a much greater variety of styles. Their main disadvantages are that they do not offer as much weather protection as the frame mounted types, rarely offer any storage space, and, if poorly fitted or naturally incompatible, can have an adverse effect on the stability of the motorcycle.

The second group, the frame mounted fairings, are secured so rigidly to the main frame of the motorcycle that they can offer a substantial amount of protection to motorcycle and rider in the event of a crash. They offer almost complete protection from the weather and, if double-skinned in construction, can provide a great deal of useful storage space. The feeling of peace, quiet and complete relaxation encountered when riding behind a good full fairing has to be experienced to be believed. For this reason full fairings are considered essential by most touring motorcyclists and by many people who ride all year round. The main disadvantages of this type are that fitting can take a long time, often involving removal or modification of standard motorcycle components, they restrict the steering lock and they can add up to about 40 lb to the weight of the machine. They do not usually affect the stability of the machine to any great extent once the front tyre pressure and suspension have been adjusted to compensate for the extra weight, but can be affected by sidewinds.

The first thing to look for when purchasing a fairing is the quality of the fittings. A good fairing will have strong, substantial brackets constructed from heavy-gauge tubing; the brackets must be shaped to fit the frame or forks evenly so that the minimum of stress is imposed on the assembly when it is bolted down. The brackets should be properly painted or finished – a nylon coating being the favourite of the better manufacturers – the nuts and bolts provided should be of the same thread and size standard as is used on the motorcycle and be properly plated. Look also for shakeproof locking nuts or locking washers to ensure that everything remains securely tightened down. The fairing shell is generally made from one of two materials: fibreglass or ABS plastic. Both have their advantages and disadvantages, but the main consideration for the owner is that fibreglass is much easier to repair in the event of damage occurring to the fairing. Whichever material is used, check that it is properly finished inside as well as out, that the edges are protected by beading and that the fairing shell is insulated from vibration by the use of rubber grommets at all mounting points. Also be careful to check that the windscreen is retained by plastic bolts which will snap on impact so that the windscreen will break away and not cause personal injury in the event of an accident.

Having purchased your fairing or windscreen, read the manufacturer's fitting instructions very carefully and check that you have all the necessary brackets and fittings. Ensure that the mounting brackets are located correctly and bolted down securely. Note that some manufacturers use hose clamps to retain the mounting brackets; these should be discarded as they are convenient to use but not strong enough for the task. Stronger clamps should be substituted; car exhaust pipe clamps of suitable size would be a good alternative. Ensure that the front forks can turn through the full steering lock available without fouling the fairing. With many types of frame-mounted fairing the handlebars will have to be altered or a different type fitted and the steering lock will be restricted by stops provided with the fittings. Also check that the fairing does not foul the front wheel or mudguard, in any steering position, under full fork compression. Re-route any cables, brake pipes or electrical wiring which may snag on the fairing and take great care to protect all electrical connections, using insulating tape. If the manufacturer's instructions are followed carefully at every stage no serious problems should be encountered. Remember that hydraulic pipes that have been disconnected must be carefully re-tightened and the hydraulic system purged of air bubbles by bleeding.

Two things will become immediately apparent when taking a motorcycle on the road for the first time with a fairing – the first is the tendency to underestimate the road speed because of the lack of wind pressure on the body. This must be very carefully watched until one has grown accustomed to riding behind the fairing. The second thing is the alarming increase in engine noise which is an unfortunate but inevitable by-product of fitting any type of fairing or windscreen, and is caused by normal engine noise being reflected, and in some cases amplified, by the flat surface of the fairing.

## Luggage racks or carriers

Carriers are possibly the commonest item to be fitted to modern motorcycles. They vary enormously in size, carrying capacity, and durability. When selecting a carrier, always look for one which is made specifically for your machine and which is bolted on with as few separate brackets as possible. The universal-type carrier, with its mass of brackets and adaptor pieces, will generally prove too weak to be of any real use. A good carrier should bolt to the main frame, generally using the two suspension unit top mountings and a mudguard mounting bolt as attachment points, and have its luggage platform as low and as far forward as possible to minimise the effect of any load on the machine's stability. Look for good quality, heavy gauge tubing, good welding and good finish. Also ensure that the carrier does not prevent opening of the seat, sidepanels or tail compartment, as appropriate. When using a carrier, be very careful not to overload it. Excessive weight placed so high and so far to the rear of any motorcycle will have an adverse effect on the machine's steering and stability.

## Luggage

Motorcycle luggage can be grouped under two headings: soft and hard. Both types are available in many sizes and styles and have advantages and disadvantages in use.

Soft luggage is now becoming very popular because of its lower cost and its versatility. Whether in the form of tankbags, panniers, or strap-on bags, soft luggage requires in general no brackets and no modification to the motorcycle. Equipment can be swapped easily from one motorcycle to another and can be fitted and removed in seconds. Awkwardly shaped loads can easily be carried. The disadvantages of soft luggage are that the contents cannot be secure against the casual thief, very little protection is afforded in the event of a crash, and waterproofing is generally poor. Also, in the case of panniers, carrying capacity is restricted to approximately 10 lb, although this amount will vary considerably depending on the manufacturer's recommendation. When purchasing soft luggage, look for good quality material, generally vinyl or nylon, with strong, well-stitched attachment points. It is always useful to have separate pockets, especially on tank bags, for items which will be needed on the journey. When purchasing a tank bag, look for one which has a separate, well-padded, base. This will protect the tank's paintwork and permit easy access to the filler cap at petrol stations.

Hard luggage is confined to two types: panniers, and top boxes or tail trunks. Most hard luggage manufacturers produce matching sets of these items, the basis of which is generally that manufacturer's own heavy-duty luggage rack. Variations on this theme occur in the form of separate frames for the better quality panniers, fixed or quickly-detachable luggage, and in size and carrying capacity. Hard luggage offers a reasonable degree of security against theft and good protection against weather and accident damage. Carrying capacity is greater than that of soft luggage, around 15 – 20 lb in the case of panniers, although top boxes should never be loaded as much as their apparent capacity might imply. A top box should only be used for lightweight items, because one that is heavily laden can have a serious effect on the stability of the machine. When purchasing hard luggage look for the same good points as mentioned under fairings and windscreens, ie good quality mounting brackets and fittings, and well-finished fibreglass or ABS plastic cases. Again as with fairings, always purchase luggage made specifically for your motorcycle, using as few separate brackets as possible, to ensure that everything remains securely bolted in place. When fitting hard luggage, be careful to check that the rear suspension and brake operation will not be impaired in any way and remember that many pannier kits require re-siting of the indicators. Remember also that a non-standard exhaust system may make fitting extremely difficult.

## Handlebars

The occupation of fitting alternative types of handlebar is extremely popular with modern motorcyclists, whose motives may vary from the purely practical, wishing to improve the comfort of their machines, to the purely aesthetic, where form is more important than function. Whatever the reason, there are several considerations to be borne in mind when changing the handlebars of your machine. If fitting lower bars, check carefully that the switches and cables do not foul the petrol tank on full lock and that the surplus length of cable, brake pipe,

and electrical wiring are smoothly and tidily disposed of. Avoid tight kinks in cable or brake pipes which will produce stiff controls or the premature and disastrous failure of an overstressed component. If necessary, remove the petrol tank and re-route the cable from the engine/gearbox unit upwards, ensuring smooth gentle curves are produced. In extreme cases, it will be necessary to purchase a shorter brake pipe to overcome this problem. In the case of higher handlebars than standard it will almost certainly be necessary to purchase extended cables and brake pipes. Fortunately, many standard motorcycles have a custom version which will be equipped with higher handlebars and, therefore, factory-built extended components will be available from your local dealer. It is not usually necessary to extend electrical wiring, as switch clusters may be used on several different motorcycles, some being custom versions. This point should be borne in mind however when fitting extremely high or wide handlebars.

When fitting different types of handlebar, ensure that the mounting clamps are correctly tightened to the manufacturer's specifications and that cables and wiring, as previously mentioned, have smooth easy runs and do not snag on any part of the motorcycle throughout the full steering lock. Ensure that the fluid level in the front brake master cylinder remains level to avoid any chance of air entering the hydraulic system. Also check that the cables are adjusted correctly and that all handlebar controls operate correctly and can be easily reached when riding.

## Crashbars

Crashbars, also known as engine protector bars, engine guards, or case savers, are extremely useful items of equipment which can contribute protection to the machine's structure if a crash occurs. They do not, as has been inferred in the US, prevent the rider from crashing, or necessarily prevent rider injury should a crash occur.

It is recommended that only the smaller, neater, engine protector type of crashbar is considered. This type will offer protection while restricting, as little as is possible, access to the engine and the machine's ground clearance. The crashbars should be designed for use specifically on your machine, and should be constructed of heavy-gauge tubing with strong, integral mounting brackets. Where possible, they should bolt to a strong lug on the frame, usually at the engine mounting bolts.

The alternative type of crashbar is the larger cage type. This type is not recommended in spite of their appearance which promises some protection to the rider as well as to the machine. The larger amount of leverage imposed by the size of this type of crashbar increases the risk of severe frame damage in the event of an accident. This type also decreases the machine's ground clearance and restricts access to the engine. The amount of protection afforded the rider is open to some doubt as the design is based on the premise that the rider will stay in the normally seated position during an accident, and the crash bar structure will not itself fail. Neither result can in any way be guaranteed.

As a general rule, always purchase the best, ie usually the most expensive, set of crashbars you an afford. The investment will be repaid by minimising the amount of damage incurred, should the machine be involved in an accident. Finally, avoid the universal type of crashbar. This should be regarded only as a last resort to be used if no alternative exists. With its usual multitude of separate brackets and spacers, the universal crashbar is far too weak in design and construction to be of any practical value.

## Exhaust systems

The fitting of aftermarket exhaust systems is another extremely popular pastime amongst motorcyclists. The usual motive is to gain more performance from the engine but other considerations are to gain more ground clearance, to lose weight from the motorcycle, to obtain a more distinctive exhaust note or to find a cheaper alternative to the manufacturer's original equipment exhaust system. Original equipment exhaust systems often cost more and may well have a relatively short life. It should be noted that it is rare for an aftermarket exhaust system alone to give a noticeable increase in the engine's power output. Modern motorcycles are designed to give the highest power output possible allowing for factors such as quietness, fuel economy, spread of power, and long-term reliability. If there were a magic formula which allowed the exhaust system to produce more power without affecting these other considerations you can be sure that the manufacturers, with their large research and development facilities, would have found it and made use of it. Performance increases of a worthwhile and noticeable nature only come from well-tried and properly matched modifications to the entire engine, from the air filter, through the carburettors, port timing or camshaft and valve design, combustion chamber shape, compression ratio, and the exhaust system. Such modifications are well outside the scope of this manual but interested owners might refer to specialist books produced by the publisher of this manual which go into the whole subject in great detail.

Whatever your motive for wishing to fit an alternative exhaust system, be sure to seek expert advice before doing so. Changes to the carburettor jetting will almost certainly be required for which you must consult the exhaust system manufacturer. If he cannot supply adequately specific information it is reasonable to assume that insufficient development work has been carried out, and that particular make should be avoided. Other factors to be borne in mind are whether the exhaust system allows the use of both centre and side stands, whether it allows sufficient access to permit oil and filter changing and whether modifications are necessary to the standard exhaust system. Many two-stroke expansion chamber systems require the use of the standard exhaust pipe; this is all very well if the standard exhaust pipe and silencer are separate units but can cause problems if the two, as with so many modern two-strokes, are a one-piece unit. While the exhaust pipe can be removed easily by means of a hacksaw it is not so easy to refit the original silencer should you at any time wish to return the machine to standard trim. The same applies to several four-stroke systems.

On the subject of the finish of aftermarket exhausts, avoid black-painted systems unless you enjoy painting. As any trail-bike owner will tell you, rust has a great affinity for black exhausts and re-painting or rust removal becomes a task which must be carried out with monotonous regularity. A bright chrome finish is, as a general rule, a far better proposition as it is much easier to keep clean and to prevent rusting. Although the general finish of aftermarket exhaust systems is not always up to the standard of the original equipment the lower cost of such systems does at least reflect this fact.

When fitting an alternative system always purchase a full set of new exhaust gaskets, to prevent leaks. Fit the exhaust first to the cylinder head or barrel, as appropriate, tightening the retaining nuts or bolts by hand only and then line up the exhaust rear mountings. If the new system is a one-piece unit and the rear mountings do not line up exactly, spacers must be fabricated to take up the difference. Do not force the system into place as the stress thus imposed will rapidly cause cracks and splits to appear. Once all the mountings are loosely fixed, tighten the retaining nuts or bolts securely, being careful not to overtighten them. Where the motorcycle manufacturer's torque settings are available, these should be used. Do not forget to carry out any carburation changes recommended by the exhaust system's manufacturer.

## Electrical equipment

The vast range of electrical equipment available to motorcyclists is so large and so diverse that only the most general outline can be given here. Electrical accessories vary from electronic ignition kits fitted to replace contact breaker points, to additional lighting at the front and rear, more powerful horns, various instruments and gauges, clocks, anti-theft systems, heated clothing, CB radios, radio-cassette players, and intercom systems, to name but a few of the more popular items of equipment.

As will be evident, it would require a separate manual to cover this subject alone and this section is therefore restricted to outlining a few basic rules which must be borne in mind when fitting electrical equipment. The first consideration is whether your machine's electrical system has enough reserve capacity to cope with the added demand of the accessories you wish to fit. The motorcycle's manufacturer or importer should be able to furnish this sort of information and may also be able to offer advice on uprating the electrical system. Failing this, a good dealer or the accessory manufacturer may be able to help. In some cases, more powerful generator components may be available, perhaps from another motorcycle in the manufacturer's range. The second consideration is the legal requirements in force in your area. The local police may be prepared to help with this point. In the UK for example, there are strict regulations governing the position and use of auxiliary riding lamps and fog lamps.

When fitting electrical equipment always disconnect the battery first to prevent the risk of a short-circuit, and be careful to ensure that all connections are properly made and that they are waterproof. Remember that many electrical accessories are designed primarily for use in cars and that they cannot easily withstand the exposure to vibration and to the weather. Delicate components must be rubber-mounted to insulate them from vibration, and sealed carefully to prevent the entry of rainwater and dirt. Be careful to follow exactly the accessory manufacturer's instructions in conjunction with the wiring diagram at the back of this manual.

## Accessories – general

Accessories fitted to your motorcycle will rapidly deteriorate if not cared for. Regular washing and polishing will maintain the finish and will provide an opportunity to check that all mounting bolts and nuts are securely fastened. Any signs of chafing or wear should be watched for, and the cause cured as soon as possible before serious damage occurs.

As a general rule, do not expect the re-sale value of your motorcycle to increase by an amount proportional to the amount of money and effort put into fitting accessories. It is usually the case that an absolutely standard motorcycle will sell more easily at a better price than one that has been modified. If you are in the habit of exchanging your machine for another at frequent intervals, this factor should be borne in mind to avoid loss of money.

# Fault diagnosis

## Contents

## 1 Introduction

This Section provides an easy reference-guide to the more common ailments that are likely to afflict your machine. Obviously, the opportunities are almost limitless for faults to occur as a result of obscure failures, and to try and cover all eventualities would require a book. Indeed, a number have been written on the subject.

Successful fault diagnosis is not a mysterious 'black art' but the application of a bit of knowledge combined with a systematic and logical approach to the problem. Approach any fault diagnosis by first accurately identifying the symptom and then checking through the list of possible causes, starting with the simplest or most obvious and progressing in stages to the most complex. Take nothing for granted, but above all apply liberal quantities of common sense.

The main symptom of a fault is given in the text as a major heading below which are listed, as Section headings, the various systems or areas which may contain the fault. Details of each possible cause for a fault and the remedial action to be taken are given, in brief, in the paragraphs below each Section heading. Further information should be sought in the relevant Chapter.

## *Starter motor problems*

## 2 Starter motor not rotating

Engine stop switch off.

Fuse blown. Check the main fuse located behind the battery side cover.

Battery voltage low. Switching on the headlamp and operating the horn will give a good indication of the charge level. If necessary recharge the battery from an external source.

Neutral gear not selected. Where a neutral indicator switch is fitted.

Faulty neutral indicator switch or clutch interlock switch. Check the switch wiring and switches for correct operation.

Ignition switch defective. Check switch for continuity and connections for security.

Engine stop switch defective. Check switch for continuity in 'Run' position. Fault will be caused by broken, wet or corroded switch contacts. Clean or renew as necessary.

Starter button switch faulty. Check continuity of switch. Faults as for engine stop switch.

Starter relay (solenoid) faulty. If the switch is functioning correctly a pronounced click should be heard when the starter button is depressed. This presupposes that current is flowing to the solenoid when the button is depressed.

Wiring open or shorted. Check first that the battery terminal connections are tight and corrosion free. Follow this by checking that all wiring connections are dry, tight and corrosion free. Check also for frayed or broken wiring. Occasionally a wire may become trapped between two moving components, particularly in the vicinity of the steering head, leading to breakage of the internal core but leaving the softer but more resilient outer cover intact. This can cause mysterious intermittent or total power loss.

Starter motor defective. A badly worn starter motor may cause high current drain from a battery without the motor rotating. If current is found to be reaching the motor, after checking the starter button and starter relay, suspect a damaged motor. The motor should be removed for inspection.

## 3 Starter motor rotates but engine does not turn over

Starter motor clutch defective. Suspect jammed or worn engagement rollers, plungers and springs.

Damaged starter motor drive train. Inspect and renew component where necessary. Failure in this area is unlikely.

## 4 Starter motor and clutch function but engine will not turn over

Engine seized. Seizure of the engine is always a result of damage to internal components due to lubrication failure, or component breakage resulting from abuse, neglect or old age. A seizing or partially seized component may go un-noticed until the engine has cooled down and an attempt is made to restart the engine. Suspect first seizure of the valves, valve gear and the pistons. Instantaneous seizure whilst the engine is running indicates component breakage. In either case major dismantling and inspection will be required.

## *Engine does not start when turned over*

## 5 No fuel flow to carburettor

No fuel or insufficient fuel in tank.

Fuel tap lever position incorrectly selected.

Float chambers require priming after running dry.

Tank filler cap air vent obstructed. Usually caused by dirt or water. Clean the vent orifice.

Fuel tap or filter blocked. Blockage may be due to accumulation of rust or paint flakes from the tank's inner surface or of foreign matter from contaminated fuel. Remove the tap and clean it and the filter. Look also for water droplets in the fuel.

Fuel line blocked. Blockage of the fuel line is more likely to result from a kink in the line rather than the accumulation of debris.

## 6 Fuel not reaching cylinder

Float chamber not filling. Caused by float needle or floats sticking in up position. This may occur after the machine has been left standing for an extended length of time allowing the fuel to evaporate. When this occurs a gummy residue is often left which hardens to a varnish-like substance. This condition may be worsened by corrosion and crystaline deposits produced prior to the total evaporation of contaminated fuel. Sticking of the float needle may also be caused by wear. In any case removal of the float chamber will be necessary for inspection and cleaning.

Blockage in starting circuit, slow running circuit or jets. Blockage of these items may be attributable to debris from the fuel tank by-passing the filter system or to gumming up as described in paragraph 1. Water droplets in the fuel will also block jets and passages. The carburettor should be dismantled for cleaning.

Fuel level too low. The fuel level in the float chamber is controlled by float height. The float height may increase with wear or damage but will never reduce, thus a low float height is an inherent rather than a developing condition. Check the float height and make any necessary adjustment.

## 7 Engine flooding

Float valve needle worn or stuck open. A piece of rust or other debris can prevent correct seating of the needle against the valve seat thereby permitting an uncontrolled flow of fuel. Similarly, a worn needle or needle seat will prevent valve closure. Dismantle the carburettor float bowl for cleaning and, if necessary, renewal of the worn components.

Fuel level too high. The fuel level is controlled by the float height which may increase due to wear of the float needle, pivot pin or operating tang. Check the float height, and make any necessary adjustment. A leaking float will cause an increase in fuel level, and thus should be renewed.

Cold starting mechanism. Check the choke (starter mechanism) for correct operation. If the mechanism jams in the 'On' position subsequent starting of a hot engine will be difficult.

Blocked air filter. A badly restricted air filter will cause flooding. Check the filter and clean or renew as required. A collapsed inlet hose will have a similar effect.

## 8 No spark at plug

Ignition switch not on.

Engine stop switch off.

Fuse blown. Check fuse for ignition circuit. See wiring diagram.

Battery voltage low. The current draw required by a starter motor is sufficiently high that an under-charged battery may not have enough spare capacity to provide power for the ignition circuit during starting.

Starter motor inefficient. A starter motor with worn brushes and a worn or dirty commutator will draw excessive amounts of current causing power starvation in the ignition system. See the preceding paragraph. Starter motor overhaul will be required.

Spark plug failure. Clean the spark plug thoroughly and reset the electrode gap. Refer to the spark plug section ~~and the colour condition guide~~ in Chapter 3. If the spark plug shorts internally or has sustained visible damage to the electrodes, core or ceramic insulator it should be renewed. On rare occasions a plug that appears to spark vigorously will fail to do so when refitted to the engine and subjected to the compression pressure in the cylinder.

Spark plug cap or high tension (HT) lead faulty. Check condition and security. Replace if deterioration is evident.

Spark plug cap loose. Check that the spark plug cap fits securely over the plug and, where fitted, the screwed terminal on the plug end is secure.

Shorting due to moisture. Certain parts of the ignition system are susceptible to shorting when the machine is ridden or parked in wet weather. Check particularly the area from the spark plug cap back to the ignition coil. A water dispersant spray may be used to dry out waterlogged components. Recurrence of the problem can be prevented by using an ignition sealant spray after drying out and cleaning.

Ignition or stop switch shorted. May be caused by water, corrosion or wear. Water dispersant and contact cleaning sprays may be used. If this fails to overcome the problem dismantling and visual inspection of the switches will be required.

Shorting or open circuit in wiring. Failure in any wire connecting any of the ignition components will cause ignition malfunction. Check also that all connections are clean, dry and tight.

Ignition coil failure. Check the coil, referring to Chapter 3.

Failure of ignition pickup (pulser coils) or spark unit. See Chapter 3.

## 9 Weak spark at plug

Feeble sparking at the plug may be caused by any of the faults mentioned in the preceding Section other than those items in paragraphs 1, 2 and 3. Check first the spark plug, this being the most likely culprit.

## 10 Compression low

Spark plug loose. This will be self-evident on inspection, and may be accompanied by a hissing noise when the engine is turned over. Remove the plug and check that the threads in the cylinder head are not damaged. Check also that the plug sealing washer is in good condition.

Cylinder head gasket leaking. This condition is often accompanied by a high pitched squeak from around the cylinder head and oil loss, and may be caused by insufficiently tightened cylinder head fasteners, a warped cylinder head or mechanical failure of the gasket material. Re-torqueing the fasteners to the correct specification may seal the leak in some instances but if damage has occurred this course of action will provide, at best, only a temporary cure.

Valve not seating correctly. The failure of a valve to seat may be caused by insufficient valve clearance, pitting of the valve seat or face, carbon deposits on the valve seat or seizure of the valve stem or valve gear components. Valve spring breakage will also prevent correct valve closure. The valve clearances should be checked first and then, if these are found to be in order, further dismantling will be required to inspect the relevant components for failure.

Cylinder, piston and ring wear. Compression pressure will be lost if any of these components are badly worn. Wear in one component is invariably accompanied by wear in another. A top end overhaul will be required.

Piston rings sticking or broken. Sticking of the piston rings may be caused by seizure due to lack of lubrication or heating as a result of poor carburation or incorrect fuel type. Gumming of the rings may result from lack of use, or carbon deposits in the ring grooves. Broken rings result from over-revving, overheating or general wear. In either case a top-end overhaul will be required.

## *Engine stalls after starting*

## 11 General causes

Improper cold start mechanism operation. Check that the operating controls function smoothly and, where applicable, are correctly adjusted. A cold engine may not require application of an enriched mixture to start initially but may baulk without choke once firing. Likewise a hot engine may start with an enriched mixture but will stop almost immediately if the choke is inadvertently in operation.

Ignition malfunction. See Section 9, 'Weak spark at plug'.

Carburettor incorrectly adjusted. Maladjustment of the mixture strength or idle speed may cause the engine to stop immediately after starting. See Chapter 2.

Fuel contamination. Check for filter blockage by debris or water which reduces, but does not completely stop, fuel flow or blockage of the slow speed circuit in the carburettor by the same agents. If water is present it can often be seen as droplets in the bottom of the float bowl. Clean the filter and, where water is in evidence, drain and flush the fuel tank and float bowl.

Intake air leak. Check for security of the carburettor mounting and hose connections, and for cracks or splits in the hoses. Check also that the carburettor top is secure and that the vacuum gauge adaptor plug (where fitted) is tight.

Air filter blocked or omitted. A blocked filter will cause an over-rich mixture; the omission of a filter will cause an excessively weak mixture. Both conditions will have a detrimental affect on carburation. Clean or renew the filter as necessary.

Fuel filler cap air vent blocked. Usually caused by dirt or water. Clean the vent orifice.

## *Poor running at idle and low speed*

## 12 Weak spark at plug or erratic firing

Battery voltage low. In certain conditions low battery charge, especially when coupled with a badly sulphated battery, may result in misfiring. If the battery is in good general condition it should be recharged; an old battery suffering from sulphated plates should be renewed.

Spark plug fouled, faulty or incorrectly adjusted. See Section 8 or refer to Chapter 3.

Spark plug cap or high tension lead shorting. Check the condition of both these items ensuring that they are in good condition and dry and that the cap is fitted correctly.

Spark plug type incorrect. Fit plug of correct type and heat range as given in Specifications. In certain conditions a plug of hotter or colder type may be required for normal running.

Igniting timing incorrect. Check the ignition timing statically and dynamically, ensuring that the advance is functioning correctly.

Faulty ignition coil. Partial failure of the coil internal insulation will diminish the performance of the coil. No repair is possible, a new component must be fitted.

Faulty pickup (pulser) coil or spark unit. The former is the more likely cause of a partial failure. See Chapter 3 for test procedures.

## 13 Fuel/air mixture incorrect

Intake air leak. See Section 11.

Mixture strength incorrect. Adjust slow running mixture strength using pilot adjustment screw.

Carburettor synchronisation.

Pilot jet or slow running circuit blocked. The carburettor should be removed and dismantled for thorough cleaning. Blow through all jets and air passages with compressed air to clear obstructions.

Air cleaner clogged or omitted. Clean or fit air cleaner element as necessary. Check also that the element and air filter cover are correctly seated.

Cold start mechanism in operation. Check that the choke has not been left on inadvertently and the operation is correct. Where applicable check the operating cable free play.

Fuel level too high or too low. Check the float height and adjust as necessary. See Section 7.

Fuel tank air vent obstructed. Obstruction usually caused by dirt or water. Clean vent orifice.

Valve clearance incorrect. Check, and if necessary, adjust, the clearances.

## 14 Compression low

See Section 10.

## *Acceleration poor*

## 15 General causes

All items as for previous Section.

Timing not advancing. This is caused by a failure of the advance control circuit in the spark unit. The unit should be renewed; repair is not practical.

Sticking throttle vacuum piston.

Brakes binding. Usually caused by maladjustment or partial seizure of the operating mechanism due to poor maintenance. Check brake adjustment (where applicable). A bent wheel spindle or warped brake disc can produce similar symptoms.

## *Poor running or lack of power at high speeds*

## 16 Weak spark at plug or erratic firing

All items as for Section 12.

HT lead insulation failure. Insulation failure of the HT lead and spark plug cap due to old age or damage can cause shorting when the engine is driven hard. This condition may be less noticeable, or not noticeable at all at lower engine speeds.

## 17 Fuel/air mixture incorrect

All items as for Section 13, with the exception of items 2 and 4.

Main jet blocked. Debris from contaminated fuel, or from the fuel tank, and water in the fuel can block the main jet. Clean the fuel filter, the float bowl area, and if water is present, flush and refill the fuel tank.

Main jet is the wrong size. The standard carburettor jetting is for sea level atmospheric pressure. For high altitudes, usually above 5000 ft, a smaller main jet will be required.

Jet needle and needle jet worn. These can be renewed individually but should be renewed as a pair. Renewal of both items requires partial dismantling of the carburettor.

Air bleed holes blocked. Dismantle carburettor and use compressed air to blow out all air passages.

Reduced fuel flow. A reduction in the maximum fuel flow from the fuel tank to the carburettor will cause fuel starvation, proportionate to the engine speed. Check for blockages through debris or a kinked fuel line.

Vacuum diaphragm split. Renew.

## 18 Compression low

See Section 10.

## *Knocking or pinking*

## 19 General causes

Carbon build-up in combustion chamber. After high mileages have been covered large accumulation of carbon may occur. This may glow red hot and cause premature ignition of the fuel/air mixture, in advance of normal firing by the spark plug. Cylinder head removal will be required to allow inspection and cleaning.

Fuel incorrect. A low grade fuel, or one of poor quality may result in compression induced detonation of the fuel resulting in knocking and pinking noises. Old fuel can cause similar problems. A too highly leaded fuel will reduce detonation but will accelerate deposit formation in the combustion chamber and may lead to early pre-ignition as described in item 1.

Spark plug heat range incorrect. Uncontrolled pre-ignition can result from the use of a spark plug the heat range of which is too hot.

Weak mixture. Overheating of the engine due to a weak mixture can result in pre-ignition occurring where it would not occur when engine temperature was within normal limits. Maladjustment, blocked jets or passages and air leaks can cause this condition.

## *Overheating*

## 20 Firing incorrect

Spark plug fouled, defective or maladjusted. See Section 6.

Spark plug type incorrect. Refer to the Specifications and ensure that the correct plug type is fitted.

Incorrect ignition timing. Timing that is far too much advanced or far too much retarded will cause overheating. Check the ignition timing is correct and that the advance mechanism is functioning.

## 21 Fuel/air mixture incorrect

Slow speed mixture strength incorrect. Adjust pilot air screw.

Main jet wrong size. The carburettor is jetted for sea level atmospheric conditions. For high altitudes, usually above 5000 ft, a smaller main jet will be required.

Air filter badly fitted or omitted. Check that the filter element is in place and that it and the air filter box cover are sealing correctly. Any leaks will cause a weak mixture.

Induction air leaks. Check the security of the carburettor mountings and hose connections, and for cracks and splits in the hoses. Check also that the carburettor top is secure and that the vacuum gauge adaptor plug (where fitted) is tight.

Fuel level too low. See Section 6.

Fuel tank filler cap air vent obstructed. Clear blockage.

## 22 Lubrication inadequate

Engine oil too low. Not only does the oil serve as a lubricant by preventing friction between moving components, but it also acts as a coolant. Check the oil level and replenish.

Engine oil overworked. The lubricating properties of oil are lost slowly during use as a result of changes resulting from heat and also contamination. Always change the oil at the recommended interval.

Engine oil of incorrect viscosity or poor quality. Always use the recommended viscosity and type of oil.

Oil filter and filter by-pass valve blocked. Renew filter and clean the by-pass valve.

## 23 Miscellaneous causes

Engine fins clogged. A build-up of mud in the cylinder head and cylinder barrel cooling fins will decrease the cooling capabilities of the fins. Clean the fins as required.

## *Clutch operating problems*

## 24 Clutch slip

No clutch lever play. Adjust clutch lever end play according to the procedure in Chapter

Friction plates worn or warped. Overhaul clutch assembly, replacing plates out of specification.

Steel plates worn or warped. Overhaul clutch assembly, replacing plates out of specification.

Clutch springs broken or wear. Old or heat-damaged (from slipping clutch) springs should be replaced with new ones.

Clutch inner cable snagging. Caused by a frayed cable or kinked outer cable. Replace the cable with a new one. Repair of a frayed cable is not advised.

Clutch release mechanism defective. Worn or damaged parts in the clutch release mechanism could include the shaft, cam, actuating arm or pivot. Replace parts as necessary.

Clutch hub and outer drum worn. Severe indentation by the clutch plate tangs of the channels in the hub and drum will cause snagging of the plates preventing correct engagement. If this damage occurs, renewal of the worn components is required.

Lubricant incorrect. Use of a transmission lubricant other than that specified may allow the plates to slip.

## 25 Clutch drag

Clutch lever play excessive. Adjust lever at bars or at cable end if necessary.

Clutch plates warped or damaged. This will cause a drag on the clutch, causing the machine to creep. Overhaul clutch assembly.

Clutch spring tension uneven. Usually caused by a sagged or broken spring. Check and replace springs.

Engine oil deteriorated. Badly contaminated engine oil and a heavy deposit of oil sludge and carbon on the plates will cause plate sticking. The oil recommended for this machine is of the detergent type, therefore it is unlikely that this problem will arise unless regular oil changes are neglected.

Engine oil viscosity too high. Drag in the plates will result from the use of an oil with too high a viscosity. In very cold weather clutch drag may occur until the engine has reached operating temperature.

Clutch hub and outer drum worn. Indentation by the clutch plate tangs of the channels in the hub and drum will prevent easy plate disengagement. If the damage is light the affected areas may be dressed with a fine file. More pronounced damage will necessitate renewal of the components.

Clutch housing seized to shaft. Lack of lubrication, severe wear or damage can cause the housing to seize to the shaft. Overhaul of the clutch, and perhaps the transmission, may be necessary to repair damage.

Clutch release mechanism defective. Worn or damaged release mechanism parts can stick and fail to provide leverage. Overhaul clutch cover components.

Loose clutch hub nut. Causes drum and hub misalignment, putting a drag on the engine. Engagement adjustment continually varies. Overhaul clutch assembly.

## Gear selection problems

## 26 Gear lever does not return

Weak or broken centraliser spring. Renew the spring.

Gearchange shaft bent or seized. Distortion of the gearchange shaft often occurs if the machine is dropped heavily on the gear lever. Provided that damage is not severe straightening of the shaft is permissible.

## 27 Gear selection difficult or impossible

Clutch not disengaging fully. See Section 25.

Gearchange shaft bent. This often occurs if the machine is dropped heavily on the gear lever. Straightening of the shaft is permissible if the damage is not too great.

Gearchange arms, pawls or pins worn or damaged. Wear or breakage of any of these items may cause difficulty in selecting one or more gears. Overhaul the selector mechanism.

Gearchange shaft centraliser spring maladjusted. This is often characterised by difficulties in changing up or down, but rarely in both directions. Adjust the centraliser anchor bolt as described in Chapter 1.

Gearchange arm spring broken. Renew spring.

Gearchange drum stopper cam damaged. Failure, rather than wear, of these items may jam the drum thereby preventing gearchanging. The damaged items must be renewed.

Selector forks bent or seized. This can be caused by dropping the machine heavily on the gearchange lever or as a result of lack of lubrication. Though rare, bending of a shaft can result from a missed gearchange or false selection at high speed.

Selector fork end and pin wear. Pronounced wear of these items and the grooves in the gearchange drum can lead to imprecise selection and, eventually, no selection. Renewal of the worn components will be required.

Structural failure. Failure of any one component of the selector rod and change mechanism will result in improper or fouled gear selection.

## 28 Jumping out of gear

Stopper arm assembly worn or damaged. Wear of the roller and the cam with which it locates and breakage of the detent spring can cause imprecise gear selection resulting in jumping out of gear. Renew the damaged components.

Gear pinion dogs worn or damaged. Rounding off the dog edges and the mating recesses in adjacent pinion can lead to jumping out of gear when under load. The gears should be inspected and renewed. Attempting to reprofile the dogs is not recommended.

Selector forks, gearchange drum and pinion grooves worn. Extreme wear of these interconnected items can occur after high mileages especially when lubrication has been neglected. The worn components must be renewed.

Gear pinions, bushes and shafts worn. Renew the worn components.

Bent gearchange shaft. Often caused by dropping the machine on the gear lever.

Gear pinion tooth broken. Chipped teeth are unlikely to cause jumping out of gear once the gear has been selected fully; a tooth which is completely broken off, however, may cause problems in this respect and in any event will cause transmission noise.

## 29 Overselection

Pawl spring weak or broken. Renew the spring.

Stopper arm spring worn or broken. Renew the spring.

Gearchange arm stop pads worn. Repairs can be made by welding and reprofiling with a file.

Selector limiter claw components (where fitted) worn or damaged. Renew the damaged items.

## Abnormal engine noise.

## 30 Knocking or pinking

See Section 19.

## 31 Piston slap or rattling from cylinder

Cylinder bore/piston clearance excessive. Resulting from wear, partial seizure or improper boring during overhaul. This condition can often be heard as a high, rapid tapping noise when the engine is under little or no load, particularly when power is just beginning to be applied. Reboring to the next correct oversize should be carried out and a new oversize piston fitted.

Connecting rod bent. This can be caused by over-revving, trying to start a very badly flooded engine (resulting in a hydraulic lock in the cylinder) or by earlier mechanical failure such as a dropped valve. Attempts at straightening a bent connecting rod from a high performance engine are not recommended. Careful inspection of the crankshaft should be made before renewing the damaged connecting rod.

Gudgeon pin, piston boss bore or small-end bearing wear or seizure. Excess clearance or partial seizure between normal moving parts of these items can cause continuous or intermittent tapping noises. Rapid wear or seizure is caused by lubrication starvation resulting from an insufficient engine oil level or oilway blockage.

Piston rings worn, broken or sticking. Renew the rings after careful inspection of the piston and bore.

## 32 Valve noise or tapping from the cylinder head

Valve clearance incorrect. Adjust the clearances with the engine cold.

Valve spring broken or weak. Renew the spring set.

Camshaft or crankshaft head worn or damaged. The camshaft lobes are the most highly stressed of all components in the engine and are subject to high wear if lubrication becomes inadequate. The bearing surfaces on the camshaft and cylinder head are also sensitive to a lack of lubrication. Lubrication failure due to blocked oilways can occur, but over-enthusiastic revving before engine warm-up is complete is the usual cause.

Worn camshaft drive components. A rustling noise or light tapping which is not improved by correct re-adjustment of the cam chain tension can be emitted by a worn cam chain or worn sprockets and chain. If uncorrected, subsequent cam chain breakage may cause extensive damage. The worn components must be renewed before wear becomes too far advanced.

## 33 Other noises

Big-end bearing wear. A pronounced knock from within the crankcase which worstens rapidly is indicative of big-end bearing failure as a result of extreme normal wear or lubrication failure. Remedial action in the form of a bottom end overhaul should be taken; continuing to run the engine will lead to further damage including the possibility of connecting rod breakage.

Main bearing failure. Extreme normal wear or failure of the main bearings is characteristically accompanied by a rumble from the crankcase and vibration felt through the frame and footrests. Renew the worn bearings and carry out a very careful examination of the crankshaft.

Crankshaft excessively out of true. A bent crank may result from over-revving or damage from an upper cylinder component or gearbox failure. Damage can also result from dropping the machine on either crankshaft end. Straightening of the crankshaft is not possible in normal circumstances; a replacement item should be fitted.

Engine mounting loose. Tighten all the engine mounting nuts and bolts.

Cylinder head gasket leaking. The noise most often associated with a leaking head gasket is a high pitched squeaking, although any other noise consistent with gas being forced out under pressure from a small orifice can also be emitted. Gasket leakage is often accompanied by oil seepage from around the mating joint or from the cylinder head holding down bolts and nuts. Leakage into the cam chain tunnel or oil return passages will increase crankcase pressure and may cause oil leakage at joints and oil seals. Also, oil contamination will be accelerated. Leakage results from insufficient or uneven tightening of the cylinder head fasteners, or from random mechanical failure. Retightening to the correct torque figure will, at best, only provide a temporary cure. The gasket should be renewed at the earliest opportunity.

Exhaust system leakage. Popping or crackling in the exhaust system, particularly when it occurs with the engine on the overrun, indicates a poor joint either at the cylinder port or at the exhaust pipe/silencer connection. Failure of the gasket or looseness of the clamp should be looked for.

## *Abnormal transmission noise*

## 34 Clutch noise

Clutch outer drum/friction plate tang clearance excessive.
Clutch outer drum/thrust washer clearance excessive.
Primary drive gear teeth worn or damaged.
Clutch shock absorber assembly worn or damaged.

## 35 Transmission noise

Bearing or bushes worn or damaged. Renew the affected components.

Gear pinions worn or chipped. Renew the gear pinions.

Metal chips jams in gear teeth. This can occur when pieces of metal from any failed component are picked up by a meshing pinion. The condition will lead to rapid bearing wear or early gear failure.

Engine/transmission oil level too low. Top up immediately to prevent damage to gearbox and engine.

Gearchange mechanism worn or damaged. Wear or failure of certain items in the selection and change components can induce mis-selection of gears (see Section 27) where incipient engagement of more than one gear set is promoted. Remedial action, by the overhaul of the gearbox, should be taken without delay.

Loose gearbox chain sprocket. Remove the sprocket and check for impact damage to the splines of the sprocket and shaft. Excessive slack between the splines will promote loosening of the securing nut; renewal of the worn components is required. When retightening the nut ensure that it is tightened fully and that, where fitted, the lock washer is bent up against one flat of the nut.

Chain snagging on cases or cycle parts. A badly worn chain or one that is excessively loose may snag or smack against adjacent components.

## *Exhaust smokes excessively*

## 36 White/blue smoke (caused by oil burning)

Piston rings worn or broken. Breakage or wear of any ring, but particularly the oil control ring, will allow engine oil past the piston into the combustion chamber. Overhaul the cylinder barrel and piston.

Cylinder cracked, worn or scored. These conditions may be caused by overheating, lack of lubrication, component failure or advanced normal wear. The cylinder barrel should be renewed or rebored and the next oversize piston fitted.

Valve oil seal damages or worn. This can occur as a result of valve guide failure or old age. The emission of smoke is likely to occur when the throttle is closed rapidly after acceleration, for instance, when changing gear. Renew the valve oil seals and, if necessary, the valve guides.

Valve guides worn. See the preceding paragraph.

Engine oil level too high. This increases the crankcase pressure and allows oil to be forced pass the piston rings. Often accompanied by seepage of oil at joints and oil seals.

Cylinder head gasket blown between cam chain tunnel or oil return passage. Renew the cylinder head gasket.

Abnormal crankcase pressure. This may be caused by blocked breather passages or hoses causing back-pressure at high engine revolutions.

## 37 Black smoke (caused by over-rich mixture)

Air filter element clogged. Clean or renew the element.

Main jet loose or too large. Remove the float chamber to check for tightness of the jet. If the machine is used at high altitudes rejetting will be required to compensate for the lower atmospheric pressure.

Cold start mechanism jammed on. Check that the mechanism works smoothly and correctly and that, where fitted, the operating cable is lubricated and not snagged.

Fuel level too high. The fuel level is controlled by the float height which can increase as a result of wear or damage. Remove the float bowl and check the float height. Check also that floats have not punctured; a punctured float will loose buoyancy and allow an increased fuel level.

Float valve needle stuck open. Caused by dirt or a worn valve. Clean the float chamber or renew the needle and, if necessary, the valve seat.

## Oil pressure indicator lamp goes on

### 38 Engine lubrication system failure

Engine oil defective. Oil pump shaft or locating pin sheared off from ingesting debris or seizing from lack of lubrication (low oil level).

Engine oil screen clogged. Change oil and filter and service pickup screen.

Engine oil level too low. Inspect for leak or other problem causing low oil level and add recommended lubricant.

Engine oil viscosity too low. Very old, thin oil, or an improper weight of oil used in engine. Change to correct lubricant.

Camshaft or journals worn. High wear causing drop in oil pressure. Replace cam and/or head. Abnormal wear could be caused by oil starvation at high rpm from low oil level, improper oil weight or type, or loose oil fitting on upper cylinder oil line.

Crankshaft and/or bearings worn. Same problems as paragraph 5. Overhaul lower end

Relief valve stuck open. This causes the oil to be dumped back into the sump. Repair or replace.

### 39 Electrical system failure

Oil pressure switch defective. Check switch according to the procedures in Chapter ?? . Replace if defective.

Oil pressure indicator lamp wiring system defective. Check for pinched, shorted, disconnected or damaged wiring.

## Poor handling or roadholding

### 40 Directional instability

Steering head bearing adjustment too tight. This will cause rolling or weaving at low speeds. Re-adjust the bearings.

Steering head bearing worn or damaged. Correct adjustment of the bearing will prove impossible to achieve if wear or damage has occurred. Inconsistent handling will occur including rolling or weaving at low speed and poor directional control at indeterminate higher speeds. The steering head bearing should be dismantled for inspection and renewed if required. Lubrication should also be carried out.

Bearing races pitted or dented. Impact damage caused, perhaps, by an accident or riding over a pot-hole can cause indentation of the bearing, usually in one position. This should be noted as notchiness when the handlebars are turned. Renew and lubricate the bearings.

Steering stem bent. This will occur only if the machine is subjected to a high impact such as hitting a curb or a pot-hole. The lower yoke/stem should be renewed; do not attempt to straighten the stem.

Front or rear tyre pressures too low.

Front or rear tyre worn. General instability, high speed wobbles and skipping over white lines indicates that tyre renewal may be required. Tyre induced problems, in some machine/tyre combinations, can occur even when the tyre in question is by no means fully worn.

Swinging arm bearings worn. Difficulties in holding line, particularly when cornering or when changing power settings indicates wear in the swinging arm bearings. The swinging arm should be removed from the machine and the bearings renewed.

Swinging arm flexing. The symptoms given in the preceding paragraph will also occur if the swinging arm fork flexes badly. This can be caused by structural weakness as a result of corrosion, fatigue or impact damage, or because the rear wheel spindle is slack.

Wheel bearings worn. Renew the worn bearings.

Tyres unsuitable for machine. Not all available tyres will suit the characteristics of the frame and suspension, indeed, some tyres or tyre combinations may cause a transformation in the handling characteristics. If handling problems occur immediately after changing to a new tyre type or make, revert to the original tyres to see whether an improvement can be noted. In some instances a change to what are, in fact, suitable tyres may give rise to handling deficiences. In this case a thorough check should be made of all frame and suspension items which affect stability.

### 41 Steering bias to left or right

Rear wheel out of alignment. Caused by uneven adjustment of chain tensioner adjusters allowing the wheel to be askew in the fork ends. A bent rear wheel spindle will also misalign the wheel in the swinging arm.

Wheels out of alignment. This can be caused by impact damage to the frame, swinging arm, wheel spindles or front forks. Although occasionally a result of material failure or corrosion it is usually as a result of a crash.

Front forks twisted in the steering yokes. A light impact, for instance with a pot-hole or low curb, can twist the fork legs in the steering yokes without causing structural damage to the fork legs or the yokes themselves. Re-alignment can be made by loosening the yoke pinch bolts, wheel spindle and mudguard bolts. Re-align the wheel with the handlebars and tighten the bolts working upwards from the wheel spindle. This action should be carried out only when there is no chance that structural damage has occurred.

### 42 Handlebar vibrates or oscillates

Tyres worn or out of balance. Either condition, particularly in the front tyre, will promote shaking of the fork assembly and thus the handlebars. A sudden onset of shaking can result if a balance weight is displaced during use.

Tyres badly positioned on the wheel rims. A moulded line on each wall of a tyre is provided to allow visual verification that the tyre is correctly positioned on the rim. A check can be made by rotating the tyre; any misalignment will be immediately obvious.

Wheels rims warped or damaged. Inspect the wheels for runout as described in Chapter .

Swinging arm bearings worn. Renew the bearings.

Wheel bearings worn. Renew the bearings.

Steering head bearings incorrectly adjusted. Vibration is more likely to result from bearings which are too loose rather than too tight. Re-adjust the bearings.

Loosen fork component fasteners. Loose nuts and bolts holding the fork legs, wheel spindle, mudguards or steering stem can promote shaking at the handlebars. Fasteners on running gear such as the forks and suspension should be check tightened occasionally to prevent dangerous looseness of components occurring.

Engine mounting bolts loose. Tighten all fasteners.

### 43 Poor front fork performance

Air pressure adjustment outside limits. Damping fluid level incorrect. If the fluid level is too low poor suspension control will occur resulting in a general impairment of roadholding and early loss of tyre adhesion when cornering and braking. Too much oil is unlikely to change the fork characteristics unless severe overfilling occurs when the fork action will become stiffer and oil seal failure may occur.

Damping oil viscosity incorrect. The damping action of the fork is directly related to the viscosity of the damping oil. The lighter the oil used, the less will be the damping action imparted. For general use, use the recommended viscosity of oil, changing to a slightly higher or heavier oil only when a change in damping characteristic is required. Overworked oil, or oil contaminated with water which has found its way past the seals, should be renewed to restore the correct damping performance and to prevent bottoming of the forks.

Damping components worn or corroded. Advanced normal wear of the fork internals is unlikely to ocur until a very high mileage has been covered. Continual use of the machine with damaged oil seals which allows the ingress of water, or neglect, will lead to rapid corrosion and wear. Dismantle the forks for inspection and overhaul. See Chapter 4.

Weak fork springs. Progressive fatigue of the fork springs, resulting in a reduced spring free length, will occur after extensive use. This condition will promote excessive fork dive under braking, and in its advanced form will reduce the at-rest extended length of the forks and thus the fork geometry. Renewal of the springs as a pair is the only satisfactory course of action.

Bent stanchions or corroded stanchions. Both conditions will prevent correct telescoping of the fork legs, and in an advanced state can cause sticking of the fork in one position. In a mild form corrosion will cause stiction of the fork thereby increasing the time the suspension takes to react to an uneven road surface. Bent fork stanchions should be attended to immediately because they indicate that impact damage has occurred, and there is a danger that the forks will fail with disastrous consequences.

## 44 Front fork judder when braking (see also Section 52)

Wear between the fork stanchions and the fork legs. Renewal of the affected components is required.

Slack steering head bearings. Re-adjust the bearings.

Warped brake disc. If irregular braking action occurs fork judder can be induced in what are normally serviceable forks. Renew the damaged brake components.

## 45 Poor rear suspension performance

Air pressure adjustment outside limits. Rear suspension unit damper worn out or leaking. The damping performance of most rear suspension units falls off with age. This is a gradual process, and thus may not be immediately obvious. Indications of poor damping include hopping of the rear end when cornering or braking, and a general loss of positive stability. See Chapter 4.

Weak rear springs. If the suspension unit springs fatigue they will promote excessive pitching of the machine and reduce the ground clearance when cornering. Although replacement springs are available separately from the rear suspension damper unit it is probable that if spring fatigue has occurred the damper units will also require renewal.

Swinging arm flexing or bearings worn. See Sections 40 and 41.

Bent suspension unit damper rod. This is likely to occur only if the machine is dropped or if seizure of the piston occurs. If either happens the suspension units should be renewed as a pair.

## Abnormal frame and suspension noise

## 46 Front end noise

Oil level low or too thin. This can cause a 'spurting' sound and is usually accompanied by irregular fork action.

Spring weak or broken. Makes a clicking or scraping sound. Fork oil will have a lot of metal particles in it.

Steering head bearings loose or damaged. Clicks when braking. Check, adjust or replace.

Fork clamps loose. Make sure all fork clamp pinch bolts are tight.

Fork stanchion bent. Good possibility if machine has been dropped. Repair or replace tube.

## 47 Rear suspension noise

Fluid level too low. Leakage of a suspension unit, usually evident by oil on the outer surfaces, can cause a spurting noise. The suspension units should be renewed as a pair.

Defective rear suspension unit with internal damage. Renew the suspension units as a pair.

## Brake problems

## 48 Brakes are spongy or ineffective

Air in brake circuit. This is only likely to happen in service due to neglect in checking the fluid level or because a leak has developed. The problem should be identified and the brake system bled of air.

Pad worn. Check the pad wear against the wear lines provided and renew the pads if necessary.

Contaminated pads. Cleaning pads which have been contaminated with oil, grease or brake fluid is unlikely to prove successful; the pads should be renewed.

Pads glazed. This is usually caused by overheating. The surface of the pads may be roughened using glass-paper or a fine file.

Brake fluid deterioration. A brake which on initial operation is firm but rapidly becomes spongy in use may be failing due to water contamination of the fluid. The fluid should be drained and then the system refilled and bled.

Master cylinder seal failure. Wear or damage of master cylinder internal parts will prevent pressurisation of the brake fluid. Overhaul the master cylinder unit.

Caliper seal failure. This will almost certainly be obvious by loss of fluid, a lowering of fluid in the master cylinder reservoir and contamination of the brake pads and caliper. Overhaul the caliper assembly.

Brake lever or pedal improperly adjusted. Adjust the clearance between the lever end and master cylinder plunger to take up lost motion, as recommended in Routine maintenance.

## 49 Brakes drag

Disc warped. The disc must be renewed.

Caliper piston, caliper or pads corroded. The brake caliper assembly is vulnerable to corrosion due to water and dirt, and unless cleaned at regular intervals and lubricated in the recommended manner, will become sticky in operation.

Piston seal deteriorated. The seal is designed to return the piston in the caliper to the retracted position when the brake is released. Wear or old age can affect this function. The caliper should be overhauled if this occurs.

Brake pad damaged. Pad material separating from the backing plate due to wear or faulty manufacture. Renew the pads. Faulty installation of a pad also will cause dragging.

Wheel spindle bent. The spindle may be straightened if no structural damage has occurred.

Brake lever or pedal not returning. Check that the lever or pedal works smoothly throughout its operating range and does not snag on any adjacent cycle parts. Lubricate the pivot if necessary.

Twisted caliper support bracket. This is likely to occur only after impact in an accident. No attempt should be made to re-align the caliper; the bracket should be renewed.

## 50 Brake lever or pedal pulsates in operation

Disc warped or irregularly worn. The disc must be renewed.

Wheel spindle bent. The spindle may be straightened provided no structural damage has occurred.

## 51 Disc brake noise

Brake squeal. This can be caused by the omission or incorrect installation of the anti-squeal shim fitted to the rear of one pad. The arrow on the shim should face the direction of wheel normal rotation. Squealing can also be caused by dust on the pads, usually in combination with glazed pads, or other contamination from oil, grease, brake fluid or corrosion. Persistent squealing which cannot be traced to any of the normal causes can often be cured by applying a thin layer of high temperature silicone grease to the rear of the pads. Make absolutely certain that no grease is allowed to contaminate the braking surface of the pads.

Glazed pads. This is usually caused by high temperatures or contamination. The pad surfaces may be roughened using glass-paper or a fine file. If this approach does not effect a cure the pads should be renewed.

Disc warped. This can cause a chattering, clicking or intermittent squeal and is usually accompanied by a pulsating brake lever or pedal or uneven braking. The disc must be renewed.

Brake pads fitted incorrectly or undersize. Longitudinal play in the pads due to omission of the locating springs (where fitted) or because pads of the wrong size have been fitted will cause a single tapping noise every time the brake is operated. Inspect the pads for correct installation and security.

## 52 Brake induced fork judder

Worn front fork stanchions and legs, or worn or badly adjusted steering head bearings. These conditions, combined with uneven or pulsating braking as described in Sections 50 and 54 will induce more or less judder when the brakes are applied, dependent on the degree of wear and poor brake operation. Attention should be given to both areas of malfunction. See the relevant Sections.

## *Electrical problems*

## 53 Battery dead or weak

Battery faulty. Battery life should not be expected to exceed 3 to 4 years, particularly where a starter motor is used regularly. Gradual sulphation of the plates and sediment deposits will reduce the battery performance. Plate and insulator damage can often occur as a result of vibration. Complete power failure, or intermittent failure, may be due to a broken battery terminal. Lack of electrolyte will prevent the battery maintaining charge.

Battery leads making poor contact. Remove the battery leads and clean them and the terminals, removing all traces of corrosion and tarnish. Reconnect the leads and apply a coating of petroleum jelly to the terminals.

Load excessive. If additional items such as spot lamps, are fitted, which increase the total electrical load above the maximum alternator output, the battery will fail to maintain full charge. Reduce the electrical load to suit the electrical capacity.

Regulator/rectifier failure.

Alternator generating coils open-circuit or shorted.

Charging circuit shorting or open circuit. This may be caused by frayed or broken wiring, dirty connectors or a faulty ignition switch. The system should be tested in a logical manner. See Section 56.

## 54 Battery overcharged

Rectifier/regulator faulty. Overcharging is indicated if the battery becomes hot or it is noticed that the electrolyte level falls repeatedly between checks. In extreme cases the battery will boil causing corrosive gases and electrolyte to be emitted through the vent pipes.

Battery wrongly matched to the electrical circuit. Ensure that the specified battery is fitted to the machine.

## 55 Total electrical failure

Fuse blown. Check the main fuse. If a fault has occurred, it must be rectified before a new fuse is fitted.

Battery faulty. See Section 53.

Earth failure. Check that the frame main earth strap from the battery is securely affixed to the frame and is making a good contact.

Ignition switch or power circuit failure. Check for current flow through the battery positive lead (red) to the ignition switch. Check the ignition switch for continuity.

## 56 Circuit failure

Cable failure. Refer to the machine's wiring diagram and check the circuit for continuity. Open circuits are a result of loose or corroded connections, either at terminals or in-line connectors, or because of broken wires. Occasionally, the core of a wire will break without there being any apparent damage to the outer plastic cover.

Switch failure. All switches may be checked for continuity in each switch position, after referring to the switch position boxes incorporated in the wiring diagram for the machine. Switch failure may be a result of mechanical breakage, corrosion or water.

Fuse blown. Refer to the wiring diagram to check whether or not a circuit fuse is fitted. Replace the fuse, if blown, only after the fault has been identified and rectified.

## 57 Bulbs blowing repeatedly

Vibration failure. This is often an inherent fault related to the natural vibration characteristics of the engine and frame and is, thus, difficult to resolve. Modifications of the lamp mounting, to change the damping characteristics may help.

Intermittent earth. Repeated failure of one bulb, particularly where the bulb is fed directly from the generator, indicates that a poor earth exists somewhere in the circuit. Check that a good contact is available at each earthing point in the circuit.

Reduced voltage. Where a quartz-halogen bulb is fitted the voltage to the bulb should be maintained or early failure of the bulb will occur. Do not overload the system with additional electrical equipment in excess of the system's power capacity and ensure that all circuit connections are maintained clean and tight.

# SUZUKI GSX/GS 1000 & 1100 4-VALVE FOURS

## Check list – UK

**Every 600 miles (1000 km)**

1 Check the battery electrolyte level
2 Check tighten all engine fasteners
3 Check the valve clearances
4 Check the engine compression
5 Examine and clean the spark plugs
6 Check the ignition timing
7 Check carburettor adjustment and synchronisation
8 Renew the engine oil and filter
9 Check the clutch adjustment
10 Clean, adjust and lubricate the final drive chain
11 Check the brake fluid level and pad wear
12 Check the tyre pressures and inspect for wear and damage
13 Check the steering head bearing adjustment
14 Renew the front fork oil

**Every 2000 miles (3000 km)**

1 Clean the air filter element

**Every 3000 miles (5000 km)**

1 Check the engine oil pressure
2 Check the front fork air pressure
3 Lubricate the throttle, clutch and choke cables
4 Lubricate the brake pedal pivot

**Every 6000 miles (10 000 km)**

1 Renew the spark plugs
2 Clean the sump oil strainer
3 Grease the throttle twistgrip
4 Grease the speedometer and tachometer cables
5 Grease the ATU sliding surfaces

**Every year**

1 Renew the hydraulic brake fluid

**Every 8000 miles (12 000 km)**

1 Renew the air filter element

**Every two years or 12 000 miles (20 000 km)**

1 Renew the fuel and vacuum pipes
2 Renew the brake hoses
3 Grease the steering head bearings
4 Grease the swinging arm bearings

## Check list – US

**Every two months or 600 miles (1000 km)**

1 Check tighten the exhaust system fasteners and all frame fasteners
2 Check the valve clearances
3 Check the engine idle speed
4 Renew the engine oil and filter
5 Check the clutch adjustment
6 Inspect, clean and lubricate the final drive chain
7 Check the brake fluid level and pad wear
8 Check the tyre pressures and inspect for wear and damage
9 Check the steering head bearing adjustment

**Every 2000 miles (3000 km)**

1 Clean the air filter element

**Every six months**

1 Check the front fork air pressure

**Every year or 4000 miles (6000 km)**

1 Check the battery electrolyte level
2 Clean the spark plugs
3 Lubricate the handlebar lever pivots
4 Lubricate the throttle and clutch cables
5 Lubricate the brake pedal pivot
6 Lubricate the stand pivots

**Every two years**

1 Renew the fuel and vacuum pipes
2 Renew the brake hoses

**Every two years or 7500 miles (12 000 km)**

1 Renew the spark plugs
2 Renew the air filter element
3 Renew the front fork oil
4 Grease the throttle twistgrip
5 Grease the clutch release mechanism
6 Grease the speedometer and tachometer cables
7 Grease the ATU sliding surfaces

**Every two years or 15 000 miles (24 000 km)**

1 Grease the steering head bearings
2 Grease the swinging arm bearings

## Adjustment data

| | |
|---|---|
| Valve clearances | 0.07 – 0.12 mm (0.003 – 0.005 in) |
| Engine idle speed | 950 – 1150 rpm |

**Spark plug type**
UK models   NGK DR8EA or ND X24ESR-U
US models   NGK D8EA or ND X24ES-U

Spark plug gap   0.6 – 0.7 mm (0.024 – 0.028 in)

**Front fork air pressure**
T, ET, EX and EZ models   7.11 psi (0.5 kg/cm²)
LT model   11.8 psi (0.8 kg/cm²)

| Tyre pressures | Front | Rear |
|---|---|---|
| Normal riding – solo | 25 psi | 28 psi |
| Normal riding – dual: | | |
|   All models except LT | 28 psi | 36 psi |
|   LT model | 28 psi | 32 psi |
| Continuous high speed – solo: | | |
|   All models except LT | 28 psi | 36 psi |
|   LT model | 28 psi | 32 psi |
| Continuous high speed – dual | 32 psi | 40 psi |

## Recommended lubricants

| | Component | Quantity | Type/Viscosity |
|---|---|---|---|
| 1 | Engine/transmission: | | SAE 10W/40 motor oil SE or SF type |
| | At oil change | 3.2 lit (6.8/5.6 US/ Imp pt) | |
| | At oil and filter change | 3.6 lit (7.6/6.4 US/ Imp pt) | |
| 2 | Front forks – see section 20 of Routine Maintenance in manual | | |
| 3 | Final drive chain | As required | Gear oil or aerosol lubricant suitable for O-ring chains |
| 4 | Brakes | As required | DOT 3, 4 or SAE J1703 hydraulic fluid |
| 5 | Wheel bearings | As required | High melting point grease |
| 6 | Steering head bearings | As required | General purpose grease |
| 7 | Swinging arm bearings | As required | General purpose grease |
| 8 | Instrument drive cables | As required | General purpose grease |
| 9 | Control cables | As required | Engine oil or WD40 |
| 10 | Throttle twistgrip | As required | General purpose grease |
| 11 | Pivot points | As required | Engine oil or WD40 |

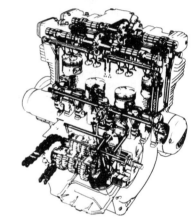

The engine lubrication system

# ROUTINE MAINTENANCE GUIDE

# Routine Maintenance Schedule – UK models

| Operation | Initial 600 mi (1000 km) | Every 3000 mi (5000 km) | Every 6000 mi (10 000 km) | See Section No. |
|---|---|---|---|---|
| **Engine** | | | | |
| Air filter | Clean every 2000 miles (3000 km) and renew every 8000 miles (12 000 km) | | | 1 |
| Battery | Check | Check | – | 2 |
| Engine fasteners | Check | Check | – | 3 |
| Valve clearances | Check | Check | – | 4 |
| Compression | Check | Check | – | 5 |
| Spark plug | Check | Check | Renew | 6 |
| Ignition timing | Check | Check | – | 7 |
| Carburettor | Check | Check | – | 8 |
| Fuel pipes | Renew every two years | | | 8 |
| Engine oil | Change | Change | – | 9 |
| Engine oil filter | Renew | Renew | – | 9 |
| Sump strainer | – | – | Clean | 10 |
| Oil pressure | – | Check | – | 11 |
| Clutch | Check | Check | – | 12 |
| **Chassis** | | | | |
| Drive chain | Clean & adjust every 600 miles (1000 km) | | | 14 |
| Brakes | Check | Check | – | 15 |
| Brake hoses | Renew every two years | | | 16 |
| Brake fluid | Renew every year | | | 17 |
| Tyres | Check | Check | – | 18 |
| Steering | Check | Check | – | 19 |
| Front fork oil | Change | – | Change | 20 |
| Front fork air | Check every 3000 miles (5000 km) or 6 monthly | | | 20 |

## Lubrication

| | Every 3000 miles (5000 km) | Every 6000 miles (10 000 km) | See Section No. |
|---|---|---|---|
| Throttle cable | Motor oil | – | 21 |
| Throttle twistgrip | – | Grease | 21 |
| Clutch cable | Motor oil | – | 21 |
| Choke cable | Motor oil | – | 21 |
| Speedometer cable | – | Grease | 21 |
| Tachometer cable | – | Grease | 21 |
| Drive chain | Motor oil every 600 miles (1000 km) | | 21 |
| Brake pedal pivot | Grease or oil | – | 21 |
| Automatic timing unit | – | Grease | 21 |
| Steering bearings | Grease every 2 years or 12 000 miles (20 000 km) | | 21 |
| Swinging arm bushes | Grease every 2 years or 12 000 miles (20 000 km) | | 21 |

## Routine Maintenance Schedule – US models

| Operation | *Interval: mile | 600 | 4000 | 7500 | 11 000 | 15 000 | See |
|---|---|---|---|---|---|---|---|
| | : km | 1000 | 6000 | 12 000 | 18 000 | 24 000 | Sec |
| | : month | 2 | 12 | 24 | 36 | 48 | No. |
| Battery | | – | I | I | I | I | 2 |
| Exhaust system fasteners | | T | T | T | T | T | 13 |
| Air filter | | Clean every 2000 miles (3000 km). Renew every 7500 miles (12 000 km) | | | | | 1 |
| Valve clearances | | I | I | I | I | I | 4 |
| Spark plugs | | – | C | R | C | R | 6 |
| Idle speed | | I | I | I | I | I | 8 |
| Fuel pipes | | Renew every 2 years | | | | | 8 |
| Engine oil and filter | | R | R | R | R | R | 9 |
| Clutch | | I | I | I | I | I | 12 |
| Drive chain | | I | I | I | I | I | 14 |
| | | Clean & lubricate every 600 miles | | | | | |
| Brakes | | I | I | I | I | I | 15 |
| Brake hoses | | Renew every 2 years | | | | | 16 |
| Tyres | | I | I | I | I | I | 18 |
| Steering | | I | I | I | I | I | 19 |
| Front fork | | – | – | I | – | I | 20 |
| Front fork air pressure | | Check every 6 months | | | | | 20 |
| Chassis fasteners | | T | T | T | T | T | – |

*Based on calendar or mileage readings, whichever comes first.

Note:
    T = Tighten
    I = Inspect
    R = Renew
    C = Clean

### Lubrication (see Section 21)

| | Every 4000 mi (6000 km) | Every 7500 mi (12 000 km) |
|---|---|---|
| Clutch/brake levers | Motor oil | – |
| Throttle cable | Motor oil | – |
| Throttle twistgrip | – | Grease |
| Clutch cable | Motor oil | – |
| Clutch release | – | Grease |
| Speedometer cable | – | Grease |
| Tachometer cable | – | Grease |
| Drive chain | Motor oil every 600 miles (1000 km) | |
| Brake pedal/linkage | Grease or oil | – |
| Stand pivots | Motor oil | – |
| Automatic timing unit | – | Grease |
| Steering head bearings | Grease every 2 years/15 000 mi | |
| Swinging arm bushes | Grease every 2 years/15 000 mi | |

# Routine maintenance

*Refer to Chapter 7 for information relating to the 1984-on 1135cc-engined GSX1100 and GS1150 models*

Regular users of Haynes Owners Workshop Manuals may notice that the Routine Maintenance details do not follow our normal practice where the various operations are given under calendar and mileage headings. This is because the schedules for the UK and US models vary considerably, so to avoid confusion two separate schedules are shown on pages 25 and 26, together with reference to the appropriate section number.

Periodical routine maintenance is essential to keep the motorcycle in a peak and safe condition. Routine maintenance also saves money because it provides the opportunity to detect and remedy a fault before it develops further and causes more damage. Maintenance should be undertaken on either a calendar or mileage basis depending on whichever comes sooner. The period between maintenance tasks serves only as a guide since there are many variables eg: age of machine, riding technique and adverse conditions.

The maintenance instructions are generally those recommended by the manufacturer but are supplemented by additional tasks which, through practical experience, the author recommends should be carried out at intervals suggested. The additional tasks are primarily of a preventative nature, which will assist in eliminating unexpected failure of a component or system, due to wear and tear, and increase safety margins when riding.

All the maintenance tasks are described together with the procedures required for accomplishing them. If necessary, more general information on each topic can be found in the relevant Chapter within the main text.

Although no special tools are required for routine maintenance, a good selection of general workshop tools is essential. Included in the tools must be a range of metric ring or combination spanners, a selection of crosshead screwdrivers, and two pairs of circlip pliers, one

external opening and the other internal opening. Additionally, owing to the extreme tightness of most casing screws on Japanese machines, an impact screwdriver, together with a choice of large or small crosshead screws bits, is absolutely indispensable. This is particularly so if the engine has not been dismantled since leaving the factory.

## 1 Cleaning the air filter

Unlock and open the seat to gain access to the air filter casing. The casing lid is secured by a screw(s). On some models it will be necessary first to remove the seat locating bracket before it can be opened. To release the element, grasp the spring retainer and pull it upwards and out of the casing. The element can now be lifted away.

The pleated paper element can be cleaned by blowing compressed air through from the inside face to dislodge the accumulated dust. On no account blow air through from the outside because this will force the dust deeper into the pores. Do not attempt to clean the element with solvents.

The element must be renewed at the specified intervals. If this is ignored the increased resistance to airflow will upset the carburation, causing poor performance and fuel economy. If the machine is used in unusually dusty conditions, it is advisable to halve the cleaning and renewal intervals. The element should be renewed at once if it becomes contaminated with oil or is holed or torn. Unfiltered air entering the engine will cause rapid and expensive wear to take place.

Refit the cleaned or new element by reversing the removal sequence, ensuring that the spring retainer is located correctly in the casing. For further information, see Chapter 2, Section 10.

Tank and locating bracket may have to be raised to permit removal of air filter lid

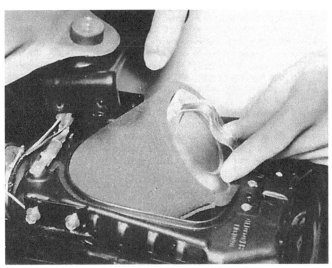

Withdraw spring retainer and pull element out of casing

Air filter must be removed to gain access to battery

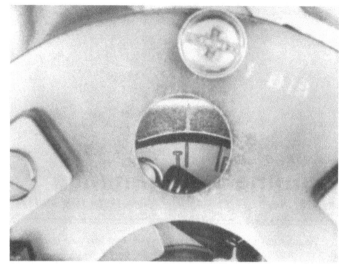

Rotate crankshaft to align the "T" mark ...

## 2 Checking the battery

The battery electrolyte level is monitored by the level sensor, where fitted. To gain access for a physical check, or to remove the battery for charging, remove the side panels, rear master cylinder reservoir and the air cleaner casing. Details of this procedure and a full description of maintenance procedures will be found in Chapter 6, Section 8.

## 3 Checking and tightening the engine fasteners

This operation is particularly important on new machines or where the engine has recently been removed and reconditioned. In addition, it provides a valuable general check, often highlighting problems which might otherwise go unnoticed. Refer to Chapter 1, Section 42 for details of the appropriate tightening sequence and torque settings.

... with camshaft notches in required position (see text)

## 4 Checking the valve clearances

The valve clearances must be checked with the engine cold, preferably after it has been left overnight. Remove the seat and fuel tank to gain access to the cylinder head cover. Remove the cover retaining bolts, noting the positions of those fitted with sealing washers, then lift the cover away. If it is stuck to the gasket, tap around the joint face with a soft-faced mallet to break the seal. Remove the inspection cover at the right-hand end of the crankshaft.

The clearances are checked by measuring the gap between the adjuster and the end of the valve stem. It is important that the rocker is correctly positioned when the measurement is taken, and to simplify the process the following procedure should be followed.

Using the large hexagon on the crankshaft end, turn it until the 'T' mark is aligned with the index mark and the notches on the right-hand ends of the camshafts face outwards. With the crankshaft set in this position, check and adjust the following valves:

Cylinder No. 1 Inlet and exhaust
Cylinder No. 2 Exhaust only
Cylinder No. 3 Inlet only

The clearance for both inlet and exhaust valves is 0.07-0.12 mm (0.003-0.005 in). If any valve is outside this range, slacken the locknut using a ring spanner, then turn the adjuster head to obtain the correct clearance. Hold the adjuster firmly and secure the locknut, then recheck the clearance before moving on to the next valve. If required, a Suzuki adjuster tool, Part Number 09917 - 14910 can be obtained through dealers, though a small open-ended spanner will suffice.

Check the valve clearances using feeler gauges

When the above valves have been dealt with, turn the crankshaft through one complete rotation until the 'T' mark is again aligned, but this time with the camshaft notches facing inwards. The following valves can now be checked and adjusted:

Cylinder No. 2 Inlet only
Cylinder No. 3 Exhaust only
Cylinder No. 4 Inlet and exhaust

After adjustment is complete, clean the joint face of the cylinder head and cover and place a new gasket in position. A thin film of RTV sealant should be applied to both sides of the gasket to ensure an oil-tight seal. The sealing washers fitted to some of the holding bolts should also be renewed. Refit the crankcase inspection cover, then refit the tank and seat.

## 5 Checking the compression

The following check requires the use of the Suzuki compression gauge, Part Number 09915-64510, and an adaptor, Part Number 09915-63210, or an equivalent gauge arrangement. If the equipment is not available, it may be preferable to have the check carried out by a Suzuki dealer.

Before making the test the cylinder head nuts should be tightened to the specified torque (see Chapter 1, Section 39). Run the engine until normal operating temperature is reached, then remove the seat and fuel tank.

Disconnect the plug caps and remove the plugs. Connect the gauge assembly to one of the cylinders via the spark plug thread. Set the engine kill switch to the 'off' position, hold the throttle fully open, then crank the engine for a few seconds, noting the maximum reading for that cylinder. Repeat this process on the remaining cylinders, then compare the readings obtained with the following:

| | |
|---|---|
| Standard compression: | 9-12 kg cm² (128-171 psi) |
| Service limit: | 7 kg cm² (100 psi) |
| Max, difference between cylinders: | 2 kg cm² (28.4 psi) |

If the readings obtained are lower than those specified, one of the following faults is indicated.

a) Broken or leaking cylinder head gasket
b) Sticking or burnt valves
c) Gummed or broken piston rings
d) Badly worn piston rings or bores

In the case of c or d, this can be confirmed by pouring a small quantity of oil into the bore(s) and repeating the test. If the reading improves, the fault is likely to be related to a defect in the rings or bores, whilst if no difference is found the fault can probably be traced to the head gasket or valves.

## 6 Checking the spark plugs

Remove the spark plugs for examination and cleaning, noting that comparing the appearance of the electrode area with that shown in Chapter 3 can give a good indication of engine condition and mixture strength. The plugs are best cleaned using an abrasive blasting process. Many garages offer a cleaning service, or one of the inexpensive home units can be used. Be careful to ensure that all traces of abrasive are removed after cleaning to avoid bore damage. In the absence of blasting facilities, accumulated carbon can be removed by judicious scraping, using a knife or a small screwdriver. Care must be taken to avoid damage to the electrodes or insulator nose.

Examine the plug electrodes for wear. In time, the outer earth (ground) electrode will become eroded. If it is thin or badly stepped at the end, the plug should be renewed. If the centre electrode has worn to the extent that it is nearly flush with the insulator nose, renewal will again be required. Lesser degrees of wear can be corrected by careful dressing, using a fine file or abrasive paper.

Measure the electrode gap using feeler gauges, noting that the correct clearance is 0.6-0.7 mm (0.024-0.028 in). If adjustment is required, bend the outer electrode only. Any attempt to bend the centre electrode will almost invariably crack the porcelain insulator.

The recommended plug grades are shown below. It should not normally be necessary to depart from the standard type, but in some circumstances a change to a hotter or colder grade may be indicated. If there have been persistent problems of fouling, the plugs becoming wet due to very short journeys in cold climates, select a hotter grade. If on the other hand the machine is ridden hard in a hot climate, the plugs may become overheated, taking on a white, blistered appearance. In this case a colder type may be required.

### Recommended spark plug grades

| UK models: | Hot type | Standard type | Cold type |
|---|---|---|---|
| Nippon Denso | X22ESR-U | X24ESR-U | X27ESR-U |
| NGK | DR7EA | DR8EA | DR9EA |

| US models: | Hot type | Standard type | Cold type |
|---|---|---|---|
| Nippon Denso | X22ES-U | X24ES-U | X27ES-U |
| NGK | D7EA | D8EA | D9EA |

## 7 Checking the ignition timing

The ignition timing should be checked using a stroboscopic timing lamp as described in Chapter 3, Section 10. In addition, check that as engine speed rises the ignition advances to the appropriate mark on the timing unit. If the ignition advance is hesitant or erratic, remove the automatic timing unit and lubricate the pivots (see Section 21).

## 8 Checking and adjusting the carburettors

### Throttle cable free play

Check the amount of free play in the throttle cable and adjust it if necessary, using the adjuster at the lower end. The prescribed free play is 0.5-1.0 mm (0.02-0.04 in).

### Fuel level check

Check the fuel level in each carburettor using the Suzuki gauge, Part Number 09913-14540, or a home-made equivalent. The check is described in Chapter 2, Section 8.

### Carburettor synchronisation

Accurate carburettor synchronisation is important if the engine is to run smoothly and deliver full performance and fuel economy. In addition, poor synchronisation will often produce serious-sounding noises in the primary drive due to uneven idling. The synchronisation operation requires the use of vacuum gauge equipment and is described in Chapter 2, Section 9. In the absence of the above equipment, take the machine to a Suzuki dealer to have the synchronisation checked.

### Idle speed adjustment

The idle speed should be checked after synchronisation and with the engine at normal operating temperature. Using the large knurled knob at the centre underside of the carburettor bank, set the idle speed to 950-1150rpm. Note that on no account should the small pilot screws, located next to the intake hoses, be disturbed. These are pre-set during manufacture and no setting data is available if adjustment is lost.

### Fuel hose

Check the condition of the fuel and vacuum hoses, renewing them if there is any indication of splitting or leakage. Use only synthetic rubber hose as a replacement. The hoses should be renewed every two years as a precautionary measure.

## 9 Changing the engine oil and filter

It is vital that the engine oil and filter are changed at the recommended intervals. Apart from the risk of damage due to the degraded oil being unable to cope with the loads imposed on it, it should be remembered that if the filter becomes badly obstructed and the bypass valve opens, any contaminants in the oil will be carried around the engine, causing severe wear or damage if left for long.

Place the machine on its centre stand and run the engine until it reaches normal operating temperature. It is preferable to carry out the oil change immediately after a run, for obvious reasons. Place a drain tray or bowl of at least 1 gallon (5 litres) capacity beneath the drain plug, which can now be removed.

While the oil is draining, slacken and remove the five nuts which retain the filter cover at the front of the crankcase. Lift away the cover and remove the old filter. Clean out any residual oil from the housing, then fit the new filter element. Clean the filter cover and renew the O-ring if it is damaged. Offer up the cover and fit the retaining nuts, using a thread locking compound on their threads. Tighten them evenly and progressively to 0.6-0.8 kgf m (4.5-6.0 lbf ft).

Clean and refit the drain plug, tightening it securely. Remove the filler cap at the top of the clutch housing and add an API classification SD or SE type motor oil of SAE 10W/40 viscosity. Oil should be added until it reaches the upper limit line in the sight glass. This will require about 3.2 litres (6.8/5.6 US/Imp pint). Start the engine and allow it to idle for a few minutes while checking for leaks, then stop the engine and allow the oil to settle. The level will probably have fallen now that the filter chamber has filled, which will necessitate topping up to the correct level.

Top up with specified oil and recheck after running engine

## 10 Cleaning the sump strainer

In addition to the normal oil change, the oil pickup strainer in the sump should be removed and cleaned at the specified interval. Drain the engine oil, then remove the exhaust system to gain access to the sump (oil pan). For more information refer to Chapter 1, Section 4. Slacken evenly and progressively the sump retaining bolts, noting the position of the wiring clips, then lift the sump away.

Clean out the inside of the sump and place it to one side to await reassembly. Remove the three bolts which retain the sump strainer to the underside of the crankcase and lift it away. Clean the strainer in a degreasing solvent and check that it is undamaged. When refitting the strainer, note that the oil inlet tubes must face forward.

## 11 Checking the oil pressure

The engine oil pressure should be checked with the engine at normal operating temperature, using a Suzuki oil pressure gauge, Part Number 09915-74510, or equivalent. Remove the end plug from the right-hand end of the main oil gallery and attach the gauge. Run the engine at 3000 rpm and note the indicated pressure. The minimum pressure is 0.1 kg cm$^2$ (1.42 psi), maximum pressure being 0.5 kg cm$^2$ (7.11 psi). If pressure is below the minimum figure, the oil pump or engine components may be badly worn.

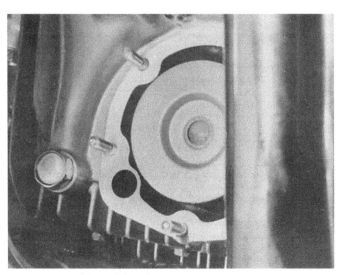

Place new element in filter housing ...

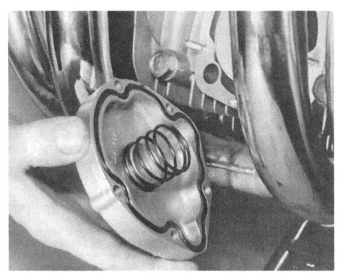

... then refit cover. Do not omit the spring

Note direction of inlet pipes on sump strainer

## 12 Adjusting the clutch

Slacken the adjuster locknut at the clutch lever and screw the adjuster fully inwards. Moving to the lower adjuster, slacken the locknut and set the adjuster to give 2-3 mm (0.08-0.12 in) measured between the lever stock and blade. Tighten the locknut and slide the rubber boot back into place. Subsequent fine adjustment can be carried out at the lever adjuster.

Set clutch cable free play to 2-3 mm at lever

## 13 Checking the exhaust system fasteners

Check the security of the nuts and bolts which retain the exhaust system to the cylinder head and the frame. If tightening is required, start with the exhaust port fasteners, then the silencer mountings.

## 14 Checking, adjusting and lubricating the drive chain

The final drive chain should be examined, cleaned and adjusted as described in Chapter 5, Section 17.

Fluid level must be kept above 'lower' mark

## 15 Checking the braking system

It is important that the hydraulic fluid level is checked regularly. Although the change in level due to pad wear will be very gradual, any sudden level change can be indicative of a leak or seal failure in the system. The front reservoir level can be checked via the sight window in its side. The rear reservoir is comparatively hidden behind the right-hand side panel, and so incorporates a level sensor to warn if the fluid level has dropped to the minimum mark. The side panel should be removed periodically and the level checked physically. If topping up is necessary, use only fresh hydraulic fluid conforming to DOT 3, 4 or SAE J1703 specification.

Check the condition of the brake pads, noting that they should be renewed as pairs if worn to the limit line. For further information, see Chapter 5, Section 8.

## 16 Checking the brake hoses

Inspect the brake hoses, looking for signs of cracking or splitting of the outer casing. If deterioration or leakage is suspected, the hose must be renewed without delay. It is recommended that the hoses are renewed as a matter of course every two years, in conjunction with the annual fluid change. For further details, refer to Chapter 5, Section 11.

## 17 Changing the hydraulic fluid

The hydraulic fluid in the front and rear braking systems should be changed as a precautionary measure on an annual basis. This is necessary because the fluid is hygroscopic, absorbing moisture from the surrounding air. As this occurs, the boiling point of the fluid is gradually lowered, and if left for long enough this may fall to the point where heat generated under hard braking may cause the formation of bubbles in the fluid. In addition, regular fluid changes will flush the system which in turn will prolong the life of the brake components.

For details of the fluid changing operation, refer to Chapter 5, Section 16.

## 18 Checking the tyre pressures

The tyre pressures should always be checked when the tyres are cold, preferably after the machine has been standing overnight. This ensures that the resulting reading is not affected by the significant pressure rise which takes place when the machine is ridden. It is recommended that a pressure gauge is kept with the machine's toolkit and used in preference to the sometimes unreliable gauges found at filling stations.

Remove cap and diaphragm to top up rear reservoir

Check the tyre treads and sidewalls for cuts or other damage. Remove any small stones which may have become trapped between the tread blocks. If left in place these can work into the tyre and may eventually cause a puncture.

| Tyre pressures – cold | Front | | Rear | |
|---|---|---|---|---|
| | psi | kg cm² | psi | kg cm² |
| Normal riding – solo | 25 | 1.75 | 28 | 2.00 |
| Normal riding – dual: | | | | |
| All models except LT | 28 | 2.00 | 36 | 2.50 |
| LT | 28 | 2.00 | 32 | 2.25 |
| Continuous high speed solo: | | | | |
| All models except LT | 28 | 2.00 | 36 | 2.50 |
| LT | 28 | 2.00 | 32 | 2.25 |
| Continuous high speed-dual | 32 | 2.25 | 40 | 2.80 |

### 19  Checking and adjusting the steering head bearings

The machines covered by this manual are equipped with tapered roller steering head bearings. These can be expected to last for a very long time if adjustment is maintained correctly. The procedure is described in Chapter 4, Section 7.

### 20  Changing the front fork oil

The front fork oil will gradually deteriorate in use, allowing the damping performance to fall off. To prevent this, the damping oil should be changed at the recommended interval. Start by removing the fork air valve cap(s) and depress the valve insert(s) to release the fork air pressure. Place a jack or wooden blocks beneath the crankcase to raise the front wheel clear of the ground.

On most models the fork top bolts are obstructed by the handlebars, and it will be necessary to release the clamp bolts and displace the handlebar assembly slightly to gain access. Slacken and remove the top bolts. Where a floating piston is fitted above the fork springs, compress the forks slightly to displace them. Remove the fork springs and place them to one side.

Unscrew the fork drain plugs at the bottom of the lower legs and allow the oil to drain. This can be assisted by 'pumping' the forks up and down to expel the oil. Once all the old oil is drained, clean and refit the drain plugs. Top up the fork legs with the prescribed grade and quantity of oil as shown below. Check that the oil comes to the prescribed level below the top of the stanchions, noting that the measurement is made with the forks compressed.

When the forks have been filled to the correct level, refit the fork springs and the floating piston (where fitted) and secure the top bolts. On models featuring air assistance, the fork air pressure should be set to the figure shown below. When adding air to the forks note that an air line must not be used. Use only a manual pump designed for use with forks. Where two separate air valves are fitted, ensure that each leg is set to the same pressure. For further information on fork adjustment and maintenance, refer to Chapter 5.

### Fork oil grades

| | |
|---|---|
| T, ET, EX | 50% SAE 10W/30, 50% ATF (automatic transmission fluid) |
| LT | SAE 10W/20 |
| EZ, SZ | SAE 15W fork oil |
| SD, ESD | SE or SF type SAE 10W/40 |

### Fork oil quantity per leg

**UK 1100 models**

| | |
|---|---|
| T | 302 ml (10.63/10.21 Imp/US fl oz) |
| ET | 252 ml (8.87/8.52 Imp/US fl oz) |
| LT | 245 ml (8.63/8.28 Imp/US fl oz) |
| EX | 232 ml (7.84/8.17 Imp/US fl oz) |
| EZ | 246 ml (8.66/8.32 Imp/US fl oz) |
| SZ, SD, ESD | 227 ml (7.99/7.67 Imp/US fl oz) |

**US 1100 models**

| | |
|---|---|
| T | 290 ml (9.80 US fl oz) |
| ET, EX | 238 ml (8.04 US fl oz) |
| LT | 245 ml (8.28 US fl oz) |
| EZ | 246 ml (8.31 US fl oz) |

**UK 1000 models**

| | |
|---|---|
| SD, SZ | 227 ml (7.99/7.67 Imp/US fl oz) |

**US 1000 models**

| | |
|---|---|
| SZ | 227 ml (7.99/7.67 Imp/US fl oz) |

### Fork oil level

**UK 1100 models**

| | |
|---|---|
| T | 152 mm (6.0 in) |
| ET | 193 mm (7.6 in) |
| LT | 260 mm (10.2 in) |
| EX | 226 mm (8.9 in) |
| EZ | 195 mm (7.7 in) |
| SZ, SD, ESD | 221 mm (8.7 in) |

Release handlebars, top bolt and piston (where fitted)

Top up to specified level with correct grade of oil

**US 1100 models**

|   |   |
|---|---|
| T, ET, EX | 216 mm (8.5 in) |
| LT | 260 mm (10.2 in) |
| EZ | 195 mm (7.7 in) |

**UK 1000 models**

|   |   |
|---|---|
| SZ, SD | 221 mm (8.7 in) |

**US 1000 models**

|   |   |
|---|---|
| SZ | 221 mm (8.7 in) |

## Fork air pressure (where applicable)

**UK and US 1100 models**

|   |   |
|---|---|
| T, ET, EX, EZ | 0.5 kg cm² (7.11 psi) |
| LT | 0.8 kg cm² (11.38 psi) |

## 21 General lubrication

Regular cleaning of the controls, cables and stand pivots will prolong the life of these components and will help to draw attention to loose or damaged fasteners. It is recommended that general lubrication should be carried out immediately after the machine has been cleaned. This will make the task more pleasant and will help prevent corrosion where degreasing solvents have been used.

### Control levers, pedals and gearchange linkage

Lubricate the moving parts using engine oil applied with an oil can or one of the popular aerosol maintenance sprays such as WD 40. Check that pivot and retaining bolts are secure and that the control operates smoothly.

### Throttle and clutch cables

Disconnect the cable at its upper end and lubricate it by introducing engine oil and allowing it to work through the cable overnight. This can be accomplished using the improvised funnel arrangement shown in the accompanying line drawing. A faster alternative is to lubricate the cable using one of the proprietary cable lubricators available from most motor cycle dealers.

While the cables are disconnected check the ends of the inner cable for signs of fraying, renewing the cable complete if damage is noted. Before the throttle cable is refitted, grease the throttle twistgrip as described below. Remember to adjust the cables after refitting.

### Throttle twistgrip

The throttle twistgrip housing halves can be separated after the retaining screws have been removed, allowing access for throttle cable maintenance and twistgrip lubrication. Remove all traces of old grease, then apply a film of grease to the sliding surfaces and to the cable groove.

### Clutch release mechanism

Apply a smear of grease to the external link of the clutch release arm to ensure smooth operation. The internal part of the mechanism is lubricated by the engine oil and should require no attention between engine overhauls.

### Speedometer and tachometer drive cables

Disconnect the upper end of the speedometer cable from the underside of the instrument panel and withdraw the inner cable. Wipe off any old grease, then apply a film of fresh grease along its length. Slide the inner cable back into place, noting that it is advisable to wipe off the grease from the uppermost six inches or so to prevent the instrument head from becoming contaminated. On those machines fitted with a mechanical tachometer, the cable can be lubricated in the same way.

### Final drive chain

The procedure for lubricating the final drive chain is given in Chapter 5, Section 17. It should be noted that a sealed O-ring chain is fitted as standard. These can be expected to provide good service if properly maintained, but it is emphasised that many of the proprietary aerosol chain lubricants contain solvents which will destroy the O-

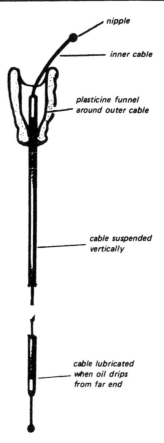

*nipple*

*inner cable*

*plasticine funnel around outer cable*

*cable suspended vertically*

*cable lubricated when oil drips from far end*

**Oiling a control cable**

rings and thus the chain. It follows that these must be avoided, but there are now one or two lubricants designed specifically for use on O-ring chains, notably PJ1 Blue Label. If the can is not clearly marked as being suitable for use on O-ring chains, stick to a heavy motor oil such as gear oil.

### Stand pivots

Clean and lubricate the pivots using motor oil or an aerosol lubricant. Check that the pivot bolts and springs are secure and in good condition.

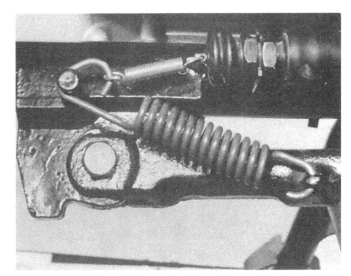

Lubricate side stand pivot and check condition of bolt and spring

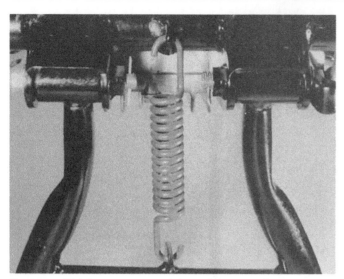

Remove and lubricate centre stand pivots. Do not omit R-pins in ends of pivot bolts

Remove ATU and grease moving parts

## Automatic timing unit (ATU)

Remove the pickup inspection cover to expose the timing unit. Mark the position of the pickup backplate, then remove it. Hold the crankshaft with the larger of the two hexagons and remove the central retaining bolt. The ATU can now be lifted away. Using a small screwdriver or a piece of wire apply a trace of grease to the sliding surfaces. Check that the unit operates smoothly. If excessive play is noted, renew the ATU to preserve the accuracy of the ignition timing.

When fitting the unit ensure that the crankshaft turning hexagon is fitted correctly and tighten the retaining bolt to 1.3-2.3 kgf m (9.5-16.5 lbf ft).

## Steering head and swinging arm

This operation requires the dismantling of the respective assemblies so that a full examination can be made prior to lubrication. Refer to Chapter 4 for details.

# Chapter 1 Engine, clutch and gearbox

*Refer to Chapter 7 for information relating to the 1984-on 1135cc-engined GSX1100 and GS1150 models*

## Contents

## Specifications

### GSX 1100 T, ET, LT, EX, EZ, SZ, SD and ESD

#### Engine

| | |
|---|---|
| Type ................................................................................ | Four cylinder, dohc 16-valve, air-cooled four-stroke |
| Bore ................................................................................ | 72.0 mm (2.835 in) |
| Stroke .............................................................................. | 66.0 mm (2.598 in) |
| Capacity .......................................................................... | 1075 cc (65.6 cu in) |
| Compression ratio: | |
|     T, ET ......................................................................... | 9.2 : 1 |
|     LT, EX, EZ, SZ, SD, ESD ......................................... | 9.5 : 1 |
| Max power ...................................................................... | N/Av |
| Max torque ..................................................................... | N/Av |

#### Cylinder head

| | |
|---|---|
| Max warpage .................................................................. | 0.2 mm (0.008 in) |

## Camshafts and rockers

| | |
|---|---|
| Cam height (inlet): | |
|    T, ET, LT, EX, SZ, SD, ESD ........................................... | 34.650 – 34.690 mm (1.3642 – 1.3657 in) |
|    Service limit ......................................................... | 34.350 mm (1.3524 in) |
| Cam height (exhaust): | |
|    T, ET, LT, EX ....................................................... | 34.650 – 43.690 mm (1.3642 – 1.3657 in) |
|    Service limit ......................................................... | 34.350 mm (1.3524 in) |
|    EZ, SZ, SD, ESD ................................................... | 34.360 – 34.400 mm (1.3528 – 1.3543 in) |
|    Service limit ......................................................... | 34.060 mm (1.3409 in) |
| Camshaft/journal oil clearance: | |
|    T, ET, LT, EX ....................................................... | 0.020 – 0.054 mm (0.0008 – 0.0021 in) |
|    Service limit ......................................................... | 0.150 mm (0.0059 in) |
|    EZ, SZ, SD, ESD ................................................... | 0.032 – 0.066 mm (0.0013 – 0.0026 in) |
|    Service limit ......................................................... | 0.150 mm (0.0059 in) |
| Camshaft journal ID: | |
|    T, ET, LT, EX ....................................................... | 22.000 – 22.013 mm (0.8661 – 0.8667 in) |
|    Service limit ......................................................... | not available |
|    EZ, SZ, SD, ESD ................................................... | 22.012 – 22.025 mm (0.8666 – 0.8671 in) |
|    Service limit ......................................................... | not available |
| Camshaft journal OD ..................................................... | 21.959 – 21.980 mm (0.8645 – 0.8654 in) |
| Service limit .............................................................. | not available |
| Camshaft runout – service limit ...................................... | 0.10 mm (0.004 in) |
| Cam chain length – service limit (20 links) ....................... | 157.80 mm (6.213 in) |
| Rocker arm ID ............................................................ | 12.000 – 12.018 mm (0.4724 – 0.4731 in) |
| Service limit .............................................................. | not available |
| Rocker shaft OD .......................................................... | 11.973 – 11.984 mm (0.4714 – 0.4718 in) |
| Service limit .............................................................. | not available |

## Valves and guides

| | |
|---|---|
| Valve head diameter: | |
|    Inlet ................................................................ | 27.0 mm (1.06 in) |
|    Exhaust ............................................................ | 23.0 mm (0.91 in) |
| Valve lift, inlet – all models .......................................... | 7.0 mm (0.28 in) |
| Valve lift, exhaust: | |
|    T, ET, LT, EX ....................................................... | 7.0 mm (0.28 in) |
|    EZ, SZ, SD, ESD ................................................... | 6.5 mm (0.26 in) |
| Valve clearance (cold) – inlet and exhaust ....................... | 0.07 – 0.12 mm (0.003 – 0.005 in) |
| Guide to stem clearance: | |
|    Inlet ................................................................ | 0.025 – 0.052 mm (0.0010 – 0.0020 in) |
|    Service limit ....................................................... | 0.35 mm (0.014 in) |
|    Exhaust ............................................................ | 0.040 – 0.067 mm (0.0016 – 0.0026 in) |
|    Service limit ....................................................... | 0.35 mm (0.014 in) |
| Valve guide ID ........................................................... | 5.500 – 5.512 mm (0.2165 – 0.2170 in) |
| Valve stem OD: | |
|    Inlet ................................................................ | 5.460 – 5.475 mm (0.2165 – 0.2170 in) |
|    Exhaust ............................................................ | 5.445 – 5.460 mm (0.2144 – 0.2150 in) |
| Valve stem runout (In/Ex) – service limit ......................... | 0.05 mm (0.002 in) |
| Valve head thickness (In/Ex) – service limit ..................... | 0.5 mm (0.02 in) |
| Valve stem end length (In/Ex) – service limit ................... | 3.6 mm (0.14 in) |
| Valve seat width (In/Ex) .............................................. | 0.9 – 1.1 mm (0.035 – 0.043 in) |
| Valve head radial runout .............................................. | 0.03 mm (0.001 in) |
| Valve spring free length: | |
|    Inner (service limit) ............................................. | 31.9 mm (1.26 in) |
|    Outer (service limit) ............................................. | 35.6 mm (1.40 in) |
| Valve spring pressure: | |
|    Inner ............................................................... | 4.4 – 6.4 kg (9.7 – 14.1 lb) @ 28.5 mm (1.12 in) |
|    Outer ............................................................... | 6.5 – 8.9 kg (14.3 – 19.6 lb) @ 32.0 mm (1.26 in) |

## Cylinder bores

| | |
|---|---|
| Compression pressure .................................................. | 9 – 12 kg cm² (128 – 171 psi) |
| Service limit .............................................................. | 7 kg cm² (100 psi) |
| Pressure difference between cylinders – service limit ......... | 2 kg cm² (28.4 psi) |
| Bore size (Std) .......................................................... | 72.000 – 72.015 mm (2.8346 – 2.8352 in) |
| Service limit .............................................................. | 72.080 mm (2.8378 in) |
| Bore ovality – service limit ........................................... | 0.2 mm (0.008 in) |

## Pistons

| | |
|---|---|
| Piston/cylinder clearance ............................................. | 0.050 – 0.060 mm (0.0020 – 0.0024 in) |
| Service limit .............................................................. | 0.120 mm (0.0047 in) |
| Piston diameter (Std) .................................................. | 71.945 – 71.960 mm (2.8325 – 2.8331 in) |
| Service limit .............................................................. | 71.880 mm (2.8299 in) |
| Ring end gap (free:) | |
|    Top ................................................................. | About 9.5 mm (0.37 in) |
|    Service limit ....................................................... | 7.6 mm (0.30 in) |
|    2nd ................................................................. | About 11.0 mm (0.43 in) |
|    Service limit ....................................................... | 8.8 mm (0.35 in) |

Ring end gap (installed):
    Top and 2nd .................................................................. 0.10 – 0.30 mm (0.004 – 0.012 in)
    Service limit ................................................................... 0.7 mm (0.03 in)
Ring to groove clearance:
    Top (service limit) ......................................................... 0.180 mm (0.0071 in)
    2nd (service limit) ......................................................... 0.150 mm (0.0059 in)
Ring groove width:
    Top (except SD and ESD) ............................................... 1.03 – 1.05 mm (0.040 – 0.041 in)
    Top (SD, ESD) ............................................................... 0.10 – 0.30 mm (0.004 – 0.012 in)
    2nd ................................................................................ 1.21 – 1.23 mm (0.047 – 0.048 in)
    Oil:
        T, ET ....................................................................... 1.170 – 1.190 mm (0.0461 – 0.0469 in)
        LT, EX, EZ, SZ, SD, ESD ........................................ 2.510 – 2.530 mm (0.099 – 0.100 in)
Gudgeon pin bore ID ........................................................... 18.001 – 18.006 mm (0.7087 – 0.7089 in)
Service limit ........................................................................ 18.030 mm (0.7098 in)
Gudgeon pin OD .................................................................. 17.996 – 18.000 mm (0.7085 – 0.7086 in)
Service limit ........................................................................ 17.980 mm (0.7079 in)

## Crankshaft and connecting rods

Crankshaft runout – service limit ....................................... 0.1 mm (0.004 in)
Small end ID ........................................................................ 18.006 – 18.014 mm (0.7089 – 0.7092 in)
Service limit ........................................................................ 18.040 mm (0.7102 in)
Connecting rod deflection – service limit .......................... 3.0 mm (0.12 in)
Big-end side clearance ....................................................... 0.10 – 0.65 mm (0.004 – 0.026 in)
Service limit ........................................................................ 1.00 mm (0.040 in)

## Clutch

Type .................................................................................... Wet, multiplate
Friction plate thickness:
    All models except SD, ESD ........................................... 2.9 – 3.1 mm (0.11 – 0.12 in)
    Service limit ................................................................... 2.6 mm (0.10 in)
    SD, ESD ......................................................................... 2.15 – 2.35 mm (0.085 – 0.093 in)
Friction plate tang width .................................................... 15.6 – 15.8 mm (0.61 – 0.62 in)
Service limit ........................................................................ 14.8 mm (0.58 in)
Plain plate thickness .......................................................... 2.00 ± 0.06 mm (0.008 ± 0.002 in)
Service limit ........................................................................ not available
Plain plate warpage – service limit .................................... 0.1 mm (0.004 in)
Clutch spring free length – service limit ............................ 38.5 mm (1.52 in)
Clutch/primary drive pinion:
    Backlash ......................................................................... 0 – 0.02 mm (0.0008 in)
    Service limit ................................................................... 0.08 mm (0.003 in)
Clutch cable free play ........................................................ 2 – 3 mm (0.08 – 0.12 in)

## Transmission

Primary reduction ............................................................... 1.775 : 1 (87/49 T)
Final reduction:
    T, ET, EX, EZ, SZ, SD, ESD .......................................... 2.800 : 1 (45/15 T)
    LT ................................................................................... 2.666 : 1 (40/15 T)
Final drive chain:
    Type ............................................................................... Daido D.I.D. 630YL or Takasago RK630GSV
    Length ............................................................................ 96 links
    20 links length (max) ..................................................... 383.0 mm (15.08 in)
    Free play ........................................................................ 20 – 30 mm (0.8 – 1.2 in)
Gearbox type ...................................................................... 5-speed constant mesh
Gearbox ratios:
    1st .................................................................................. 2.500 : 1 (35/14 T)
    2nd ................................................................................ 1.777 : 1 (32/18 T)
    3rd ................................................................................. 1.380 : 1 (29/21 T)
    4th ................................................................................. 1.125 : 1 (27/24 T)
    Top ................................................................................ 0.961 : 1 (25/26 T)
Gear backlash:
    1st, 2nd, 3rd .................................................................. 0.03 mm (0.001 in)
    Service limit ................................................................... 0.08 mm (0.003 in)
    4th, Top ......................................................................... 0.10 mm (0.004 in)
    Service limit ................................................................... 0.15 mm (0.006 in)
Output shaft length – 1st to 2nd ...................................... 111.4 – 111.5 mm (4.386 – 4.390 in)
Selector fork to groove clearance ..................................... 0.40 – 0.60 mm (0.016 – 0.024 in)
Service limit ........................................................................ 0.8 mm (0.031 in)
Fork groove width .............................................................. 5.45 – 5.55 mm (0.215 – 0.219 in)
Fork thickness .................................................................... 4.95 – 5.05 mm (0.195 – 0.199 in)

## Torque wrench settings

| Component | kgf m | lbf ft |
|---|---|---|
| Cylinder head cover bolt | 0.9 – 1.0 | 6.5 – 7.0 |
| Cylinder head bolt | 0.7 – 1.1 | 5.0 – 8.0 |
| Cylinder head nut | 3.5 – 4.0 | 25.5 – 29.0 |
| Rocker shaft stopper bolt | 0.8 – 1.0 | 6.0 – 7.0 |
| Valve clearance locknut | 0.9 – 1.1 | 6.5 – 8.0 |
| Camshaft cap bolt | 0.8 – 1.2 | 6.0 – 8.5 |
| Camshaft sprocket bolt: | | |
| T, ET, LT, EZ, SZ | 0.9 – 1.2 | 6.5 – 8.5 |
| EX, SD, ESD | 2.4 – 2.6 | 17.5 – 19.0 |
| Cam chain tensioner holding bolt | 0.6 – 0.8 | 4.5 – 6.0 |
| Cam chain tensioner shaft | 3.1 – 3.5 | 22.0 – 25.5 |
| Cam chain tensioner locknut | 0.9 – 1.4 | 6.5 – 10.0 |
| Cam chain guide bolt | 0.9 – 1.4 | 6.5 – 10.0 |
| Alternator rotor nut: | | |
| T, ET, LT, EX, SZ | 13.0 – 14.0 | 94.0 – 101.5 |
| EZ, SD, ESD | 16.0 – 17.0 | 115.7 – 123.0 |
| Starter clutch Allen bolt: | | |
| All models except SD, ESD | 1.5 – 2.0 | 11.0 – 14.5 |
| SD, ESD | 2.3 – 2.8 | 16.5 – 20.0 |
| ATU holding bolt | 1.3 – 2.3 | 9.5 – 16.5 |
| Crankcase bolt (6 mm) | 0.9 – 1.3 | 6.5 – 9.5 |
| Crankcase bolt (8 mm) | 2.0 – 2.4 | 14.5 – 17.0 |
| Starter motor mounting bolt | 0.4 – 0.7 | 3.0 – 5.0 |
| Sump bolt | 1.0 | 7.0 |
| Oil filter cover nut | 0.6 – 0.8 | 4.5 – 6.0 |
| Neutral stopper | 1.8 – 2.8 | 13.0 – 20.0 |
| Gearchange stopper | 1.5 – 2.3 | 11.0 – 16.5 |
| Clutch centre nut | 5.0 – 7.0 | 36.0 – 50.5 |
| Clutch spring bolt | 1.1 – 1.3 | 8.0 – 9.5 |
| Gearbox sprocket nut: | | |
| T, ET, LT, EX, SD, ESD | 10.0 – 15.0 | 72.0 – 108.5 |
| EZ, SZ | 9.0 – 10.0 | 65.0 – 72.0 |
| Engine mounting bolts: | | |
| (A) 10 mm | 4.5 – 5.5 | 32.5 – 40.0 |
| (B) 10 mm | 3.0 – 3.7 | 21.5 – 27.0 |
| (C) 8 mm | 2.0 – 3.0 | 14.5 – 21.5 |
| Gearchange pedal pinch bolt | 1.3 – 2.3 | 9.5 – 16.5 |
| Clutch arm bolt | 0.6 – 1.0 | 4.5 – 7.0 |

# GSX 1000 SZ, SD

## Engine

| | |
|---|---|
| Type | Four cylinder, dohc 16-valve, air-cooled four-stroke |
| Bore | 69.4 mm (2.732 in) |
| Stroke | 66.0 mm (2.598 in) |
| Capacity | 998 cc (60.9 cu in) |
| Compression ratio | 9.5 : 1 |
| Max power | N/Av |
| Max torque | N/Av |

## Cylinder head

| | |
|---|---|
| Max warpage | 0.2 mm (0.008 in) |

## Camshafts and rockers

| | |
|---|---|
| Cam height (inlet) | 34.650 – 34.690 mm (1.3642 – 1.3657 in) |
| Service limit | 34.350 mm (1.3524 in) |
| Cam height (exhaust) | 34.360 – 34.400 mm (1.3528 – 1.3543 in) |
| Service limit | 34.060 mm (1.3409 in) |
| Camshaft/journal oil: | |
| Clearance | 0.032 – 0.066 mm (0.0013 – 0.0026 in) |
| Service limit | 0.150 mm (0.0059 in) |
| Camshaft journal ID | 22.012 – 22.025 mm (0.8666 – 0.8671 in) |
| Service limit | not available |
| Camshaft journal OD | 21.959 – 21.980 mm (0.8645 – 0.8654 in) |
| Service limit | not available |
| Camshaft runout – service limit | 0.10 mm (0.004 in) |
| Cam chain length – service limit (20 links) | 157.80 mm (6.213 in) |
| Rocker arm ID | 12.000 – 12.018 mm (0.4724 – 0.4731 in) |
| Service limit | not available |
| Rocker shaft OD | 11.973 – 11.984 mm (0.4714 – 0.4718 in) |
| Service limit | not available |

## Valves and guides

| | |
|---|---|
| Valve head diameter: | |
|     Inlet | 27.0 mm (1.06 in) |
|     Exhaust | 23.0 mm (0.91 in) |
| Valve lift: | |
|     Inlet | 7.0 mm (0.28 in) |
|     Exhaust, SZ | 7.0 mm (0.28 in) |
|     Exhaust, SD | 6.5 mm (0.26 in) |
| Valve clearance (cold) – inlet and exhaust | 0.07 – 0.12 mm (0.003 – 0.005 in) |
| Guide to stem clearance: | |
|     Inlet | 0.025 – 0.052 mm (0.0010 – 0.0020 in) |
|     Service limit | 0.35 mm (0.014 in) |
|     Exhaust | 0.040 – 0.067 mm (0.0016 – 0.0026 in) |
|     Service limit | 0.35 mm (0.014 in) |
| Valve guide ID | 5.500 – 5.512 mm (0.2165 – 0.2170 in) |
| Valve stem OD: | |
|     Inlet | 5.460 – 5.475 mm (0.2165 – 0.2170 in) |
|     Exhaust | 5.445 – 5.460 mm (0.2144 – 0.2150 in) |
| Valve stem runout (In/Ex) – service limit | 0.05 mm (0.002 in) |
| Valve head thickness (In/Ex) – service limit | 0.5 mm (0.02 in) |
| Valve stem end length (In/Ex) – service limit | 3.6 mm (0.14 in) |
| Valve seat width (In/Ex) | 0.9 – 1.1 mm (0.035 – 0.043 in) |
| Valve head radial runout | 0.03 mm (0.001 in) |
| Valve spring free length: | |
|     Inner (service limit) | 31.9 mm (1.26 in) |
|     Outer (service limit) | 35.6 mm (1.40 in) |
| Valve spring pressure: | |
|     Inner | 4.4 – 6.4 kg (9.7 – 14.1 lb) @ 28.5 mm (1.12 in) |
|     Outer | 6.5 – 8.9 kg (14.3 – 19.6 lb @ 32.0 mm (1.26 in) |

## Cylinder bores

| | |
|---|---|
| Compression pressure | 11 – 14 kg cm$^2$ (156 – 199 psi) |
| Service limit | 9 kg cm$^2$ (128 psi) |
| Pressure difference between cylinders – service limit | 2 kg cm$^2$ (28.4 psi) |
| Bore size (Std) | 69.400 – 69.415 mm (2.7323 – 2.7329 in) |
| Service limit | 69.480 mm (2.7354 in) |
| Bore ovality – service limit | 0.2 mm (0.008 in) |

## Pistons

| | |
|---|---|
| Piston/cylinder clearance | 0.050 – 0.060 mm (0.0020 – 0.0024 in) |
| Service limit | 0.120 mm (0.0047 in) |
| Piston diameter (Std) | 69.345 – 69.360 mm (2.7301 – 2.7307 in) |
| Service limit | 69.280 mm (2.7276 in) |
| Ring end gap (free:) | |
|     Top | About 8.5 mm (0.33 in) |
|     Service limit | 6.8 mm (0.27 in) |
|     2nd | About 10.0 mm (0.39 in) |
|     Service limit | 8.0 mm (0.31 in) |
| Ring end gap (installed): | |
|     Top | 0.10 – 0.25 mm (0.004 – 0.010 in) |
|     2nd, SZ | 0.10 – 0.25 mm (0.004 – 0.010 in) |
|     2nd, SD | 0.10 – 0.30 mm (0.004 – 0.012 in) |
|     Service limit | 0.7 mm (0.03 in) |
| Ring to groove clearance: | |
|     Top (service limit) | 0.180 mm (0.0071 in) |
|     2nd (service limit) | 0.150 mm (0.0059 in) |
| Ring groove width: | |
|     Top | 1.01 – 1.03 mm (0.0398 – 0.0406 in) |
|     2nd | 1.21 – 1.25 mm (0.047 – 0.048 in) |
|     Oil | 2.51 – 2.53 mm (0.099 – 0.100 in) |
| Gudgeon pin bore ID | 18.001 – 18.006 mm (0.7087 – 0.7089 in) |
| Service limit | 18.030 mm (0.7098 in) |
| Gudgeon pin OD | 17.996 – 18.000 mm (0.7085 – 0.7086 in) |
| Service limit | 17.980 mm (0.7079 in) |

## Crankshaft and connecting rods

| | |
|---|---|
| Crankshaft runout – service limit | 0.1 mm (0.004 in) |
| Small end ID | 18.006 – 18.014 mm (0.7089 – 0.7092 in) |
| Service limit | 18.040 mm (0.7102 in) |
| Connecting rod deflection – service limit | 3.0 mm (0.12 in) |
| Big-end side clearance | 0.10 – 0.65 mm (0.004 – 0.026 in) |
| Service limit | 1.0 mm (0.040 in) |

## Clutch

| | |
|---|---|
| Type | Wet, multiplate |
| Friction plate thickness | 2.9 – 3.1 mm (0.11 – 0.12 in) |
| Service limit | 2.6 mm (0.10 in) |
| Friction plate tang width | 15.6 – 15.8 mm (0.61 – 0.62 in) |
| Service limit | 14.8 mm (0.58 in) |
| Plain plate thickness | 2.00 ± 0.06 mm (0.008 ± 0.002 in) |
| Service limit | not available |
| Plain plate warpage – service limit | 0.1 mm (0.004 in) |
| Clutch spring free length – service limit | 38.5 mm (1.52 in) |
| Clutch cable free play | 2 – 3 mm (0.08 – 0.12 in) |

## Transmission

| | |
|---|---|
| Primary reduction | 1.775 : 1 (87/49T) |
| Final reduction | 2.800 : 1 (45/15T) |
| Final drive chain: | |
|     Type | Daido D.I.D. 630YL or Takasago RK630GSV |
|     Length | 96 links |
|     20 link length (max) | 383.0 mm (15.08 in) |
|     Free play | 20 – 30 mm (0.8 – 1.2 in) |
| Gearbox type | 5-speed constant mesh |
| Gearbox ratios: | |
|     1st | 2.500 : 1 (35/14 T) |
|     2nd | 1.777 : 1 (32/18 T) |
|     3rd | 1.380 : 1 (29/21 T) |
|     4th | 1.125 : 1 (27/24 T) |
|     Top | 0.961 : 1 (25/26 T) |
| Output shaft length – 1st to 2nd | 111.4 – 111.5 mm (4.386 – 4.390 in) |
| Selector fork to groove clearance | 0.40 – 0.60 mm (0.016 – 0.024 in) |
| Service limit | 0.8 mm (0.031 in) |
| Fork groove width | 5.45 – 5.55 mm (0.215 – 0.219 in) |
| Fork thickness | 4.95 – 5.05 mm (0.195 – 0.199 in) |

## Torque wrench settings

| Component | kgf m | lbf ft |
|---|---|---|
| Cylinder head cover bolt | 0.9 – 1.0 | 6.5 – 7.0 |
| Cylinder head bolt | 0.7 – 1.1 | 5.0 – 8.0 |
| Cylinder head nut | 3.5 – 4.0 | 25.5 – 29.0 |
| Rocker shaft stopper bolt | 0.8 – 1.0 | 6.0 – 7.0 |
| Valve clearance locknut | 0.9 – 1.1 | 6.5 – 8.0 |
| Camshaft cap bolt | 0.8 – 1.2 | 6.0 – 8.5 |
| Camshaft sprocket bolt, SZ | 0.9 – 1.2 | 6.5 – 8.5 |
| Camshaft sprocket bolt, SD | 2.4 – 2.6 | 17.5 – 19.0 |
| Cam chain tensioner holding bolt | 0.6 – 0.8 | 4.5 – 6.0 |
| Cam chain tensioner shaft | 3.1 – 3.5 | 22.0 – 25.5 |
| Cam chain tensioner locknut | 0.9 – 1.4 | 6.5 – 10.0 |
| Cam chain guide bolt | 0.9 – 1.4 | 6.5 – 10.0 |
| Alternator rotor nut, SZ | 13.0 – 14.0 | 94.0 – 101.5 |
| Alternator rotor nut, SD | 16.0 – 17.0 | 115.7 – 123.0 |
| Starter clutch Allen bolt, SZ | 1.5 – 2.0 | 11.0 – 14.5 |
| Starter clutch Allen bolt, SD | 2.3 – 2.8 | 16.5 – 20.0 |
| ATU holding bolt | 1.3 – 2.3 | 9.5 – 16.5 |
| Crankcase bolt (6 mm) | 0.9 – 1.3 | 6.5 – 9.5 |
| Crankcase bolt (8 mm) | 2.0 – 2.4 | 14.5 – 17.0 |
| Starter motor mounting bolt | 0.4 – 0.7 | 3.0 – 5.0 |
| Sump bolt | 1.0 | 7.0 |
| Oil filter cover nut | 0.6 – 0.8 | 4.5 – 6.0 |
| Neutral stopper | 1.8 – 2.8 | 13.0 – 20.0 |
| Gearchange stopper | 1.5 – 2.3 | 11.0 – 16.5 |
| Clutch centre nut | 5.0 – 7.0 | 36.0 – 50.5 |
| Clutch spring bolt | 1.1 – 1.3 | 8.0 – 9.5 |
| Gearbox sprocket nut, SZ | 9.0 – 10.0 | 65.0 – 72.0 |
| Gearbox sprocket nut, SD | 10.0 – 15.0 | 72.0 – 108.5 |
| Engine mounting bolts: | | |
|     (A) 10 mm | 4.5 – 5.5 | 32.5 – 40.0 |
|     (B) 10 mm | 3.0 – 3.7 | 21.5 – 27.0 |
|     (C) 8 mm | 2.0 – 3.0 | 14.5 – 21.5 |
| Gearchange pedal pinch bolt | 1.3 – 2.3 | 9.5 – 16.5 |
| Clutch arm bolt | 0.6 – 1.0 | 4.5 – 7.0 |

# GS 1000 SZ Katana (US)

## Engine
Type .................................................................................... Four cylinder, dohc 16-valve, air-cooled four-stroke
Bore ..................................................................................... 69.4 mm (2.732 in)
Stroke .................................................................................. 66.0 mm (2.598 in)
Capacity ............................................................................... 998 cc (60.9 cu in)
Compression ratio ............................................................... 9.5 : 1
Max power ........................................................................... N/Av
Max torque .......................................................................... N/Av

## Cylinder head
Max warpage ........................................................................ 0.2 mm (0.008 in)

## Camshafts and rockers
Cam height (inlet) ................................................................ 34.650 − 34.690 mm (1.3642 − 1.3657 in)
Service limit ......................................................................... 34.350 mm (1.3524 in)
Cam height (exhaust) .......................................................... 34.360 − 34.400 mm (1.3528 − 1.3543 in)
Service limit ......................................................................... 34.060 mm (1.3409 in)
Camshaft/journal oil:
    Clearance ...................................................................... 0.032 − 0.066 mm (0.0013 − 0.0026 in)
    Service limit .................................................................. 0.150 mm (0.0059 in)
Camshaft journal ID ............................................................. 22.012 − 22.025 mm (0.8666 − 0.8671 in)
Service limit ......................................................................... not available
Camshaft journal OD ........................................................... 21.959 − 21.980 mm (0.8645 − 0.8654 in)
Service limit ......................................................................... not available
Camshaft runout − service limit ......................................... 0.10 mm (0.004 in)
Cam chain length − service limit (20 links) ....................... 157.80 mm (6.213 in)
Rocker arm ID ...................................................................... 12.000 − 12.018 mm (0.4724 − 0.4731 in)
Service limit ......................................................................... not available
Rocker shaft OD ................................................................... 11.973 − 11.984 mm (0.4714 − 0.4718 in)
Service limit ......................................................................... not available

## Valves and guides
Valve head diameter:
    Inlet ............................................................................... 27.0 mm (1.06 in)
    Exhaust .......................................................................... 23.0 mm (0.91 in)
Valve lift:
    Inlet ............................................................................... 7.0 mm (0.28 in)
    Exhaust .......................................................................... 7.0 mm (0.28 in)
Valve clearance (cold) − inlet and exhaust ....................... 0.07 − 0.12 mm (0.003 − 0.005 in)
Guide to stem clearance:
    Inlet ............................................................................... 0.025 − 0.052 mm (0.0010 − 0.0020 in)
    Service limit .................................................................. 0.35 mm (0.014 in)
    Exhaust .......................................................................... 0.040 − 0.067 mm (0.0016 − 0.0026 in)
    Service limit .................................................................. 0.35 mm (0.014 in)
Valve guide ID ..................................................................... 5.500 − 5.512 mm (0.2165 − 0.2170 in)
Valve stem OD:
    Inlet ............................................................................... 5.460 − 5.475 mm (0.2165 − 0.2170 in)
    Exhaust .......................................................................... 5.445 − 5.460 mm (0.2144 − 0.2150 in)
Valve stem runout (In/Ex) − service limit .......................... 0.05 mm (0.002 in)
Valve head thickness (In/Ex) − service limit ..................... 0.5 mm (0.02 in)
Valve stem end length (In/Ex) − service limit ................... 3.6 mm (0.14 in)
Valve seat width (In/Ex) ...................................................... 0.9 − 1.1 mm (0.035 − 0.043 in)
Valve head radial runout ..................................................... 0.03 mm (0.001 in)
Valve spring free length:
    Inner (service limit) ..................................................... 31.9 mm (1.26 in)
    Outer (service limit) .................................................... 35.6 mm (1.40 in)
Valve spring pressure:
    Inner .............................................................................. 4.4 − 6.4 kg (9.7 − 14.1 lb) @ 28.5 mm (1.12 in)
    Outer ............................................................................. 6.5 − 8.9 kg (14.3 − 19.6 lb @ 32.0 mm (1.26 in)

## Cylinder bores
Compression pressure .......................................................... 11 − 14 kg·cm$^2$ (156 − 199 psi)
Service limit ......................................................................... 9 kg·cm$^2$ (128 psi)
Pressure difference between cylinders − service limit ....... 2 kg·cm$^2$ (28.4 psi)
Bore size (Std) ..................................................................... 69.400 − 69.415 mm (2.7323 − 2.7329 in)
Service limit ......................................................................... 69.480 mm (2.7354 in)
Bore ovality − service limit ................................................. 0.2 mm (0.008 in)

## Pistons
Piston/cylinder clearance .................................................... 0.050 − 0.060 mm (0.0020 − 0.0024 in)
Service limit ......................................................................... 0.120 mm (0.0047 in)
Piston diameter (Std) .......................................................... 69.345 − 69.360 mm (2.7301 − 2.7307 in)
Service limit ......................................................................... 69.280 mm (2.7276 in)

Ring end gap (free):
    Top ............................................................................ About 8.5 mm (0.33 in)
    Service limit .............................................................. 6.8 mm (0.27 in)
    2nd ........................................................................... About 10.0 mm (0.39 in)
    Service limit .............................................................. 8.0 mm (0.31 in)
Ring end gap (installed):
    Top ............................................................................ 0.10 – 0.25 mm (0.004 – 0.010 in)
    2nd ........................................................................... 0.10 – 0.30 mm (0.004 – 0.012 in)
    Service limit .............................................................. 0.7 mm (0.03 in)
Ring to groove clearance:
    Top (service limit) ................................................... 0.180 mm (0.0071 in)
    2nd (service limit) ................................................... 0.150 mm (0.0059 in)
Ring groove width:
    Top ............................................................................ 1.01 – 1.03 mm (0.0398 – 0.0406 in)
    2nd ........................................................................... 1.21 – 1.23 mm (0.047 – 0.048 in)
    Oil ............................................................................. 2.51 – 2.53 mm (0.099 – 0.100 in)
Gudgeon pin bore ID .................................................... 18.001 – 18.006 mm (0.7087 – 0.7089 in)
Service limit .................................................................. 18.030 mm (0.7098 in)
Gudgeon pin OD ........................................................... 17.996 – 18.000 mm (0.7085 – 0.7086 in)
Service limit .................................................................. 17.980 mm (0.7079 in)

## Crankshaft and connecting rods

Crankshaft runout – service limit ............................... 0.1 mm (0.004 in)
Small end ID ................................................................. 18.006 – 18.014 mm (0.7089 – 0.7092 in)
Service limit .................................................................. 18.040 mm (0.7102 in)
Connecting rod deflection – service limit ................... 3.0 mm (0.12 in)
Big-end side clearance ................................................. 0.10 – 0.65 mm (0.004 – 0.026 in)
Service limit .................................................................. 1.0 mm (0.040 in)

## Clutch

Type .............................................................................. Wet, multiplate
Friction plate thickness ............................................... 2.9 – 3.1 mm (0.11 – 0.12 in)
Service limit .................................................................. 2.6 mm (0.10 in)
Friction plate tang width ............................................. 15.6 – 15.8 mm (0.61 – 0.62 in)
Service limit .................................................................. 14.8 mm (0.58 in)
Plain plate thickness .................................................... 2.00 ± 0.06 mm (0.008 ± 0.002 in)
Service limit .................................................................. not available
Plain plate warpage – service limit ............................ 0.1 mm (0.004 in)
Clutch spring free length – service limit .................... 38.5 mm (1.52 in)
Clutch cable free play ................................................. 2 – 3 mm (0.08 – 0.12 in)

## Transmission

Primary reduction ........................................................ 1.775 : 1 (87/49T)
Final reduction ............................................................. 2.800 : 1 (45/15T)
Final drive chain:
    Type ............................................................................ Daido D.I.D. 630YL or Takasago RK630GSV
    Length ......................................................................... 96 links
    20 link length (max) .................................................. 383.0 mm (15.08 in)
    Free play ..................................................................... 20 – 30 mm (0.8 – 1.2 in)
Gearbox type ................................................................ 5-speed constant mesh
Gearbox ratios:
    1st .............................................................................. 2.500 : 1 (35/14 T)
    2nd ........................................................................... 1.777 : 1 (32/18 T)
    3rd ............................................................................. 1.380 : 1 (29/21 T)
    4th ............................................................................. 1.125 : 1 (27/24 T)
    Top ............................................................................ 0.961 : 1 (25/26 T)
Output shaft length – 1st to 2nd ............................... 111.4 – 111.5 mm (4.386 – 4.390 in)
Selector fork to groove clearance .............................. 0.40 – 0.60 mm (0.016 – 0.024 in)
Service limit .................................................................. 0.8 mm (0.031 in)
Fork groove width ........................................................ 5.45 – 5.55 mm (0.215 – 0.219 in)
Fork thickness .............................................................. 4.95 – 5.05 mm (0.195 – 0.199 in)

## Torque wrench settings

| Component | kgf m | lbf ft |
| --- | --- | --- |
| Cylinder head cover bolt | 0.9 – 1.0 | 6.5 – 7.0 |
| Cylinder head bolt | 0.7 – 1.1 | 5.0 – 8.0 |
| Cylinder head nut | 3.5 – 4.0 | 25.5 – 29.0 |
| Rocker shaft stopper bolt | 0.8 – 1.0 | 6.0 – 7.0 |
| Valve clearance locknut | 0.9 – 1.1 | 6.5 – 8.0 |
| Camshaft cap bolt | 0.8 – 1.2 | 6.0 – 8.5 |
| Camshaft sprocket bolt | 0.9 – 1.2 | 6.5 – 8.5 |
| Cam chain tensioner holding bolt | 0.6 – 0.8 | 4.5 – 6.0 |
| Cam chain tensioner shaft | 3.1 – 3.5 | 22.0 – 25.5 |
| Cam chain tensioner locknut | 0.9 – 1.4 | 6.5 – 10.0 |
| Cam chain guide bolt | 0.9 – 1.4 | 6.5 – 10.0 |

| | | |
|---|---|---|
| Alternator rotor nut | 13.0 – 14.0 | 94.0 – 101.5 |
| Starter clutch Allen bolt | 1.5 – 2.0 | 11.0 – 14.5 |
| ATU holding bolt | 1.3 – 2.3 | 9.5 – 16.5 |
| Crankcase bolt (6 mm) | 0.9 – 1.3 | 6.5 – 9.5 |
| Crankcase bolt (8 mm) | 2.0 – 2.4 | 14.5 – 17.0 |
| Starter motor mounting bolt | 0.4 – 0.7 | 3.0 – 5.0 |
| Sump bolt | 1.0 | 7.0 |
| Oil filter cover nut | 0.6 – 0.8 | 4.5 – 6.0 |
| Neutral stopper | 1.8 – 2.8 | 13.0 – 20.0 |
| Gearchange stopper | 1.5 – 2.3 | 11.0 – 16.5 |
| Clutch centre nut | 5.0 – 7.0 | 36.0 – 50.5 |
| Clutch spring bolt | 1.1 – 1.3 | 8.0 – 9.5 |
| Gearbox sprocket nut | 9.0 – 10.0 | 65.0 – 72.0 |
| Engine mounting bolts: | | |
| (A) 10 mm | 4.5 – 5.5 | 32.5 – 40.0 |
| (B) 10 mm | 3.0 – 3.7 | 21.5 – 27.0 |
| (C) 8 mm | 2.0 – 3.0 | 14.5 – 21.5 |
| Gearchange pedal pinch bolt | 1.3 – 2.3 | 9.5 – 16.5 |
| Clutch arm bolt | 0.6 – 1.0 | 4.5 – 7.0 |

# GS 1100 ET, LT, EX, EZ, ED, ESD (US)

## Engine
| | |
|---|---|
| Type | Four cylinder, dohc 16-valve, air-cooled four-stroke |
| Bore | 72.0 mm (2.835 in) |
| Stroke | 66.0 mm (2.598 in) |
| Capacity | 1075 cc (65.6 cu in) |
| Compression ratio | 9.5 : 1 |
| Max power | Not available |
| Max torque | Not available |

## Cylinder head
| | |
|---|---|
| Max warpage | 0.2 mm (0.008 in) |

## Camshafts and rockers
| | |
|---|---|
| Cam height (inlet): | |
| ET, LT, EX, EZ, ED, ESD | 34.650 – 34.690 mm (1.3642 – 1.3657 in) |
| Service limit | 34.350 mm (1.3524 in) |
| Cam height (exhaust): | |
| ET, LT, EX, EZ, ED, ESD | 34.360 – 34.400 mm (1.3528 – 1.3543 in) |
| Service limit | 34.060 mm (1.3409 in) |
| Camshaft/journal oil clearance: | |
| ET, LT, EX | 0.020 – 0.054 mm (0.0008 – 0.0021 in) |
| EZ, ED, ESD | 0.032 – 0.066 mm (0.0013 – 0.0026 in) |
| Service limit | 0.150 mm (0.0059 in) |
| Camshaft journal ID: | |
| ET, LT, EX | 22.000 – 22.013 mm (0.8661 – 0.8667 in) |
| EZ, ED, ESD | 22.012 – 22.025 mm (0.8666 – 0.8671 in) |
| Service limit | Not available |
| Camshaft journal OD | 21.959 – 21.980 mm (0.8645 – 0.8654 in) |
| Service limit | Not available |
| Camshaft runout – service limit | 0.10 mm (0.004 in) |
| Cam chain length – service limit (20 links) | 157.80 mm (6.213 in) |
| Rocker arm ID | 12.000 – 12.018 mm (0.4724 – 0.4731 in) |
| Service limit | Not available |
| Rocker shaft OD | 11.973 – 11.984 mm (0.4714 – 0.4718 in) |
| Service limit | Not available |

## Valves and guides
| | |
|---|---|
| Valve head diameter: | |
| Inlet | 26.9 – 27.1 mm (1.06 – 1.07 in) |
| Exhaust | 22.9 – 23.1 mm (0.90 – 0.91 in) |
| Valve lift, inlet | 7.0 mm (0.28 in) |
| Valve lift, exhaust – ET, LT, EX, EZ | 6.5 mm (0.26 in) |
| Valve lift, exhaust – ED, ESD | 7.0 mm (0.28 in) |
| Valve clearance (cold) – inlet and exhaust | 0.07 – 0.12 mm (0.003 – 0.005 in) |
| Guide to stem clearance: | |
| Inlet ET, LT | 0.025 – 0.12 mm (0.0010 – 0.005 in) |
| EX, EZ, ED, ESD | 0.025 – 0.052 mm (0.0010 – 0.0020 in) |
| Service limit | 0.35 mm (0.014 in) |
| Exhaust | 0.040 – 0.067 mm (0.0016 – 0.0026 in) |
| Service limit | 0.35 mm (0.014 in) |
| Valve guide ID | 5.500 – 5.512 mm (0.2165 – 0.2170 in) |

Valve stem OD:
    Inlet ........................................................................... 5.460 – 5.475 mm (0.2165 – 0.2170 in)
    Exhaust ...................................................................... 5.445 – 5.460 mm (0.2144 – 0.2150 in)
Valve stem runout (In/Ex) – service limit ................................ 0.05 mm (0.002 in)
Valve head thickness (In/Ex) – service limit ........................... 0.5 mm (0.02 in)
Valve stem end length (In/Ex) – service limit ......................... 3.6 mm (0.14 in).
Valve seat width (In/Ex) ............................................................. 0.9 – 1.1 mm (0.035 – 0.043 in)
Valve head radial runout .......................................................... 0.03 mm (0.001 in)
Valve spring free length:
    Inner (service limit) ..................................................... 31.9 mm (1.26 in)
    Outer (service limit) .................................................... 35.6 mm (1.40 in)
Valve spring pressure:
    Inner ........................................................................... 4.4 – 6.4 kg (9.7 – 14.1 lb) @ 28.5 mm (1.12 in)
    Outer .......................................................................... 6.5 – 8.9 kg (14.3 – 19.6 lb) @ 32.0 mm (1.26 in)

## Cylinder bores
Compression pressure ............................................................... $9 - 12$ kg cm$^2$ (128 – 171 psi)
Service limit .............................................................................. 7 kg cm$^2$ (100 psi)
Pressure difference between cylinders – service limit .............. 2 kg cm$^2$ (28.4 psi)
Bore size (Std) .......................................................................... 72.000 – 72.015 mm (2.8346 – 2.8352 in)
Service limit .............................................................................. 72.080 mm (2.8378 in)
Bore ovality – service limit ....................................................... 0.2 mm (0.008 in)

## Pistons
Piston/cylinder clearance ......................................................... 0.050 – 0.060 mm (0.0020 – 0.0024 in)
Service limit .............................................................................. 0.120 mm (0.0047 in)
Piston diameter (Std) ................................................................ 71.945 – 71.960 mm (2.8325 – 2.8331 in)
Service limit .............................................................................. 71.880 mm (2.8299 in)
Ring end gap (free):
    Top .............................................................................. About 9.5 mm (0.37 in)
    Service limit ................................................................ 7.6 mm (0.30 in)
    2nd .............................................................................. About 11.0 mm (0.43 in)
    Service limit ................................................................ 8.8 mm (0.35 in)
Ring end gap (installed):
    Top & 2nd ................................................................... 0.10 – 0.30 mm (0.004 – 0.012 in)
    Service limit ................................................................ 0.7 mm (0.03 in)
Ring to groove clearance:
    Top (service limit) ....................................................... 0.180 mm (0.0071 in)
    2nd (service limit) ...................................................... 0.150 mm (0.0059 in)
Ring groove width:
    Top, ET ........................................................................ 1.03 – 1.05 mm (0.040 – 0.041 in)
    2nd, ET ........................................................................ 1.21 – 1.23 mm (0.047 – 0.048 in)
    Oil, ET ......................................................................... 2.51 – 2.53 mm (0.099 – 0.100 in)
Piston ring thickness:
    Top .............................................................................. 0.975 – 0.990 mm (0.0384 – 0.0390 in)
    2nd .............................................................................. 1.170 – 1.190 mm (0.0461 – 0.0469 in)
Gudgeon pin bore ID ................................................................. 18.001 – 18.006 mm (0.7087 – 0.7089 in)
Service limit .............................................................................. 18.030 mm (0.7098 in)
Gudgeon pin OD ........................................................................ 17.996 – 18.000 mm (0.7085 – 0.7086 in)
Service limit .............................................................................. 17.980 mm (0.7079 in)

## Crankshaft and connecting rods
Crankshaft runout – service limit ............................................. 0.1 mm (0.004 in)
Small end ID ............................................................................. 18.006 – 18.014 mm (0.7089 – 0.7092 in)
Service limit .............................................................................. 18.040 mm (0.7102 in)
Connecting rod deflection – service limit ................................ 3.0 mm (0.12 in)
Big-end side clearance ............................................................. 0.10 – 0.65 mm (0.004 – 0.026 in)
Service limit .............................................................................. 1.00 mm (0.040 in)

## Clutch
Type .......................................................................................... Wet, multiplate
Friction plate thickness:
    ET, LT, EX, EZ ............................................................ 2.9 – 3.1 mm (0.11 – 0.12 in)
    Service limit ................................................................ 2.6 mm (0.10 in)
Friction plate thickness:
    ED, ESD ....................................................................... 2.15 – 2.35 mm (0.085 – 0.093 in)
    Service limit ................................................................ 1.85 mm (0.073 in)
Friction plate tang width .......................................................... 15.6 – 15.8 mm (0.61 – 0.62 in)
Service limit .............................................................................. 14.8 mm (0.58 in)
Plain plate thickness ................................................................ $2.00 \pm 0.06$ mm ($0.008 \pm 0.002$ in)
Service limit .............................................................................. Not available
Plain plate warpage – service limit .......................................... 0.1 mm (0.004 in)
Clutch spring free length – service limit .................................. 38.5 mm (1.52 in)
Clutch/primary drive pinion:
    Backlash ...................................................................... 0 – 0.02 mm (0.0008 in)
    Service limit ................................................................ 0.08 mm (0.003 in)
Clutch cable free play .............................................................. 2 – 3 mm (0.08 – 0.12 in)

## Transmission

| | |
|---|---|
| Primary reduction | 1.775 : 1 (87/49 T) |
| Final reduction: | |
| ET, EX, EZ, ED, ESD | 2.800 : 1 (45/15 T) |
| LT | 2.666 : 1 (40/15 T) |
| Final drive chain: | |
| Type | Daido D.I.D. 630YL or Takasago RK630GSV |
| Length | 96 links (98 links – ESD model) |
| 20 link length (max) | 383.0 mm (15.08 in) |
| Free play | 20 – 30 mm (0.8 – 1.2 in) |
| Gearbox type | 5-speed constant mesh |
| Gearbox ratios: | |
| 1st | 2.500 : 1 (35/14 T) |
| 2nd | 1.777 : 1 (32/18 T) |
| 3rd | 1.380 : 1 (29/21 T) |
| 4th | 1.125 : 1 (27/24 T) |
| Top | 0.961 : 1 (25/26 T) |
| Gear backlash: | |
| 1st, 2nd, 3rd | 0.03 mm (0.001 in) |
| Service limit | 0.08 mm (0.003 in) |
| 4th, Top | 0.10 mm (0.004 in) |
| Service limit | 0.15 mm (0.006 in) |
| Output shaft length – 1st to 2nd | 111.4 – 111.5 mm (4.386 – 4.390 in) |
| Selector fork to groove clearance | 0.40 – 0.60 mm (0.016 – 0.024 in) |
| Service limit | 0.8 mm (0.031 in) |
| Fork groove width | 5.45 – 5.55 mm (0.215 – 0.219 in) |
| Fork thickness | 4.95 – 5.05 mm (0.195 – 0.199 in) |

## Torque wrench settings (all models except ED/ESD)

| Component | kgf m | lbf ft |
|---|---|---|
| Cylinder head cover bolt | 0.9 – 1.0 | 6.5 – 7.0 |
| Cylinder head bolt | 0.7 – 1.1 | 5.0 – 8.0 |
| Cylinder head nut | 3.5 – 4.0 | 25.5 – 29.0 |
| Rocker shaft stopper bolt | 0.8 – 1.0 | 6.0 – 7.0 |
| Valve clearance locknut | 0.9 – 1.1 | 6.5 – 8.0 |
| Camshaft cap bolt | 0.8 – 1.2 | 6.0 – 8.5 |
| Camshaft sprocket bolt: | | |
| ET, LT, EX | 0.9 – 1.2 | 6.5 – 8.5 |
| EZ | 2.4 – 2.6 | 17.5 – 19.0 |
| Cam chain tensioner holding bolt | 0.6 – 0.8 | 4.5 – 6.0 |
| Cam chain tensioner shaft | 3.1 – 3.5 | 22.0 – 25.5 |
| Cam chain tensioner locknut | 0.9 – 1.4 | 6.5 – 10.0 |
| Cam chain guide bolt | 0.9 – 1.4 | 6.5 – 10.0 |
| Alternator rotor nut | 16.0 – 17.0 | 115.7 – 123.0 |
| Starter clutch Allen bolt | 1.5 – 2.0 | 11.0 – 14.5 |
| ATU holding bolt | 1.3 – 2.3 | 9.5 – 16.5 |
| Crankcase bolt (6 mm) | 0.9 – 1.3 | 6.5 – 9.5 |
| Crankcase bolt (8 mm) | 2.0 – 2.4 | 14.5 – 17.0 |
| Starter motor mounting bolt | 0.4 – 0.7 | 3.0 – 5.0 |
| Sump bolt | 1.0 | 7.0 |
| Oil filter cover nut | 0.6 – 0.8 | 4.5 – 6.0 |
| Neutral stopper | 1.8 – 2.8 | 13.0 – 20.0 |
| Gearchange stopper | 1.5 – 2.3 | 11.0 – 16.5 |
| Clutch centre nut | 5.0 – 7.0 | 36.0 – 50.5 |
| Clutch spring bolt | 1.1 – 1.3 | 8.0 – 9.5 |
| Gearbox sprocket nut: | | |
| ET, LT, EX | 9.0 – 10.0 | 65.0 – 72.0 |
| EZ | 10.0 – 15.0 | 72.5 – 108.5 |
| Engine mounting bolts: | | |
| (A) 10 mm | 4.5 – 5.5 | 32.5 – 40.0 |
| (B) 10 mm | 3.0 – 3.7 | 21.5 – 27.0 |
| (C) 8 mm | 2.0 – 3.0 | 14.5 – 21.5 |
| Gearchange pedal pinch bolt | 1.3 – 2.3 | 9.5 – 16.5 |
| Clutch arm bolt | 0.6 – 1.0 | 4.5 – 7.0 |

## 1 General description and engine modifications

The engine/gearbox unit is of the four-cylinder, air cooled, in-line type mounted transversely across the frame. The cylinder head is a 16 valve design, designated TSCC (Twin Swirl Combustion Chamber) by Suzuki. The valves are arranged in pairs and are driven via short rockers by the double overhead camshafts. The camshafts are driven by a central chain from the crankshaft, chain tension being controlled by a self-adjusting tensioner unit.

The horizontally separated crankcases incorporate the crankshaft, primary drive, clutch and the 5-speed constant-mesh gearbox in a common casing. The underside of the crankcase assembly is closed by a finned alloy sump (oil pan) which contains the shared lubrication supply.

In the course of its evolution, a number of modifications have been introduced, either as improvements to the original design or as the result of general uprating to provide improved performance. Brief details of the major changes are given below, but it is essential to quote the engine number in full when ordering replacement parts. This will ensure that the new parts are compatible with the associated engine components. Note that the details below are based on the UK

Fig. 1.1 Engine/gearbox unit – sectioned view

models for which information was available at the time of writing. Where this is known to apply to an equivalent US machine, the corresponding details are enclosed in brackets (thus). Where a US model is not mentioned, the modification may or may not apply; information is not available to confirm or deny this.

**GSX 1100 LT (GS 1100 LT):** The oil ring expander was changed, the new type being thicker and having rectangular, instead of round, oil holes.

The clutch centre was changed, the bosses for the spring bolts being reduced and a longer bolt and a spacer being introduced to compensate. The dished washer behind the innermost plain plate was changed along with its spring seat, the latter now having an external rather than an internal lip.

**GSX 1100 EX (GS 1100 EX):** The camshaft sprocket retaining bolts appear to have been changed twice, so that any one of three types may be found. These are as follows: A-type, a 6 mm Allen bolt, B-type, a 6 mm hexagon-headed bolt used with a double tab washer, and C-type, a 7 mm flange bolt. The three types are shown in the accompanying line drawings. Note that types A and B are interchangeable, and that each type has its own torque setting:

A-type: 0.9 – 1.2 kgf m (6.5 – 8.5 lbf ft)
B-type: 1.5 – 2.0 kgf m (11.0 – 14.5 lbf ft)
C-type: 2.4 – 2.6 kgf m (17.5 – 19.0 lbf ft)

The GSX 1100 EX model was equipped with the modified clutch centre as described for the LT model. In addition, from engine No. 113264 onwards, the back of the clutch centre was again modified and a new clutch pressure plate was fitted. (See line drawing.)

**GSX 1100 EZ (GS 1100 EZ):** The casing screws were replaced with small hexagon-headed bolts. 7 mm and 8 mm sockets or box spanners will be required, or Suzuki T-handled wrenches can be obtained, Part Numbers 09900-06711 and 09900-00303-015.

Also modified were the cylinder head cover screw positions to improve sealing (see the accompanying line drawing), meaning that the cover, head and gasket are not interchangeable with the earlier versions.

**GSX 1000/1100 SZ (GS 1000 SZ):** On these and subsequent Katana versions the diameter of the alternator rotor was reduced from 130 mm to 118 mm. This was intended to reduce crankshaft inertia, thus improving throttle response.

**GSX 1000/1100 SD and ESD:** One of the clutch cover screws was fitted with a sealing washer (see line drawing).

The clutch/primary driven gear spacer was reduced in length and a shim added, presumably to control more exactly end float in the clutch assembly. The accompanying illustration shows the new arrangement. Note that if the primary driven gear, the oil pump drive gear or their respective bearing spacers are renewed, the end float should be checked and a new shim fitted. These are available in increments of 0.05 mm from 1.05 – 1.20 mm.

A new clutch plate arrangement was introduced on the 1100 model, the previous 9 friction and 9 plain plates being replaced by 10 friction and 11 plain. Consequently, the friction plate thickness was reduced from 2.6 mm to 1.85 mm. The clutch is shown in section in the accompanying line drawing.

The oil pump retaining screws were replaced by bolts. These should be tightened to 0.7 – 0.9 kgf m.

On the GSX 1000 SD, engine No. 170134 onwards and the GSX 1100 SD engine No. 102893 onwards, the crankshaft, alternator, starter gear and starter clutch were modified as shown in the accompanying illustrations, and are incompatible with earlier arrangements.

The front engine mounting arrangement was modified as shown in the accompanying line drawing.

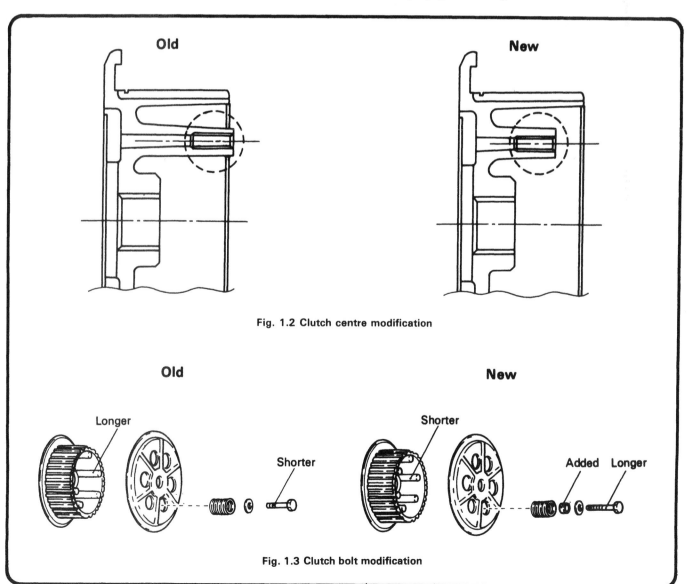

**Old**      **New**

Fig. 1.2 Clutch centre modification

**Old**      **New**

Longer    Shorter    Shorter    Added   Longer

Fig. 1.3 Clutch bolt modification

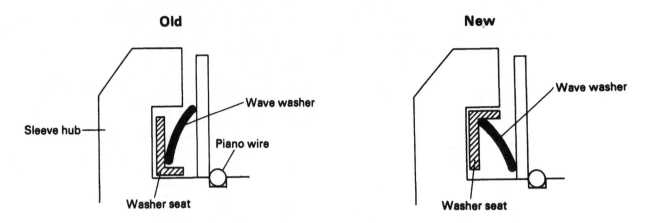

**Old**          **New**

Sleeve hub

Wave washer

Piano wire

Washer seat

Wave washer

Washer seat

Fig. 1.4 Clutch anti-snatch spring modifications

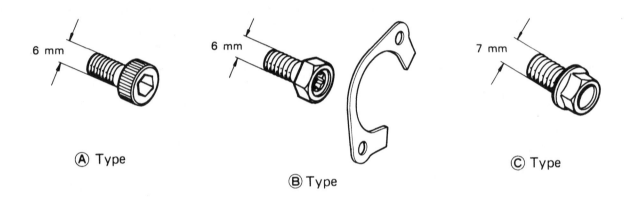

6 mm

Ⓐ Type

6 mm

Ⓑ Type

7 mm

Ⓒ Type

Fig. 1.5 Camshaft sprocket bolt modifications

A Type:  6 mm Allen bolt, torque to 0.9 – 1.2 kgf m (6.5 – 8.5 lbf ft)

B Type:  6 mm hex bolt wth double tab washer, torque to 1.5 – 2.0 kgf m (11.0 – 14.5 lbf ft)

C Type:  7 mm flange bolt, torque to 2.4 – 2.6 kgf m (17.5 – 19.0 lbf ft)

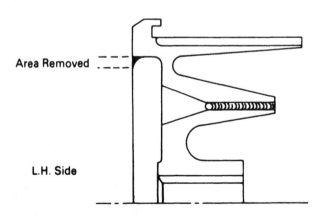

Area Removed

L.H. Side

Fig. 1.6 Clutch centre modification – 1100 engines

The oil guide area has been removed from inner face of clutch centre. Applies from engine number 113264 onwards

Old part number          21411 – 49202
New part number          21411 – 49203

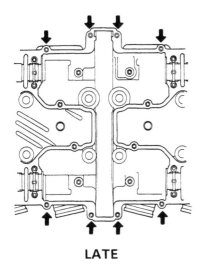

**LATE**

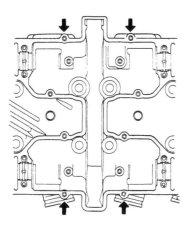

**EARLY**

**Fig. 1.7 Cylinder head cover modifications – EZ onwards**

*Late model engines have extra retaining bolts to improve sealing*

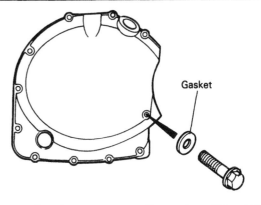

**Fig. 1.8 Clutch cover screw sealing washer – SD and ESD onwards**

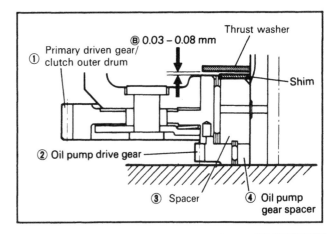

| P/No. | Thickness |
|---|---|
| 21262 – 09300 | 1.05 mm |
| 21263 – 09300 | 1.10 mm |
| 21264 – 09300 | 1.15 mm |
| 21265 – 09300 | 1.20 mm |

**Fig. 1.9 Primary driven gear/clutch outer drum spacer modification**

*On late models, the spacer (3) has been shortened and a shim added. Clearance between the shim and washer must be as shown above. Shims are available as shown in the selection chart*

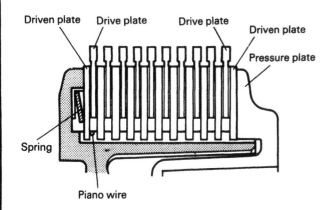

**Fig. 1.10 Section view of clutch SD and ESD onwards**

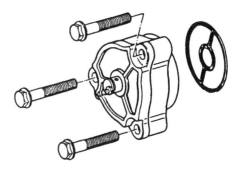

**Fig. 1.11 Oil pump modification – SD model onwards**

*Note that the three pump cover screws are deleted and cover is held by mounting bolts*

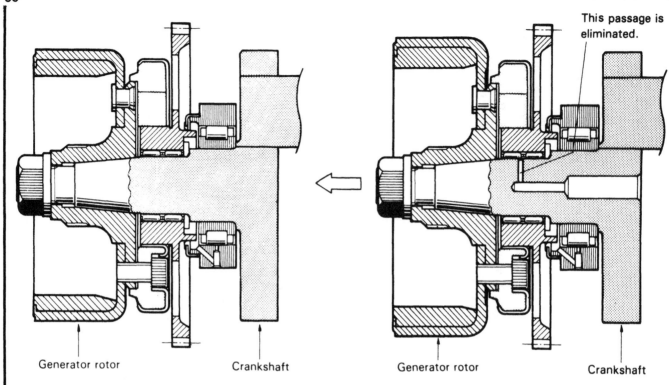

Generator rotor        Crankshaft        Generator rotor        Crankshaft

This passage is eliminated.

**Fig. 1.12 Crankshaft and alternator rotor modifications – Katana models**

*Note that starter clutch oil passage has been eliminated to strengthen crankshaft, and rotor size is reduced*

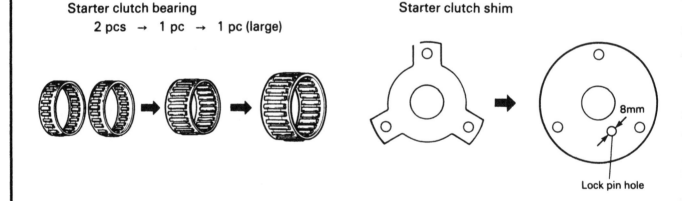

**Starter clutch bearing**
2 pcs → 1 pc → 1 pc (large)

**Starter clutch shim**

8mm

Lock pin hole

**Starter clutch allen bolt**

Knurled

|  | Old | New |
|---|---|---|
| Tightening torque | 11.0 – 14.5 lbf ft (1.5 – 2.0 kgf m) | 16.5 – 20.0 lbf ft (2.3 – 2.8 kgf m) |
| Thread lock | 1333B | 1303B |

**Fig. 1.13 Starter clutch modifications – Katana models**

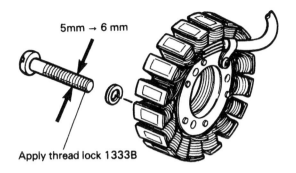

Fig. 1.14 Alternator stator mounting bolt modification

*SZ model:*   *5 mm (200W alternator)*      *SD model:*   *6 mm (250W alternator)*

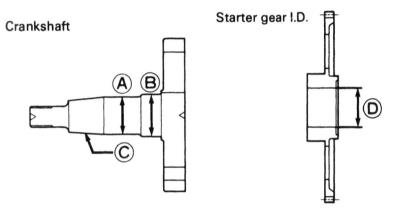

| Item | Old | New |
|------|-----|-----|
| Ⓐ | 25 mm | 29 mm |
| Ⓑ (Unchanged) | 32 mm | 32 mm |
| Ⓒ (Taper) | 1/5 | 1/7 |
| Ⓓ | 31 mm | 35 mm |
| Ⓔ | 118 mm | 128 mm |

Fig. 1.15 Crankshaft and starter gear modifications – SD model

*Dimensions are changed as shown*

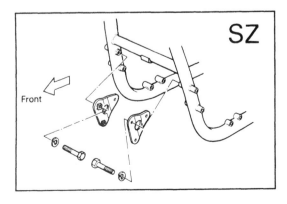
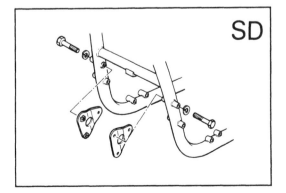

Fig. 1.16 Front engine mounting configuration – Katana models

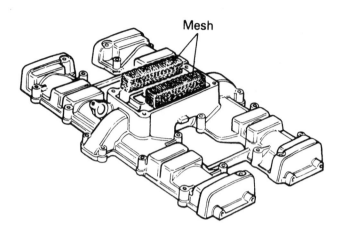

Fig. 1.17 Oil separator modification

*Late model engines have mesh inserts to improve oil separation inside breather cover*

## 2 Operations with the engine/gearbox unit in the frame

1    The following components or assemblies can be removed or dismantled without removing the engine unit from the frame. Whilst this will often be undertaken simultaneously it may prove better to remove the engine and carry out the work on the bench.

    a)   Gearchange lever and sprocket cover
    b)   Gearbox sprocket, chain and neutral switch
    c)   Alternator stator and rotor
    d)   Starter clutch and idler gear
    e)   Oil filter
    f)   Sump (oil pan) and sump filter
    g)   Tachometer drive
    h)   Cylinder head breather cover
    i)   Cam chain tensioner
    j)   Cylinder head cover
    k)   Camshafts
    l)   Cylinder head
    m)   Cylinder block and pistons
    n)   Starter motor
    o)   Oil pressure/temperature switch
    p)   Ignition pickup and automatic timing unit (ATU)
    q)   Clutch cover
    r)   Clutch assembly
    s)   Oil pump assembly
    t)   Gearchange components (except drum and forks)

## 3 Operations requiring engine/gearbox removal

1    To gain access to any of the crankcase internal components or assemblies it will first be necessary to remove the engine unit from the frame to permit crankcase separation. As a guide, engine removal will be necessary for the following operations.

    a)   Removal of the crankshaft assembly, including access to the main and big-end bearings
    b)   Removal of the gearbox shafts, or the gear selector drum and forks.

## 4 Removing the engine/gearbox unit

1    With the engine at normal operating temperature, place a container of at least 1 gal/5 litres capacity below the sump drain plug. Remove the plug and allow the oil to drain. Once the oil has drained

fully, refit the plug and then clean the engine unit thoroughly with a degreaser, or by steam or pressure cleaning to remove all road dirt and oil.

2    Place the machine securely on its centre stand. Open the seat and remove the side panels. Trace and disconnect the fuel gauge sender leads at the rear underside of the tank. Check that the fuel tap is set to the 'ON' or 'RES' position, then prise off the fuel and vacuum pipes using a small screwdriver. Remove the tank holding bolt(s), then lift the tank and pull it rearwards to free it from its front mounting rubbers.

3    Release the single bolt which secures the rear brake fluid reservoir to the frame and move it clear of the air filter casing. Where applicable, remove the tank mounting bracket. Free the bolts and hose clip which secure the rear section of the air filter casing and lift it away. Disconnect the battery leads to preclude any risk of accidental short circuits.

4    Where the horn assembly is mounted immediately above the cylinder head, pull off the horn leads and remove the horn assembly from the frame. Pull off the spark plug caps and lodge them clear of the cylinder head. Disconnect the breather hose from the breather cover.

5    Slacken the screws which retain the carburettor hose clamps between the cylinder head and the air filter trunking, leaving the first three (cylinders 1 to 3) stubs attached to the cylinder head. To obtain better working clearance, remove the No. 4 (right-hand) carburettor mounting stub completely.

6    The carburettor assembly is an extremely tight fit between the mounting stubs, removal necessitating a fair degree of patience. The task is made much easier if assistance is available. Pull the assembly back until the carburettors can be twisted free of the stubs, then slide the assembly out to the right-hand side. Once partly removed, release the throttle and starter choke cables (see photographs), disengage the drain hoses and place the assembly to one side.

7    Remove the two bolts which retain the air filter trunking front section and lift it away. Referring to the accompanying photographs, trace and disconnect the alternator output leads, the ignition pickup leads, the oil pressure and neutral indicator leads and the heavy-duty starter motor cable at the starter relay terminal.

8    Release the tachometer drive cable, (where fitted) at the cylinder head cover end. Remove the clutch operating arm pinch bolt and release the clutch cable adjuster to free the cable from the crankcase, then lodge it clear of the engine.

9    Slacken and remove the exhaust pipe retainer bolts at the cylinder head. Slacken fully the two clamp bolts which retain the inner pair of pipes at the collector below the crankcase. The inner pair of exhaust pipes (cylinders 2 and 3) can now be disengaged and removed. If corrosion makes removal difficult, soak the pipe/collector box junction with WD 40 or similar. In extreme cases, it should just be possible to remove the system complete. Remove the two nuts which retain each silencer to its respective bracket. The system can now be lowered to the ground and manoeuvred clear of the frame.

10   Remove the gearchange pedal and the left-hand footrest, noting that some models are fitted with a semi-rearset gearchange linkage secured by a pinch bolt and a circlip. Remove the sprocket cover. Straighten the sprocket nut locking tab, apply the rear brake and remove the nut. Remove the split pin from the rear wheel spindle nut, then slacken the nut and the chain adjuster bolts. Slacken the rear brake torque arm nuts. The wheel can now be pushed fully forward to give maximum chain free play.

11   Pull the chain away from the gearbox sprocket, sliding the latter off its splines. Note that with a new chain there will be barely enough clearance, in which case the rear wheel must be removed to obtain extra clearance. Disengage the chain from the shaft end and leave it to hang against the swinging arm.

12   To obtain maximum clearance during engine removal, free the rear brake pedal and the right-hand footrest. Suzuki recommend the removal of the cylinder head cover to obtain maximum clearance during engine removal and installation, the necessary removal procedure being described in Section 6 of this Chapter. Removal is just possible if the breather cover only is removed. Before proceeding further, check carefully that all leads and hoses have been disconnected.

13   The engine unit is mounted at four points by a combination of six mounting brackets, each of which is retained by bolts to the frame. The brackets, bracket retaining bolts and engine mounting bolts must be removed to obtain sufficient clearance to permit engine removal. Note that on some models the front engine plates are handed (see photograph) and must be refitted in their original position.

14 When removing the bolts, support the weight of the unit as the bolts are withdrawn, using a bar as a lever between the frame and crankcase. The front lower mounting bolts have shaped captive nuts which will drop clear as the bolts are removed. Once all the bolts have been displaced, the engine can be left to sit in the frame cradle.

15 The complete engine/gearbox unit is both heavy and bulky, and will prove difficult to extract from the frame cradle. A minimum of two persons will be necessary to ensure safe removal, and extra assistance would be desirable. To attempt single-handed removal of the unit is to invite damage to the engine and the operator, should it slip during removal. An engine hoist or jack is a distinct advantage but not essential.

16 Arrange a strong wooden crate or a similar support to the right of the engine and at the same level. Lift the engine slightly, then rotate it by a few degrees so that the various projections will clear the frame. The unit is removed by tipping it and then manoeuvring it out from the right-hand side of the frame, and onto the waiting support.

4.2a Release the tank retaining bolt(s) and lift the rear of the tank slightly

4.2b Displace the wire clip and work the fuel pipe off its stub

4.2c Trace and disconnect the fuel gauge sender leads

4.3a Release reservoir mounting bolt and lift it clear of air filter casing

4.3b Remove support bracket and air filter assembly to reveal battery

4.3c Disconnect battery leads to avoid short circuits

4.6a Push lever (a) against spring pressure and disengage the choke cable. Slacken locknut (b) and free adjuster from bracket ...

4.6b ... throttle cable inner can now be freed from pulley

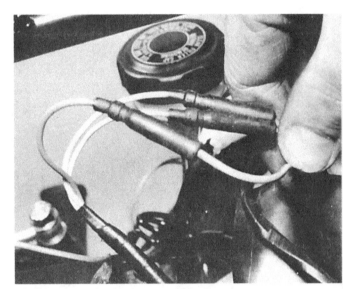

4.7a Disconnect the alternator output leads ...

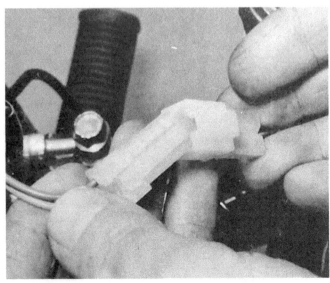

4.7b ... the ignition pickup leads ...

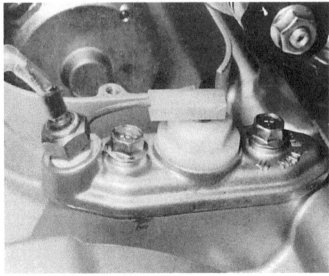

4.7c ... the oil temperature and pressure leads ...

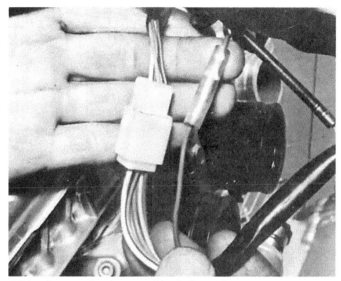

4.7d ... the neutral and gear position indicator leads ...

4.7e ... and the heavy-duty starter cable at the relay terminal

4.8 Unscrew the tachometer cable at the cylinder head end

4.9 Release the rear footrest/silencer mounting bolts

4.10a Rearset linkage (where fitted) is retained by pinch bolt at the front and by a circlip at the pedal end

4.10b Remove nut and tab washer, slide sprocket off shaft end and disengage from chain

4.12 Removal of breather cover gives minimum clearance required during engine removal

4.13 Where front mounting plates are handed, mark them as a guide during assembly

4.14 Captive nuts will drop clear as lower bolts are removed

screws. Slacken the cylinder head cover bolts evenly and progressively and lift the cover away. If it proves stubborn, tap around the joint face with a hide mallet to help break the seal. Slacken the cam chain tensioner locknut and screw the lockscrew home to hold the tensioner pushrod. Release the tensioner mounting bolts and lift the assembly away.

2    The camshaft caps are marked to indicate their position as a guide during reassembly, and should be removed in a diagonal sequence so that the shafts are kept square to the head. Slacken each bolt by about half a turn at a time until the shafts are no longer under pressure from the valve springs. Lift away the caps, taking care to retain the locating dowels.

3    Lift one of the camshafts clear of the head and disengage it from the cam chain. Place a screwdriver or a bar through the chain loop to prevent it from falling into the crankcase, then disengage the remaining shaft. The cam chain guide can now be lifted out of the tunnel.

## 5    Dismantling the engine/gearbox unit: general

1    Before commencing any dismantling work the external surfaces of the unit should be cleaned thoroughly. Plug the inlet and exhaust ports with rag and refit the cylinder head breather cover. Apply a proprietary degreaser, working it into the casting recesses with a stiff brush. Once all dirt has been loosened, wash off with water and allow to dry. Steam or hot pressure wash equipment may also be used to good effect.

2    Never use excessive force to remove a stubborn component; if a part proves difficult to remove, this is often due to corrosion or because the operation has been tackled in the wrong sequence. Where problems are likely to be encountered this will be mentioned in the text.

3    To facilitate dismantling, an improvised engine stand will prove invaluable. Failing this, collect an assortment of wooden blocks for use as props for the unit on the workbench.

## 6    Dismantling the engine/gearbox unit: removing the cylinder head cover and camshafts

1    Remove the camshaft end covers. These are made of pressed steel or cast alloy, depending on the model, and each is retained by two

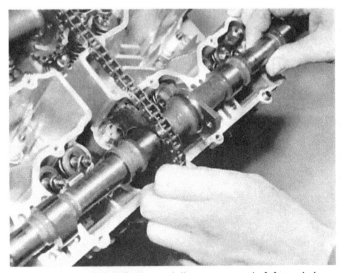

6.3 Remove the camshaft caps and disengage camshaft from chain loop

## 7 Dismantling the engine/gearbox unit: removing the cylinder head

1 The cylinder head can be removed after the camshafts as described in Section 6 above. To release the cylinder head nuts, a long 14 mm socket, part number 09911-74510 will be required. We found that a $\frac{3}{8}$ in drive thin-walled socket worked well as a substitute, but check first that there is adequate clearance in the nut recesses.

2 Release the three 6 mm bolts, these being located at the outer ends of the head and at the centre of the joint face, between the centre exhaust ports. Slacken progressively the twelve cylinder head nuts, turning each one about half a turn at a time in the reverse of the tightening sequence, which is marked in the head casting.

3 As each nut is removed it will be noted that the nut and the washer types vary. To make assembly easier, place them in marked bags or in a box or tray with compartments. The head should now lift away. If it is stuck to the gasket, try jarring it free by placing a hardwood block against the exhaust port and tapping it upwards. Once the seal has been broken, remove the head and place to one side.

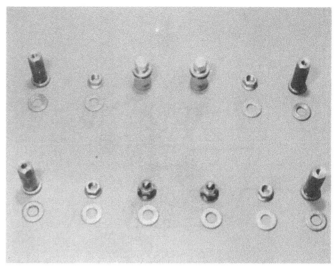

7.3 Note variety of head fasteners – keep them in order!

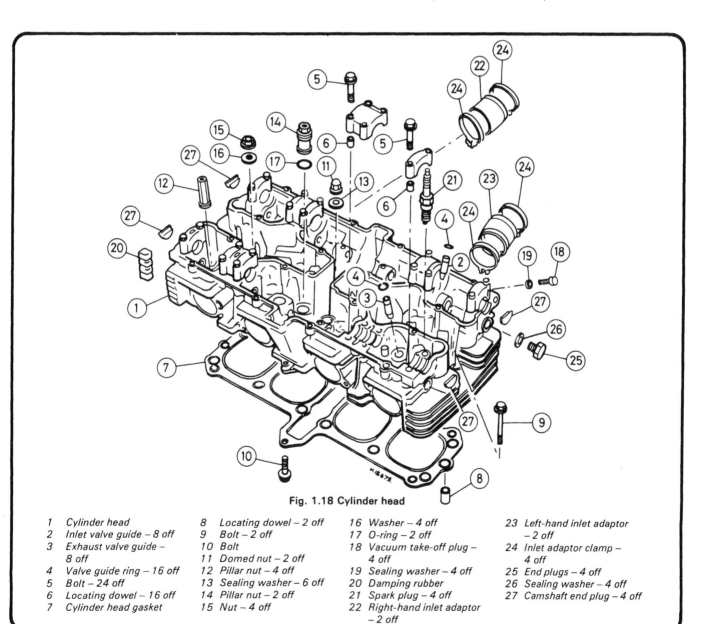

Fig. 1.18 Cylinder head

| | | | |
|---|---|---|---|
| 1 Cylinder head | 8 Locating dowel – 2 off | 16 Washer – 4 off | 23 Left-hand inlet adaptor – 2 off |
| 2 Inlet valve guide – 8 off | 9 Bolt – 2 off | 17 O-ring – 2 off | 24 Inlet adaptor clamp – 4 off |
| 3 Exhaust valve guide – 8 off | 10 Bolt | 18 Vacuum take-off plug – 4 off | 25 End plugs – 4 off |
| 4 Valve guide ring – 16 off | 11 Domed nut – 2 off | 19 Sealing washer – 4 off | 26 Sealing washer – 4 off |
| 5 Bolt – 24 off | 12 Pillar nut – 4 off | 20 Damping rubber | 27 Camshaft end plug – 4 off |
| 6 Locating dowel – 16 off | 13 Sealing washer – 6 off | 21 Spark plug – 4 off | |
| 7 Cylinder head gasket | 14 Pillar nut – 2 off | 22 Right-hand inlet adaptor – 2 off | |
| | 15 Nut – 4 off | | |

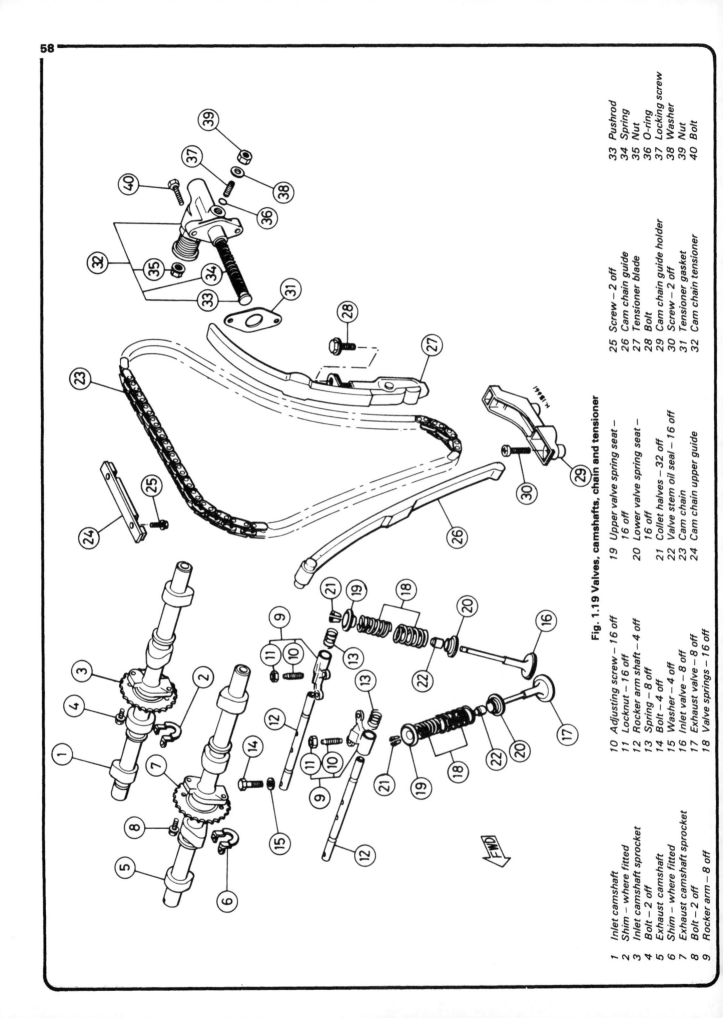

**Fig. 1.19 Valves, camshafts, chain and tensioner**

1 Inlet camshaft
2 Shim – where fitted
3 Inlet camshaft sprocket
4 Bolt – 2 off
5 Exhaust camshaft
6 Shim – where fitted
7 Exhaust camshaft sprocket
8 Bolt – 2 off
9 Rocker arm – 8 off

10 Adjusting screw – 16 off
11 Locknut – 16 off
12 Rocker arm shaft – 4 off
13 Spring – 8 off
14 Bolt – 4 off
15 Washer – 4 off
16 Inlet valve – 8 off
17 Exhaust valve – 8 off
18 Valve springs – 16 off

19 Upper valve spring seat – 16 off
20 Lower valve spring seat – 16 off
21 Collet halves – 32 off
22 Valve stem oil seal – 16 off
23 Cam chain
24 Cam chain upper guide

25 Screw – 2 off
26 Cam chain guide
27 Tensioner blade
28 Bolt
29 Cam chain guide holder
30 Screw – 2 off
31 Tensioner gasket
32 Cam chain tensioner

33 Pushrod
34 Spring
35 Nut
36 O-ring
37 Locking screw
38 Washer
39 Nut
40 Bolt

## 8   Dismantling the engine/gearbox unit: removing the cylinder block and pistons

1   Once the cylinder head has been removed, the cylinder block can be lifted off the holding studs and pistons. If the casting is firmly stuck to the base gasket, tapping around the joint face with a hide mallet will often break the seal, but take great care not to damage the rather brittle fins. A cylinder disassembling tool, Part Number 09912-34510 is available from Suzuki dealers to jack the block away from the crankcase.
2   Lift the block by about an inch or so, then pack clean rag into the crankcase mouths. This will catch any debris from the pistons when they emerge and must not be omitted. Continue lifting the block and support the pistons as they drop clear of the bores.
3   Prise out the gudgeon (piston) pin circlips. These can be retained as patterns but should **not** be reused. Small slots in the pistons are provided to allow the clips to be dislodged using a small screwdriver. Mark each piston to indicate the bore from which it came. It is normal practice to mark the piston crowns by scribing 1 to 4 working from the left-hand end.
4   The pistons can usually be freed by pushing out the gudgeon pins by hand or with a long bar. If a pin is exceptionally tight, it may prove necessary to employ a drawbolt arrangement such as that shown in the accompanying line drawing. In addition, note that warming the piston with a rag soaked in very hot water will expand the alloy, making removal much easier.
5   Remove the single bolt which retains the tensioner blade and lift it away.

8.3 Prise out the circlips using a small screwdriver

8.4a Displace and remove the gudgeon pin to free the piston

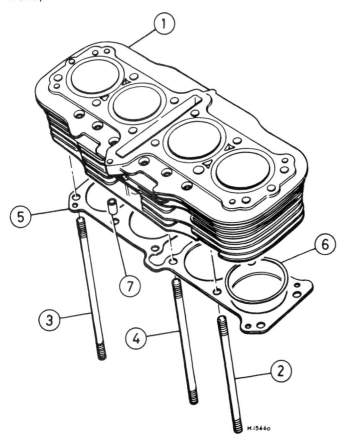

Fig. 1.20 Cylinder barrel

| 1 | Cylinder barrel | 5 | Base gasket |
|---|---|---|---|
| 2 | Stud – 4 off | 6 | O-ring – 4 off |
| 3 | Stud – 6 off | 7 | Hollow dowel – 2 off |
| 4 | Stud – 2 off | | |

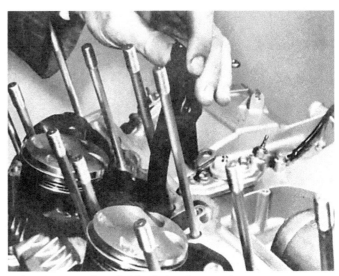

8.4b Release tensioner holding bolt and remove it

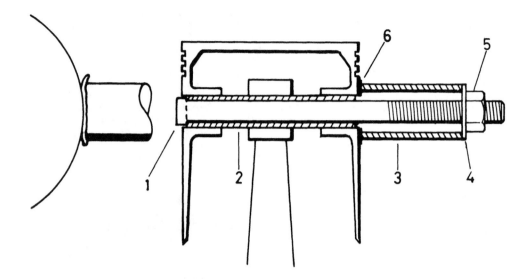

**Fig. 1.21 Drawbolt arrangement for removing a tight gudgeon pin**

| | | | | | |
|---|---|---|---|---|---|
| 1 | Extractor bolt | 3 | Tube | 5 | Nut |
| 2 | Gudgeon pin | 4 | Washer | 6 | Rubber washer |

### 9  Dismantling the engine/gearbox unit: removing the ignition pickup assembly

1   Trace the pickup wiring back along the crankcase and free it from its guide clips. Remove the three screws which retain the pickup cover and lift it away. Examine the pickup backplate, noting the position of the reference mark near the topmost screw. This should normally align with the centre of the screw, but can be marked against the crankcase for clarity, if required.

2   Release the backplate screws and remove the backplate assembly. Place a spanner on the larger of the two hexagons on the automatic timing unit (ATU), remove the holding bolt and lift the ATU away whilst preventing crankshaft rotation by holding the larger hexagon. Lift the ATU away, noting the small pin which locates it. If this is loose it can be removed from the crankshaft end.

### 10  Dismantling the engine/gearbox unit: removing the clutch

1   If the clutch is to be removed with the engine in the frame, it will first be necessary to drain the engine oil. Remove the eleven screws (early models) or bolts (late models) which retain the clutch outer cover. As the cover is pulled clear a certain amount of residual oil will be released, and some provision must be made to catch this.

2   Slacken evenly and progressively the six clutch spring bolts, then remove them together with the washers and springs. Lift out the clutch plates. The last plain plate is retained by a wire clip and should be left in position at this stage.

3   It will be necessary to hold the clutch centre to prevent rotation while the nut is removed. Suzuki dealers can supply a holding tool, Part Number 09920-53710, or a home-made equivalent can be fabricated, as shown in the accompanying line drawing and photograph, 35.4a.

4   With the clutch centre held, flatten the locking tab and remove the nut. The tab washer and the clutch centre can now be slid off the shaft, followed by the thick thrust washer.

5   To facilitate removal of the clutch drum, the central bearing and sleeve assembly must be extracted. Two threaded holes are provided for this purpose, in the sleeve. Run two 6 mm screws into these holes, using them to pull the sleeve out of the drum. The clutch drum needle roller bearings can now be removed and the drum moved to one side to permit removal. Finally, slide off the oil pump drive gear, bearing, sleeve and thrust washer.

10.2a Unscrew the clutch bolts, spacers, springs and pressure plate ...

10.2b ... then remove the clutch plates and release centre nut

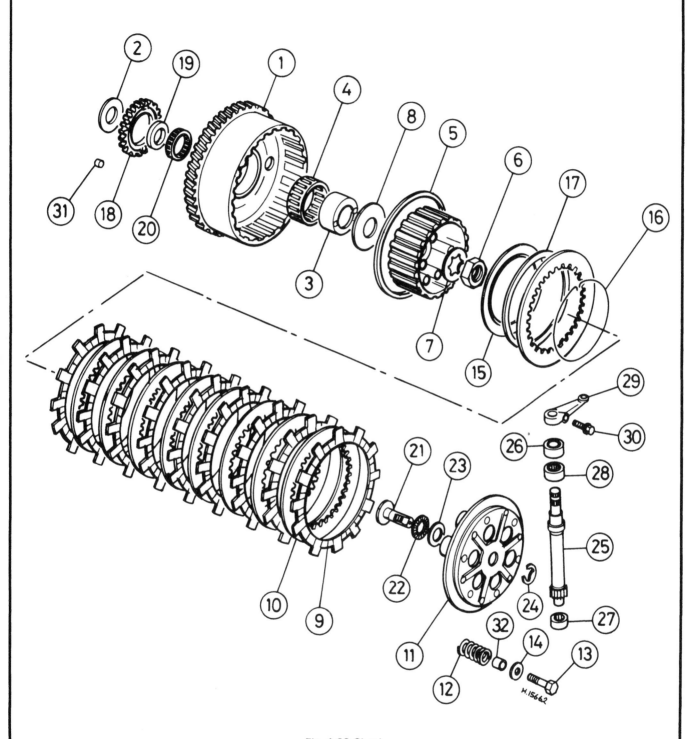

**Fig. 1.22 Clutch**

| | | | |
|---|---|---|---|
| 1 | Clutch drum | 9 | Friction plate – 9 off | 17 | Diaphragm spring | 25 | Operating shaft |
| 2 | Thrust washer | 10 | Plain plate – 9 off | 18 | Oil pump drive pinion | 26 | Oil seal |
| 3 | Sleeve | 11 | Pressure plate | 19 | Sleeve | 27 | Bearing |
| 4 | Needle roller bearing | 12 | Spring – 6 off | 20 | Needle roller bearing | 28 | Bearing |
| 5 | Clutch centre | 13 | Bolt – 6 off | 21 | Release rack | 29 | Operating lever |
| 6 | Centre nut | 14 | Washer – 6 off | 22 | Thrust bearing | 30 | Bolt |
| 7 | Splined tab washer | 15 | Spring seat | 23 | Thrust washer | 31 | Rubber peg |
| 8 | Thrust washer | 16 | Wire clip | 24 | E-clip | 32 | Spacer – 6 off |

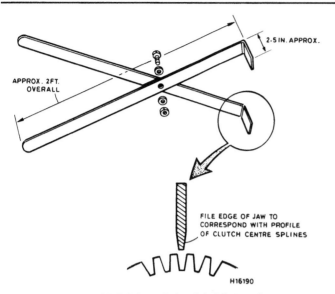

Fig. 1.23 Fabricated clutch holding tool

2·5 IN. APPROX.

APPROX. 2FT.
OVERALL

FILE EDGE OF JAW TO
CORRESPOND WITH PROFILE
OF CLUTCH CENTRE SPLINES

H16190

## 11 Dismantling the engine/gearbox unit: removing the oil pump, bearing retainers and gear indicator switch

1   With the clutch removed as described in Section 10, access to the oil pump is possible. Remove the circlip which retains the oil pump driven gear, then slide off the gear to reveal the pump body. Remove the three securing screws and pull the pump clear of the casing recess.

2   If the crankcase halves are to be separated, two bearing retainers will be found inside the clutch recess, at the ends of the gearbox shafts. These span the crankshaft joint and must be removed. The output shaft retainer is a flat plate held by four countersunk screws, whilst the adjacent input shaft retainer is a shaped component held by three screws. Using an impact driver, remove the screws and lift the retainers away. Note that it is not necessary to remove the remaining oil gallery cover.

3   Moving to the opposite side of the unit, flatten the locking tabs of the input shaft oil seal retainer, slacken the bolts and remove it. Between and below the two shafts is a single nut. This is one of the crankcase fasteners and should be removed at this stage in case it is overlooked later.

4   Release the two screws which retain the gear indicator switch to the end of the selector drum. Remove the switch, taking care not to lose the small contact and spring which will be freed as it is lifted clear.

11.1a Remove the circlip and pull off the pump pinion

11.1b The pump body is retained by three bolts

11.4a Release gear indicator switch ...

11.4b ... and remove contact and spring

## 12 Dismantling the engine/gearbox unit: removing the external gearchange components

1 Although it is not essential that the following operations are carried out prior to crankcase separation, if the crankcase components are to be removed, it is preferable to dismantle the gearchange external components now.

2 Prise off the wire retaining clip from the left-hand end of the gearchange shaft. The shaft and its claw assembly can now be displaced to the right and removed, taking care not to damage the oil seal on the shaft splines. As the shaft is displaced, the ends of the centralising spring will drop clear of their locating pin.

3 Remove the two screws which hold the selector drum bearing retainer and lift it away. The selector pawl guide can be removed in a similar manner, but take care to hold the pawls in place as they are removed; they are under spring pressure and are inclined to escape.

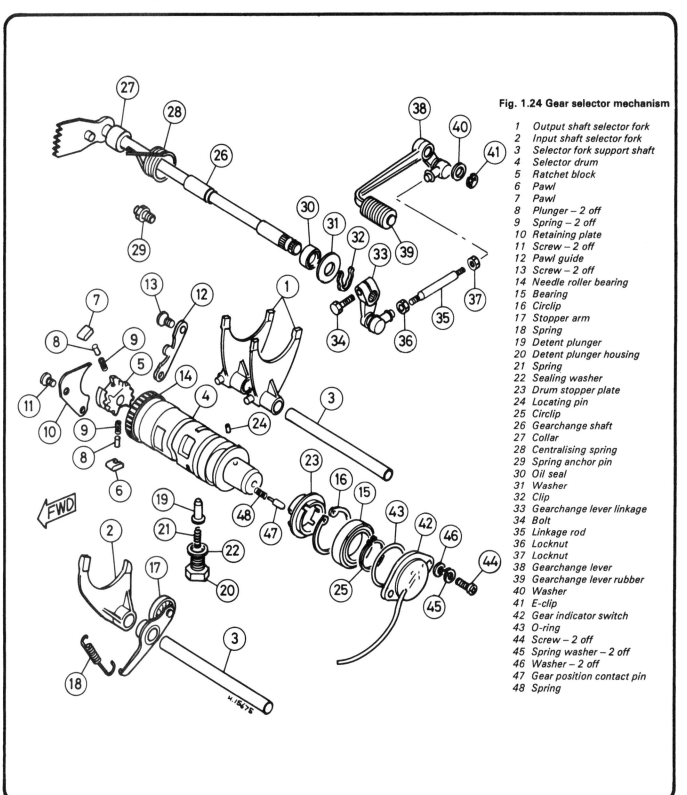

Fig. 1.24 Gear selector mechanism

1 Output shaft selector fork
2 Input shaft selector fork
3 Selector fork support shaft
4 Selector drum
5 Ratchet block
6 Pawl
7 Pawl
8 Plunger – 2 off
9 Spring – 2 off
10 Retaining plate
11 Screw – 2 off
12 Pawl guide
13 Screw – 2 off
14 Needle roller bearing
15 Bearing
16 Circlip
17 Stopper arm
18 Spring
19 Detent plunger
20 Detent plunger housing
21 Spring
22 Sealing washer
23 Drum stopper plate
24 Locating pin
25 Circlip
26 Gearchange shaft
27 Collar
28 Centralising spring
29 Spring anchor pin
30 Oil seal
31 Washer
32 Clip
33 Gearchange lever linkage
34 Bolt
35 Linkage rod
36 Locknut
37 Locknut
38 Gearchange lever
39 Gearchange lever rubber
40 Washer
41 E-clip
42 Gear indicator switch
43 O-ring
44 Screw – 2 off
45 Spring washer – 2 off
46 Washer – 2 off
47 Gear position contact pin
48 Spring

### 13 Dismantling the engine/gearbox unit: removing the starter motor

1   Remove the two screws which retain the starter motor cover and lift it away. Pull out and place to one side the small rubber block which holds the alternator wiring inside the starter motor recess. Remove the two bolts which retain the motor to the crankcase. Lever the motor body back in the recess until the drive end comes free from the casing hole, then lift the motor up and clear of the crankcase.

### 14 Dismantling the engine/gearbox unit: removing the alternator and starter drive

1   Remove the alternator cover and lift it away, together with the alternator stator. The rotor forms an assembly with the starter clutch, this being retained on the tapered crankshaft end by a central nut. To facilitate removal, it will be necessary to prevent the crankshaft from turning, and a number of methods for achieving this are described below.
2   Suzuki dealers can obtain a rotor holding tool, Part Number 09930-44910. This has a crescent shaped end with two locating bolts which engage in the large holes in the rotor edge. In the absence of this tool, a strap wrench can be used instead. If the cylinder block and pistons have been removed, a further alternative is to pass a smooth round bar through the left-hand connecting rod small end eye, supporting the bar ends on small hardwood blocks.
3   With the crankshaft suitably restrained, slacken the rotor centre nut by two or three turns. Fit the rotor extractor, Suzuki Part Number 09930-34911, to the rotor boss then tighten the extractor bolt to draw the rotor off its taper. Note that whilst clearance exists for a legged puller to be used, this method cannot be recommended because of the risk of damage to the rotor; the correct tool is considerably less expensive than a new rotor.
4   The starter idler pinion and shaft can be pulled out of the casing, taking care not to lose the thrust washer at each end of the shaft.

### 15 Dismantling the engine/gearbox unit: separating the crankcase halves

1   Remove the upper crankcase bolts, two of which are located inside the starter motor recess. Turn the unit over, placing blocks beneath the rear of the crankcase to support it horizontally. Remove the sump (oil pan) bolts and lift it away.
2   Slacken evenly and progressively the remaining crankcase bolts, and make sure that the single nut near the end of the selector drum has been removed. The crankcase halves can now be separated, lifting the lower crankcase off the inverted upper crankcase half.
3   There will usually be a certain amount of resistance caused by the joint sealant and the various locating dowels and rings. To help separate the crankcase halves, leverage points are provided. It was found that a 6 mm nut and bolt can be used to 'jack' the cases apart, the arrangement being self-explanatory if the jacking points are examined. Note also that it will help to tap around the joint, using a hide mallet.

### 16 Dismantling the engine/gearbox unit: removing the crankcase components

#### Upper crankcase half
1   Lift out the input and output shafts, placing them on a clean surface to await further attention. Remove the crankshaft assembly, which may prove somewhat reluctant to come free. Hitting the crankshaft ends upwards with the palm of the hand will usually be sufficient to dislodge it. Once the crankshaft has been lifted clear, disengage the cam chain and place it and the crankshaft to one side.
2   The crankshaft and gearbox shafts are located by various dowel pins and half-rings, and these should be removed for safe keeping only if loose. The small pins which locate the main bearings should remain in position in the crankcase.

#### Lower crankcase half
3   If still in position, remove the selector drum retainer and the pawl retainer. Slacken and remove the selector drum detent plunger assembly from the underside of the casing. Slide the gear selector fork shafts out of the casing and lift away the forks placing them on their respective shafts for safe keeping. If necessary, the shafts can be gripped with pliers to assist removal.
4   Remove the circlip from the end of the selector drum which can then be displaced and removed. Working from the inside of the casing, free the circlip which secures the selector drum bearing and tap it inwards to remove it, using a large socket as a drift.

13.1 Remove the two retaining bolts and pull motor out of casing recess

14.4 Pull starter idler assembly out of casing, taking care not to lose thrust washers

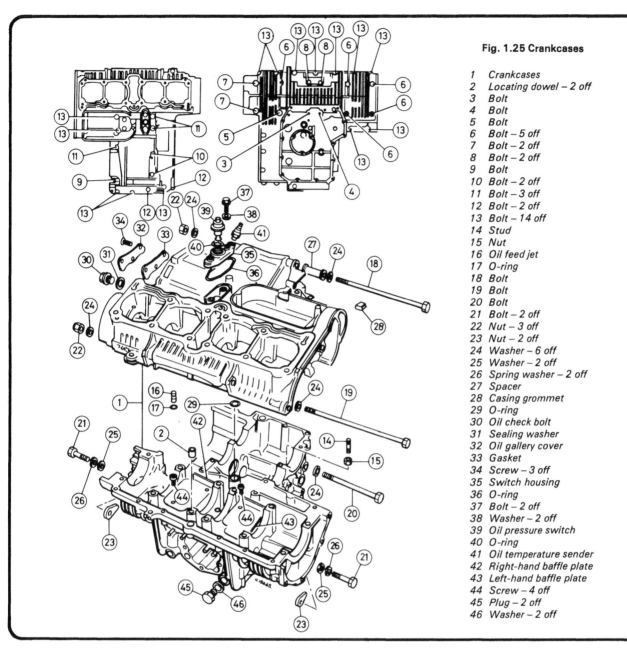

**Fig. 1.25 Crankcases**

1   Crankcases
2   Locating dowel – 2 off
3   Bolt
4   Bolt
5   Bolt
6   Bolt – 5 off
7   Bolt – 2 off
8   Bolt – 2 off
9   Bolt
10  Bolt – 2 off
11  Bolt – 3 off
12  Bolt – 2 off
13  Bolt – 14 off
14  Stud
15  Nut
16  Oil feed jet
17  O-ring
18  Bolt
19  Bolt
20  Bolt
21  Bolt – 2 off
22  Nut – 3 off
23  Nut – 2 off
24  Washer – 6 off
25  Washer – 2 off
26  Spring washer – 2 off
27  Spacer
28  Casing grommet
29  O-ring
30  Oil check bolt
31  Sealing washer
32  Oil gallery cover
33  Gasket
34  Screw – 3 off
35  Switch housing
36  O-ring
37  Bolt – 2 off
38  Washer – 2 off
39  Oil pressure switch
40  O-ring
41  Oil temperature sender
42  Right-hand baffle plate
43  Left-hand baffle plate
44  Screw – 4 off
45  Plug – 2 off
46  Washer – 2 off

16.2 Lift out the half-rings which locate the gearbox shafts

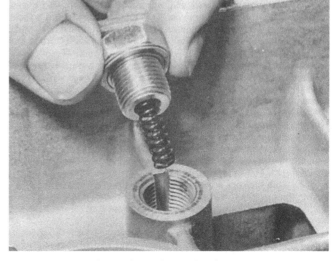

16.3a Unscrew the detent plunger bolt and spring ...

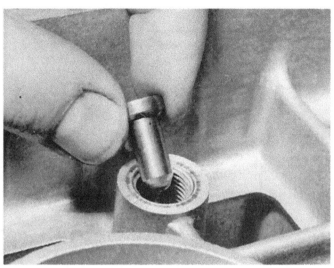

16.3b ... and remove the plunger

## 17 Examination and renovation: general

1    Before examining the parts of the dismantled engine unit for wear
it is essential that they should be cleaned thoroughly. Use a
petrol/paraffin mix or a high flash-point solvent to remove all traces of
old oil and sludge which may have accumulated within the engine.
Where petrol is included in the cleaning agent normal fire precautions
should be taken and cleaning should be carried out in a well ventilated
place.
2    Examine the crankcase castings for cracks or other signs of
damage. If a crack is discovered it will require a specialist repair.
3    Examine carefully each part to determine the extent of wear,
checking with the tolerance figures listed in the Specifications section
of this Chapter or in the main text. If there is any doubt about the
condition of a particular component, play safe and renew.
4    Use a clean lint-free rag for cleaning and drying the various
components. This will obviate the risk of small particles obstructing the
internal oilways, and causing the lubrication system to fail.

## 18 Crankshaft assembly: examination and renovation

1    The crankshaft is a pressed-up assembly, the main and big-end
bearings being of the roller type. The robust construction should
ensure that very high mileages can be covered before attention is
required, unless lubrication failure occurs. In the event of failure of one
or more components of the crankshaft it is possible to obtain
replacement parts through a Suzuki dealer, but the dismantling and
reassembly of the crankshaft should not be undertaken without the
necessary press and alignment equipment. For this reason, a damaged
crankshaft should be entrusted to a Suzuki dealer for repair.
2    Failure of one or more main bearings will be characterised by
rumbling or vibration when the engine is run. Such failure will often be
readily apparent, but each bearing can be checked after any residual oil
has been washed out. Any discernible free play or roughness will
require the renewal of the affected bearing. If in doubt, have the
condition checked by a Suzuki dealer.
3    If suitable facilities are available, set up the crankshaft ends on V-
blocks, then position a dial gauge against one of the centre main
bearings. Slowly rotate the crankshaft and check that runout does not
exceed 0.10 mm (0.004 in).
4    Big-end failure is characterised by a pronounced knocking noise,
most evident when the engine is under load and at higher engine
speeds. It should be noted that whilst a certain amount of end float
(axial clearance) is both desirable and necessary, no radial play should
be evident when the connecting rod is moved up and down.

5    Check the axial clearance (end float) of each big-end bearing using
feeler gauges. This must not exceed the service limit of 1.00 mm
(0.039 in). Radial clearance is measured using a dial gauge set up
against the small end eye of the connecting rod. Move the rod from
side to side, noting the amount of movement indicated by the gauge.
This must not exceed the 3.0 mm (0.12 in) service limit.

## 19 Connecting rods: examination and renovation

1    The connecting rods are unlikely to require attention, other than as
a result of engine damage such as seizure or a dropped valve. If any
of the rods are visibly damaged, have the crankshaft overhauled.
Remember that a stressed connecting rod which breaks in use is
extremely dangerous to the rider and other road users, and will
damage the engine unit beyond repair.
2    Examine each rod for indications of cracking, particularly where
other mechanical damage has been discovered. Renew any connecting
rod which shows signs of damage of this type.
3    The diameter of the small end eye can be checked using a bore
micrometer, and should not exceed the service limit.

18.2 Wash bearings in clean petrol and check for free play

18.5 Check big-end bearings for end float using feeler gauges

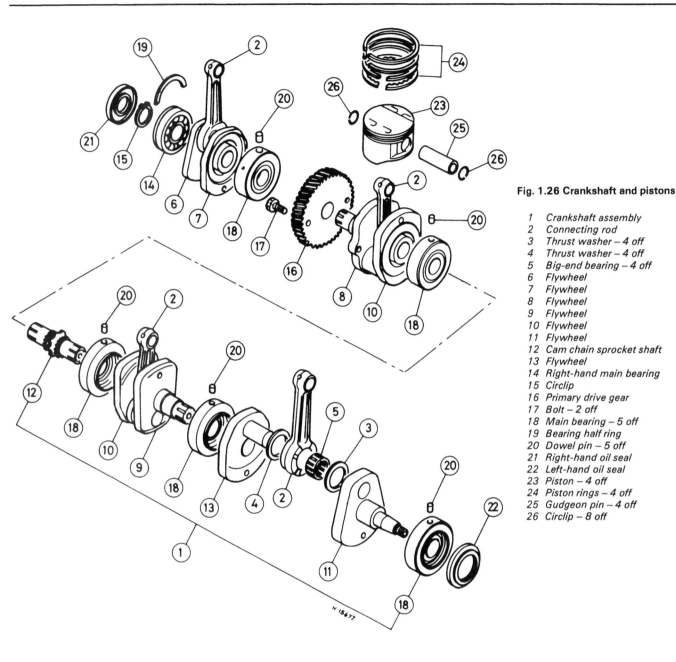

Fig. 1.26 Crankshaft and pistons

1   Crankshaft assembly
2   Connecting rod
3   Thrust washer – 4 off
4   Thrust washer – 4 off
5   Big-end bearing – 4 off
6   Flywheel
7   Flywheel
8   Flywheel
9   Flywheel
10  Flywheel
11  Flywheel
12  Cam chain sprocket shaft
13  Flywheel
14  Right-hand main bearing
15  Circlip
16  Primary drive gear
17  Bolt – 2 off
18  Main bearing – 5 off
19  Bearing half ring
20  Dowel pin – 5 off
21  Right-hand oil seal
22  Left-hand oil seal
23  Piston – 4 off
24  Piston rings – 4 off
25  Gudgeon pin – 4 off
26  Circlip – 8 off

## 20 Cylinder block: examination and renovation

1   An excessively worn cylinder block is usually indicated by smoking from the exhausts and by piston slap, a metallic rattle which occurs with little or no load on the engine. If the top of each bore is examined, a wear ridge will be found on the thrust side, the depth of which will vary according to the degree of wear present. The ridge denotes the upper limit of travel of the top ring.

2   Bore wear is measured using an internal or bore micrometer. Two measurements are made at right angles, just below the wear ridge. This is repeated about half way down the bore and again near the bottom of the bore, a total of six measurements. If any one measurement exceeds the service limit shown in the specifications, the cylinder block must be rebored and oversized pistons fitted.

3   In the absence of a bore micrometer, and if a decision cannot be made from a visual check of the bores, the block should be taken to a Suzuki dealer for checking. Needless to say, scoring or other damage to the bore surfaces will necessitate reboring, irrespective of the amount of wear.

4   Two piston oversizes are available; 0.50 mm (0.020 in) and 1.0 mm (0.040 in). It is recommended that the reboring work is entrusted to a Suzuki dealer, who will also be able to supply the correct oversize pistons.

## 21 Pistons and piston rings: examination and renovation

1   Attention to the pistons and rings can be overlooked if a rebore is required, since new components will be fitted as a matter of course.

2   Examine each piston closely, rejecting it if serious scoring or discolouration due to exhaust gas 'blow-by' is evident. Using a blunt scraper, remove carbon deposits from the piston crown, noting that the better the surface finish the slower the subsequent build up of carbon will be. If desired, a good surface finish can be achieved using metal polish.

3   Using a micrometer, measure the piston diameter at right angles to the gudgeon pin bore and about 15 mm (0.6 in) up from the bottom of the skirt. If the diameter is less than the service limit, the pistons must be renewed. Note that if this degree of wear is found it is likely that the piston/cylinder clearance will have exceeded the service limit.

If this is so, reboring and new pistons will be required.

4   Measure the ring/groove clearances of the top and 2nd rings using feeler gauges. If either exceeds the service limit, renew the piston and rings.

5   With some experience, the rings can be removed for examination and measurement by spreading the ends slightly and sliding them off the piston. It should be noted, however, that the rings are brittle and will snap if stretched too far. A safer method, and one which can be used to free gummed rings, is shown in the accompanying line drawing.

6   Carefully remove accumulated carbon from the grooves, then measure the groove width with feeler gauges. The ring thickness can be checked using a micrometer, and both readings compared with those given in the specifications.

7   The general condition of the rings is checked by measuring the free end gap using a vernier caliper. The installed end gap is measured by pushing the ring into the bottom (unworn) area of the bore, and checking the end gap with feeler gauges. This will indicate the degree of ring wear. If within tolerance in all other respects it is possible to fit new rings to compensate for normal wear.

8   Note that the wear ridge at the top of each bore must be removed, or the new rings may be broken by it. This job is done using a de-ridging tool, and is best done professionally along with glaze busting. This latter operation is essential if the new rings are to bed into the bores, and may be carried out using one of the proprietary tools designed for this work, if available.

**Fig. 1.27 Freeing gummed piston rings**

---

**22  Cylinder head and valves: examination and renovation**

---

1   Before commencing any dismantling work on the head, have ready a suitable container for the valve components. This should take the form of a box divided into 16 compartments, each marked so that the relevant valve is clearly identified; 'IN 1/L' to denote cylinder 1, inlet, left-hand, for example. On no account allow the valve components to be mixed up.

2   Pivot the rocker arms clear of the valves, then remove each one in turn, using a valve spring compressor. With the valve spring compressed, displace and remove the collet halves (a pair of tweezers or a small magnet is invaluable for this), then lift away the spring retainer, springs and spring seat. The valve can now be displaced and removed.

3   To remove the rocker arms, remove the shaft locating bolt, then unscrew the shaft end cap. Screw a 6 mm bolt or screw into the shaft end, and use this to draw the shaft out of the head. Remove the rocker components, noting their respective positions and place them in

sequence on the shaft to which they belong. Mark each shaft to ensure that it is refitted correctly.

**Cylinder head**

4   Degrease the head casting and remove accumulated carbon from the combustion chambers and ports. Check carefully for cracks, especially near the valve seats and spark plug hole. If cracking is detected, seek professional help.

5   If there has been any indication of warpage, such as a blown head gasket, check this using a straight edge and feeler gauges. If at any point the 0.2 mm (0.008 in) service limit is exceeded, the head will have to be skimmed flat or renewed.

**Valves**

6   Examine each valve, rejecting any that are badly scored or burnt across the seating face. An indication of the degree of wear is given by the width of the 'face' (or more accurately the margin); the thin parallel face between the seating face of the valve and the valve head surface. If this is worn to 0.5 mm (0.02 in) or less, the valve must be renewed.

7   The valve stem runout can be checked using a dial gauge by resting each valve stem on V-blocks. A valve should be renewed if runout exceeds 0.05 mm (0.002 in), and the cause of the bending investigated. The radial runout of the valve head is checked in a similar fashion, the service limit being 0.03 mm (0.001 in).

8   Measure each valve stem diameter with a micrometer, renewing it if worn to or beyond the service limit. Note that stem wear will contribute to stem/guide clearance, and that guide wear cannot be assessed using a worn stem.

**Valve guides**

9   Guide wear is checked by measuring the 'wobble' between the valve guide and the stem, using a dial gauge. If this exceeds the service limit and the valve stem is unworn or within limits, the guide must be renewed.

10  To renew the guides it is necessary to have to hand the various removal and fitting tools, a reamer to enlarge the bores in the cylinder head to take the oversized guides, another reamer to finish the valve guide bores and the valve seat refacing equipment so that the seats can be cut to suit the new guides. In view of the sheer cost of this equipment, the work is best entrusted to a Suzuki dealer.

**Valve springs**

11  Measure the free length of the inner and outer valve springs, renewing them as pairs if worn to or beyond the service limit. It is unlikely that the springs will have weakened without becoming shorter, but this can be checked by measuring the spring pressures when compressed to specified lengths. This is not easy to do at home, but the resourceful owner should be able to contrive a suitable method, using a spring balance and a ruler.

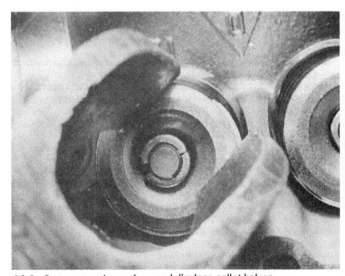

22.2a Compress valve springs and displace collet halves ...

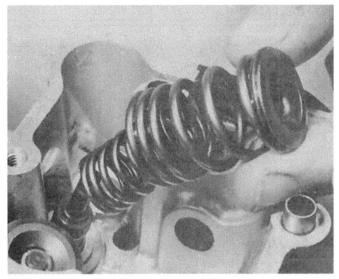

22.2b ... and lift away the springs and retainer

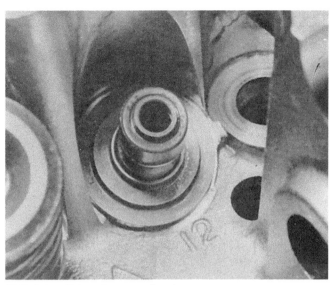

22.2c Prise off the valve stem oil seal ...

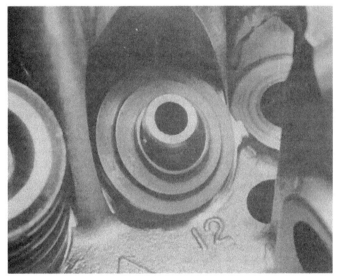

22.2d ... and lift off the lower seat

22.2e Displace and withdraw the valve

22.3a Remove the rocker shaft locating bolt ...

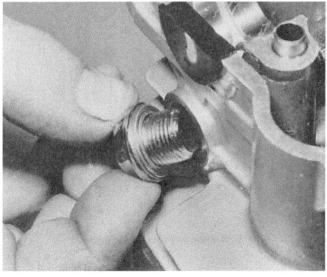

22.3b ... and unscrew the shaft end cap

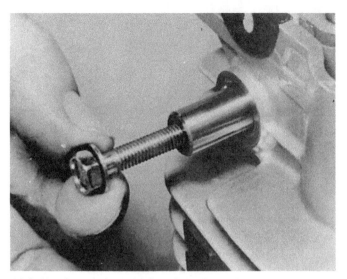

22.3c Use 6 mm bolt to withdraw the rocker shaft

22.3d Rocker arm and spring can now be lifted away

### 23 Valves and valve seats: refacing and grinding

1    If examination has shown a valve contact face to be badly pitted or eroded, it will be necessary to have it refaced by an engine reconditioner or renewed. This work requires the use of a grinding wheel and the appropriate jigs, and is not worth attempting at home. Grinding or lapping valves will not compensate for pitting, and if done to excess will pocket a valve in its seat.

2    It is likely that the valve seats will be in a similar state to the valve contact faces, and this work should be carried out at the same time as the valve refacing. Although a fairly simple operation, a good quality set of cutters is required, together with skill in their use. It is quite easy to take off too much material, rendering the seat and thus the head unserviceable. Unless experienced in this work, leave it to a professional.

3    Light wear or pitting can be removed by lapping the valve and seat, and this operation is undertaken during overhaul to restore the seal between the two. Apply a thin film of fine carborundum paste to the valve contact face, oil the valve stem and place the valve in its correct seat. Using a valve lapping tool, lap the valve using a semi-rotary motion. Lift the valve occasionally to redistribute the abrasive paste.

4    Remove the valve and wipe the compound off the valve and seat. Examine the contact faces, each of which should show an unbroken light grey ring. Any deep pits will now be highlighted, as will an uneven contact area, and a decision can be made as to whether lapping should continue or if grinding is required. The standard contact width on the valve should be 0.9 – 1.1 mm (0.035 – 0.043 in), with a corresponding band evident on the valve seat.

5    If lapping seems adequate, repeat the operation on the remaining valves, making sure that all traces of the abrasive are removed; any compound finding its way into the engine will cause serious damage.

6    It is permissible to grind flat the end of the valve stem, provided that this does not reduce the length between the stem end and the upper edge of the collet groove to less than 3.6 mm (0.14 in). If the stem is ground, it is essential that it is kept square, and when reassembled the stem end must still be proud of the collets.

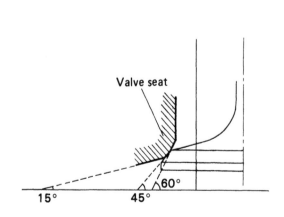

Fig. 1.28 Valve and seat grinding angles

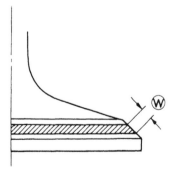

Fig. 1.29 Valve seating width

*Seat width (W) must be 0.9 – 1.1 mm (0.035 – 0.043 in)*

## 24 Cylinder head components: reassembly

1   After examination and reconditioning has been completed, check that the head and valve components are quite clean. If the valve stem seals have hardened or were removed, they must be renewed, pushing them over the guide ends using a stepped drift to avoid damage.
2   Place the valve spring seats in position, taking care not to confuse them with the upper retainers. Lubricate the valve stems with molybdenum disulphide grease, then slide them into their respective guides.
3   Assemble each set of springs in turn. Fit the springs with the closer pitch end downwards, then fit the upper retainer. Compress the springs and fit the collet halves, ensuring that they engage correctly in the groove in the valve stem. Release the compressor and check that the various components are seated correctly by tapping the end of the stem. Repeat the above sequence to fit the remaining valves.
4   Reassemble the rocker arms and shafts, ensuring that they are fitted in their original positions. Refit the shaft end plugs after securing the shaft locating bolts. If it proves necessary to turn the shaft to align the threaded hole for the locating bolt, this can be done using the 6 mm screw employed during removal.

## 25 Camshafts and drive components: examination and renovation

### Camshafts
1   Examine the cam lobes for wear or scoring. This will be most evident near the peaks of the lobes, and if present will require the renewal of the camshaft. Whilst the lobes will eventually wear down through normal use, scoring can usually be attributed to failure to change the oil at the specified interval.
2   If the lobes are undamaged, measure the overall cam lobe height to assess the degree of wear. The service limit for both the inlet and exhaust cams is given in the Specifications.
3   The camshafts run directly in bearing surfaces formed by the cylinder head casting and the camshaft caps. The clearance between the bearing surfaces and the camshaft journals can be measured using a deformable plastic strip known as Plastigage.
4   In the US, Plastigage is available from most large dealers, but availability is limited in the UK. Where the material is not available, it will be necessary to check the relevant dimensions directly, using internal and external micrometers, and comparing the readings obtained with those shown in the specifications.
5   Where Plastigage can be obtained, clean the bearing surfaces and place a strip of the material on each journal. Assemble the bearing caps, tightening the bolts to 0.8 – 1.2 kgf m (6.0 – 8.5 lbf ft), then remove them to allow the spread of the Plastigage to be read off against the scale provided. The clearance indicated at the widest point shows the clearance. If this exceeds the service limit, the camshafts, cylinder head or both may have to be renewed. Given the expense that this will entail it is suggested that the advice of an experienced Suzuki dealer is sought.
6   Camshaft runout is measured with the shaft supported on V-blocks, using a dial gauge on the centre journal. If runout exceeds the 0.1 mm (0.004 in) service limit, the camshaft must be renewed.

### Tensioner
7   The tensioner assembly is of the automatic type, an arrangement which ensures that cam chain tension is maintained as the chain wears and stretches during normal use. It is important that the mechanism operates smoothly and evenly in service, and this should be checked during overhauls.
8   Slacken by one or two turns the small locking screw which retains the tensioner plunger. Turn the knurled wheel anticlockwise against spring pressure, then check that the plunger can be moved smoothly and evenly. If resistance is noted, release the locking screw and examine and clean the plunger. If the plunger is bent or scored, it should be renewed.
9   Turn the knurled wheel fully back against spring pressure, then release it, noting whether it returns fully to its rest position. If resistance or hesitation is noted, remove the lock-shaft assembly from the tensioner body, remove the nut which retains the knurled wheel, then release the wheel and spring and displace the lock-shaft.
10   Clean and lubricate the lock-shaft components, then slide the lock-shaft into its housing. Position the spiral ramps so that the stepped ends are 10 mm apart. Place the spring over the shaft end and hook the ends into the housing and the knurled wheel. Turn the knurled wheel one turn anticlockwise, fit the knurled wheel over the lock-shaft end and fit the retaining nut, tightening it to 0.9 – 1.4 kgf m (6.5 – 10.0 lbf ft). Refit the assembled unit, tightening the body to 3.1 – 3.5 kgf m (22.5 – 25.5 lbf ft).
11   Refit the tensioner plunger, turning the knurled wheel until the rod is fully home. Hold it in place, then secure the locking screw to secure it. Note that the locking screw must align with the slot on the side of the plunger.

### Cam chain
12   Pull the cam chain taut, then measure the length of 20 links using a vernier caliper. If this exceeds the service limit of 157.80 mm (6.213 in), the chain must be renewed. The chain should be given a close visual inspection, renewing it if damage is evident.

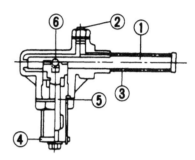

Fig. 1.30 Cam chain tensioner

| | |
|---|---|
| 1   Push rod | 4   Knurled wheel |
| 2   Lock-screw | 5   Lock-shaft |
| 3   Push rod spring | 6   Steel ball |

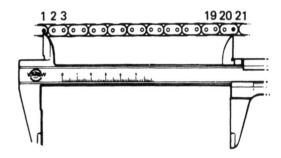

Fig. 1.31 Measuring cam chain wear

## 26 Gearbox components: examination and renovation

1   It will not be necessary to dismantle the gearbox input or output shafts unless damage to the pinions or the shafts has occurred. Reference to the accompanying line drawing and photographs shows the precise sequence in which the shafts are assembled. The input shaft cannot be dismantled or reassembled without access to a press, and should not be attempted in the absence of these facilities.
2   The shafts should be examined for indications of wear or damage. The gear teeth should be free from chipping. Each tooth face should present a polished face, with no indication of pitting. Ensure that the pinions turn smoothly with no tight spots or excessive play.
3   The general level of wear can be estimated by temporarily refitting the shafts in their casing recess and checking the backlash between the various pairs of gears, using a dial gauge. Compare the readings obtained with those given in the specifications.

### Input shaft
4   Arrange the shaft in a vice, using rag to protect the shaft surface and with the left-hand end uppermost. Remove the bearing by working

it up and off the shaft with screwdrivers or a bearing extractor. Lift away the wave washers which fit behind the bearing.

5   Assemble the press as shown in the accompanying photograph, noting that it was convenient to draw off the 5th and 2nd gears together, although it is only the latter which is pressed into position. The two pinions can now be pulled off the shaft and removed. Considerable pressure was needed to effect removal, a locking compound having been used during assembly.

6   Remove the 5th gear bush and thrust washer, if these remained on the shaft, then slide off the 3rd gear pinion. The 4th gear pinion is retained by a circlip and is removed together with its headed bush. Note that the 1st gear pinion is integral with the shaft. The remaining bearing can be left in place unless it requires renewal.

### Output shaft

7   Working from the right-hand end of the shaft, remove the bearing then slide off the large 1st gear pinion, together with the needle roller bearing which supports it, and the thrust washers which are fitted on each side. The 4th gear pinion can now be slid off the shaft.

8   Release the circlip and the splined washer and remove the 3rd gear pinion and its splined bush. The 5th gear pinion is secured by two special splined washers; the first having three locating tangs to hold the second in its locked position. Remove the tanged washer, then twist and remove the locking washer. The pinion can now be slid off the shaft.

9   Free the circlip which locates the 2nd gear pinion and remove it

together with the headed bush on which it runs. The remaining bearing can be drawn off the shaft end if it requires renewal.

### Reassembly

10   Examine the gearbox pinions for wear or damage, renewing as pairs those which show signs of chipping or burring of the teeth or the engagement dogs, or which were found to have excessive backlash. Check that all components are clean, and lubricate the shafts with molybdenum disulphide grease prior to assembly.

11   The assembly sequence is shown in the accompanying photographs, and should be followed carefully. Pay particular attention to the disposition of the washers and circlips. In the case of the latter, note that the sharper edge must face towards the direction of thrust, the curved edge facing the pinion. In the case of the output shaft 3rd gear pinion bush, ensure that the oil hole aligns with that of the shaft.

12   When fitting the input shaft 2nd gear pinion, note that it must be pressed into position until the overall distance between its outer face and that of the 1st gear pinion is 111.4 – 111.5 mm (4.386 – 4.390 in).

13   Before the 2nd gear pinion is fitted, apply a thread locking compound to its internal bore, taking care not to allow any excess to get on the 5th gear pinion. Suzuki 'thread lock super 1303B', Part Number 99000-32030, or an equivalent is recommended. Note that Suzuki advise that the pinion may be removed and refitted twice, after which the shaft should be renewed to avoid any risk of loosening in service. A non-hardening thread locking compound should be used to hold the gearbox sprocket spacer in place on the output shaft.

26.5 Draw off the 5th and 2nd gears as a pair using a press as shown

26.11a Fit the headed bush into the output shaft 2nd gear pinion ...

26.11b ... and slide into place as shown, securing it with the circlip

26.11c Fit the 5th gear pinion with the selector groove outwards

26.11d Slide the locking washer into place and turn it to retain the pinion

26.11e Fit the tanged washer as shown to secure the locking washer

26.11f Slide the splined bush into place ...

26.11g ... and fit the 3rd gear pinion ...

26.11h ... securing it with the splined washer and circlip

26.11i Fit the 4th gear pinion with selector groove inwards

26.11j Place the thrust washer and needle roller bearing

26.11k Fit the 1st gear pinion and thrust washer

26.11l Fit the sealed bearing as shown ...

26.11m ... followed by spacer and seal at other end of assembly

26.12a Fit the 4th gear pinion and its headed bush ...

26.12b ... and secure with thrust washer and circlip

26.12c Fit the 3rd gear pinion, groove inwards

26.12d Slide the thrust washer and bush against shoulder

26.12e Place 5th gear pinion in position

26.12f 2nd gear pinion must be pressed into place ...

26.12g ... until the prescribed length is reached

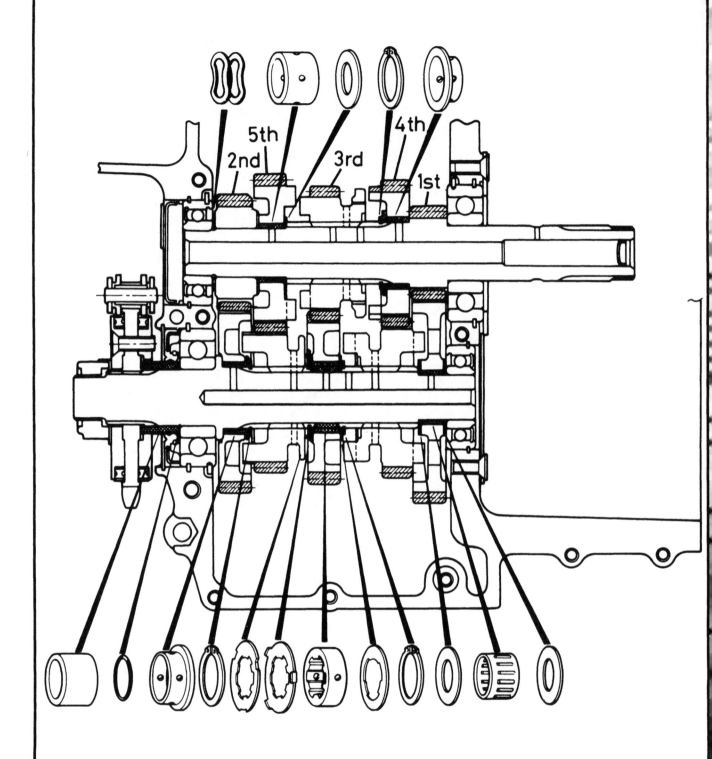

Fig. 1.32 Section view of gearbox shafts showing thrust washer, bush and circlip positions

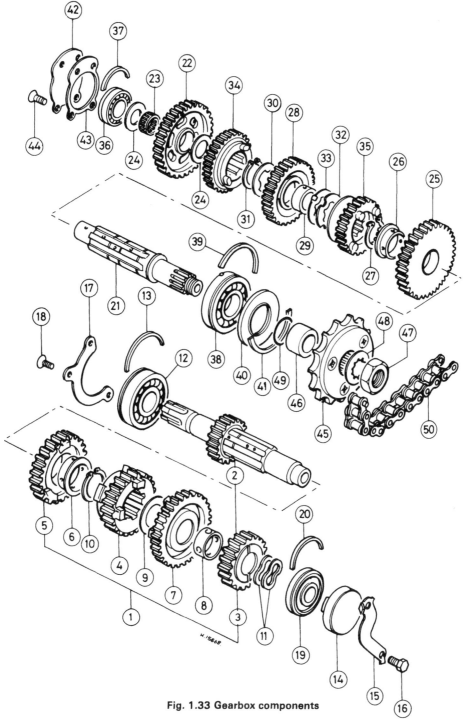

**Fig. 1.33 Gearbox components**

| | | | |
|---|---|---|---|
| 1 | Input shaft assembly | 13 | Bearing half ring |
| 2 | Input shaft | 14 | Input shaft left-hand plug |
| 3 | Input shaft 2nd gear pinion | 15 | Oil seal retaining plate |
| 4 | Input shaft 3rd gear pinion | 16 | Bolt – 2 off |
| 5 | Input shaft 4th gear pinion | 17 | Bearing retaining plate |
| 6 | Headed bush | 18 | Screw – 3 off |
| 7 | Input shaft 5th gear pinion | 19 | Input shaft left-hand bearing |
| 8 | Bush | 20 | Bearing half ring |
| 9 | Thrust washer | 21 | Output shaft |
| 10 | Circlip | 22 | Output shaft 1st gear pinion |
| 11 | Wave washer – 2 off | 23 | Needle roller bearing |
| 12 | Input shaft right-hand bearing | 24 | Thrust washer – 2 off |
| | | 25 | Output shaft 2nd gear pinion |

| | | | |
|---|---|---|---|
| 26 | Headed bush | 38 | Output shaft left-hand bearing |
| 27 | Circlip | 39 | Bearing half ring |
| 28 | Output shaft 3rd gear pinion | 40 | Output shaft left-hand oil seal |
| 29 | Splined bush | 41 | Bearing half ring |
| 30 | Splined washer | 42 | Output shaft end plate |
| 31 | Circlip | 43 | Gasket |
| 32 | Lock washer | 44 | Screw – 4 off |
| 33 | Tanged washer | 45 | Final drive sprocket |
| 34 | Output shaft 4th gear pinion | 46 | Spacer |
| 35 | Output shaft 5th gear pinion | 47 | Nut |
| 36 | Output shaft right-hand bearing | 48 | Lock washer |
| 37 | Bearing half ring | 49 | O-ring |
| | | 50 | Final drive chain |

## 27 Gearchange mechanism: examination and renovation

1    Check the selector fork shafts for straightness by rolling them on a sheet of glass. Renew the forks if bent. Examine the selector forks for wear or damage, paying particular attention to the fork ends and the pegs which engage in the selector drum grooves.
2    The tracks in the selector drum should be free from signs of wear unless lubrication has been neglected. Examine the selector pawl assembly, renewing any component showing signs of burring. Wear on the pawl ends is the most probable cause of selection problems.

## 28 Clutch assembly: examination and renovation

1    After extended use, the clutch friction plates will wear, leading to clutch slip if worn excessively. Measure the thickness of each plate, renewing them if worn to or beyond the service limit.
2    The plain steel plates are unlikely to wear, but may have warped due to overheating. This can be checked by placing each one on a surface plate or a flat glass sheet and measuring any distortion with feeler gauges. The innermost plain plate is retained on the clutch centre by a wire clip. This can be released to allow the removal of the plate, dished washer and seat.
3    The latter components provide a degree of shock absorption, making the clutch smoother in operation. It is unlikely that they will require attention, but it should be noted that the arrangement was changed after the original ET model (see Fig. 1.4).
4    The clutch springs will weaken in time, becoming shorter as they lose their elasticity. The free length should be measured and the springs renewed as a set if worn to or beyond the service limit. Note that the springs used on Katana models are made from 'a premium quality material' and are marked with white paint to denote this. Ensure that the correct springs are fitted when renewal becomes necessary.
5    Examine the clutch centre splines and also those of the clutch outer drum. In time, these will become indented by the clutch plates, causing the plates to catch on the resulting steps and making operation erratic. Light damage may be corrected by using a fine file or abrasive paper, but if wear is serious, the damaged component must be renewed.
6    The clutch drum incorporates the primary driven gear and a transmission shock absorber arrangement. If any of these items are worn it will be necessary to renew the drum complete. In the case of the shock absorber assembly, the best way of assessing wear is to compare the amount of free play with that of a new component. In extreme cases the shock absorber springs will be loose and will rattle if the clutch drum is shaken.
7    The clutch centre bearing must be free from damage or indentations, however small. It is not practicable to measure wear in needle roller bearings, but any blemish on the rollers or race surfaces indicates the need for renewal.

**Clutch modifications**
8    In the course of development, a number of modifications have been made to various clutch components, some of which have been mentioned above. In addition, the rear face of the clutch centre was changed and a modified pressure plate with twelve additional oil drainage holes was introduced. The various modifications were introduced in the middle of model runs, the exact point of introduction being dependent on engine number. For this reason it is essential that the engine number is quoted in full when ordering replacement clutch parts.

## 29 Crankcase and covers: examination and renovation

1    The crankcase halves and the various outer covers are unlikely to become damaged unless the machine is dropped or accident damage is sustained. Small cracks can be repaired by a specialist welding process, providing that care is taken to avoid distortion. Expert advice should always be sought.
2    Damaged threads can be repaired relatively easily by tapping them oversize and fitting a Helicoil thread insert. This work should be entrusted to a local dealer offering this service.
3    If the selector drum bearings require renewal they can be driven out of the casing using a large socket as a drift. Note that the journal

ball bearing is retained by a circlip on its inner face. It is advisable to warm the casing with very hot water to expand the alloy, making removal easier.
4    The oil passages and galleries should be cleaned by flushing them through with a clean solvent. This should be done outside or in a well ventilated area, and care must be taken to avoid any risk of fire. A flat cover above the input shaft boss in the upper crankcase half provides access to the main gallery and can be removed to facilitate cleaning.

## 30 Engine reassembly: general

1    Before commencing reassembly, check that each component has been cleaned and that every gasket face is free from old gaskets or jointing compound. Removal of the latter is greatly facilitated by the use of a solvent. This varies according to the composition of the compound, but methylated spirit, acetone or cellulose thinners are often effective. In some cases scraping may be the only answer, but take great care not to damage the gasket face itself.
2    Check that all gaskets, seals and O-rings are to hand and that they are of the correct type for the model in question; a dry run to check this point may save time and frustration later. Clear the workbench of all unnecessary tools or parts, and make ready an oil can filled with clean engine oil for lubrication as assembly proceeds.
3    Make sure that a torque wrench is available and refer to the torque wrench settings at the front of this Chapter. Note that failure to observe torque settings may result in oil leakage, broken fasteners or distorted castings.

## 31 Engine reassembly: refitting the lower crankcase components

1    Assemble the selector pawl components and fit the unit to the end of the selector drum. It will be noted that the notch in each pawl is offset, and should be fitted towards the drum so that it aligns with the pawl spring. Lubricate the drum bearings, then slide the drum into the crankcase. Check to ensure the circlip which locates the left-hand bearing is in position before the drum is installed. When the drum is in position, fit the circlip to its left-hand end to retain it.
2    Fit the input shaft selector fork and drum stopper arm as shown in the accompanying photograph, and slide the shaft into position to retain it. Hook the end of the stopper arm spring over the location point on the crankcase. The output shaft selector forks can now be fitted and their shaft slid into place. Check that the forks are arranged as shown and that they engage correctly in the drum.
3    Fit the retainer and the pawl guide to the right-hand end of the drum, using a thread locking compound on the retaining screws. Note that the latter are 12 mm in length and should not be confused with the 16 mm screws used elsewhere. Turn the casing over and fit the sump strainer to the casing boss, with the oil inlet pipes facing forward. A thread locking compound should be applied to the three retaining screws.

31.1a Fit springs and pawls to end of ratchet block, noting that they are offset

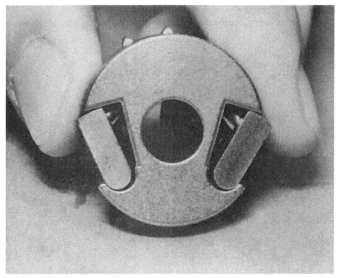

31.1b Hold pawls in position as shown ...

31.1c ... and fit assembly into end of selector drum

31.1d Ensure that drum bearing is retained by its circlip ...

31.1e ... then slide assembled drum into casing ...

31.1f ... securing its left-hand end with a circlip

31.2a Fit input shaft fork, stopper arm and support shaft as shown, noting anchor point for spring

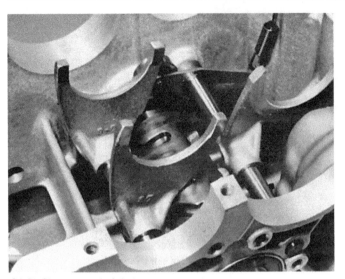

31.2b The output shaft forks can now be installed

31.3 Fit the retainer and pawl guide to the right-hand end of the selector drum

### 32  Engine reassembly: refitting the upper crankcase components

1   Check that the small oil feed jet in the right-hand main bearing recess is unobstructed; the supply of oil to the bearing is dependent on this, so do not overlook it. Fit the main bearing location pins, if removed, and fit the half-ring to the right-hand main bearing recess.

2   Place the cam chain around the crankshaft sprocket and fit a new oil seal to its right-hand end. Lower the crankshaft into position, feeding the cam chain through the opening in the crankcase.

3   Arrange the right-hand main bearing so that it locates over the half-ring and rotate it until the small locating pin sits in the cutout to the rear of the bearing recess, parallel to the gasket face.

4   Position the remaining bearings so that they engage their respective location pins. Each bearing has a small punch mark diametrically opposite the location hole as an aid to alignment. When all the bearings are aligned, the crankshaft will drop into place.

5   Fit the five half rings to their respective grooves in the gearbox shaft bearing recesses. Fit the gearbox shafts, noting that all but the output shaft right-hand bearing have small location pins which must align with the cutouts provided. Note also that the input shaft left-hand bearing must be preceded by the two wave washers removed during dismantling. Finally, fit a new O-ring to the oil feed passage between the crankshaft and gearbox shafts.

32.1a Oil feed nozzle must be unobstructed

32.1b Check that the main bearing locating pins are in place ...

32.1c ... and fit the half-ring retainer

32.2 Fit a new oil seal to the crankshaft and lower it into position

32.3 Check that half-ring locates in groove and that the small pin (arrowed) sits in cutout

32.4 Dots on main bearing outer races should be opposite locating holes

32.5a Fit half-rings and install gearbox shafts

32.5b Do not omit wave washers behind input shaft left-hand bearing

32.5c Note the location pins on all except output shaft right-hand bearing

32.5d Remember to fit new O-ring to oil passage recess

33.1 If cam chain guide holder was removed, refit it using a thread locking compound on screws

## 33 Engine reassembly: joining the crankcase halves

1    If the cam chain guide holder was removed from the lower crankcase half, it should be refitted, using a thread locking compound on the screws. Check that the crankcase mating surfaces are clean and free from oil, then apply a thin, even film of sealant. Suzuki Bond No.4 or any good quality silicone rubber RTV sealant can be used. Leave the sealant for about ten minutes to allow it to cure.

2    Offer up the lower casing half, placing it over the inverted upper half. Check that the forks engage correctly in the grooves in the gearbox pinions. Check that the casing halves seat fully and that the oil seals are located correctly. If necessary, the joint can be closed by tapping the casing with the palm of the hand, but if excessive force is needed, remove the lower half and trace the cause of the problem.

3    Fit the 6 mm and 8 mm crankcase bolts. Following the tightening sequence indicated by the numbers cast into the crankcase next to each bolt, tighten them evenly and progressively to the initial torque figure. Reset the torque wrench, then tighten to the final torque setting as shown below.

| Crankcase torque settings | kgf m | lbf ft |
| --- | --- | --- |
| 6 mm bolts: | | |
|    Initial | 0.6 | 4.5 |
|    Final | 1.3 | 9.5 |
| 8 mm bolts: | | |
|    Initial | 1.3 | 9.5 |
|    Final | 2.4 | 17.0 |

33.3 Bolts should be tightened in the sequence indicated by the numbers cast into the crankcase

4    When all of the lower crankcase bolts have been secured, check that the gearbox shafts and crankshaft turn freely. If there are any problems in this respect, resolve them now before reassembly progresses further. If all is well, fit a new O-ring to the oil feed passage in the sump area. Place a new sump gasket in position and refit the sump. Note the position of the wiring guide clips (see accompanying photograph) and secure the retaining bolts to a torque setting of 1.0 kgf m (7.0 lbf ft).

5    Invert the unit on the workbench and fit the upper crankcase securing bolts, tightening them evenly and progressively in numerical sequence to the above torque figures. Note the position and direction of the engine earth (ground) lead shown in the accompanying photograph.

6    Fit the oil gallery cover, using a new gasket, the input shaft bearing retainer and the output shaft end plate in the clutch recess, using locking compound on the retaining screws. On the right-hand side of the unit, fit the single crankcase retaining nut. Offer up the gear position indicator switch and secure the lead with the input shaft end seal retainer.

33.4a Refit the sump, using a new gasket

33.4b Note the position of the wiring clips along sump edge

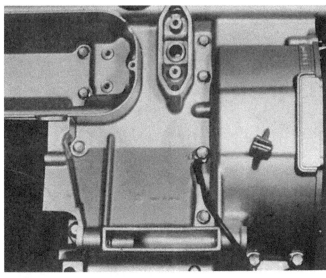

33.5 Fit the upper crankcase bolts, noting the two bolts fitted in the starter motor recess and the position of the engine earth lead

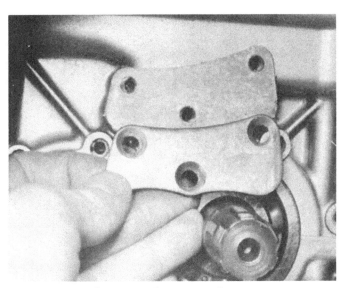

33.6a Fit the oil gallery cover using a new gasket

33.6b Secure the input shaft bearing retainer ...

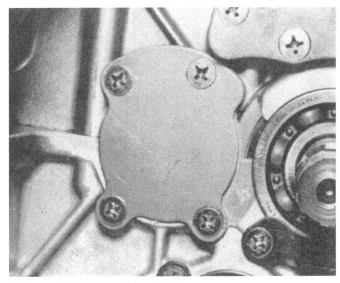

33.6c ... and the output shaft end plate

33.6d Fit and tighten the single crankcase retaining nut ...

33.6e ... then fit the gear position switch, noting wiring routing and position of end seal retainer

34.1a Clean the oil pump boss area ...

## 34 Engine reassembly: refitting the oil pump and gear selector shaft

1   Fit a new O-ring to the pump body and install it in its recess. Apply a thread locking compound to the three retaining bolts, then tighten them evenly and securely.

2   Place the driving pin through the pump spindle. Offer up the pump pinion, noting that the slot on its inner face must locate over the driving pin. Once in place, secure the pinion with its circlip.

3   Check that the gear selector pawl assembly is in position, then slide the selector shaft into the casing hole. Care should be taken to avoid damage to the shaft seal as it emerges from the left-hand side of the crankcase.

4   Arrange the claw assembly as shown in the accompanying photograph. Once the mechanism is aligned correctly, secure the end of the shaft with its plain washer and circlip. The ends of the centralising spring should be hooked over the locating pin as shown.

34.1b ... and fit a new O-ring to the pump

34.1c Install the pump and fit the drive pin as shown

34.2 Fit the pump pinion and secure with circlip

34.4a Slide claw assembly into place and position as shown

34.4b Secure end of shaft with plain washer and clip

## 35 Engine reassembly: refitting the clutch

*Note: on later models a shim is fitted to the clutch assembly between the outer drum/primary driven gear spacer and the thrust washer (items 3 and 8 in Fig. 1.22). Prior to reassembly, it is important to measure the thrust clearance and if necessary adjust this by changing the size of shim. Refer to Chapter 7, Section 4 for further details.*

1    Place the large plain thrust washer over the end of the gearbox input shaft, followed by the oil pump drive pinion spacer. Lubricate and fit the needle roller bearing, then slide the pinion into position with the engagement dogs outermost.

2    Check that the small rubber plug is in place in the back of the clutch drum (see photograph) noting that if it is omitted, a significant amount of clutch noise will result. Offer up the drum, ensuring that the dogs on the pump drive pinion engage correctly. Hold the drum in this position and slide the clutch needle roller bearing and the spacer into place. Fit the thrust washer with the grooved face innermost.

3    If the inner plain plate was removed from the clutch centre, this should now be refitted. As has been mentioned, the design of the dished washer and L-section seat which precede the inner plate have been changed on later models, but this has little effect on removal or reassembly. Fit the washer and its seat, then place the inner plate against them. Work the wire retaining clip into place, noting that the ends pass through a hole in the clutch centre (see photograph).

4    If new friction plates are to be fitted, soak them in oil prior to installation. Meanwhile, fit the locking washer and the clutch centre nut on the shaft end. Hold the clutch centre and tighten the nut to 5.0 – 7.0 kgf m (36.0 – 50.5 lbf ft). Install the plain and friction plates alternately.

5    Fit the clutch release rack, bearing and thrust washer through the clutch pressure plate, noting that the washer fits between the bearing and the pressure plate, and secure them with the E-clip. Fit the clutch springs, washers and bolts, tightening them evenly and in a diagonal sequence to the recommended torque setting of 1.1 – 1.3 kgf m (8.0 – 9.5 lbf ft).

6    Apply a thin film of Suzuki bond No.4 or an RTV silicone sealant, 1 in each side of both crankcase joints. Fit a new clutch cover gasket. Offer up the cover, taking care to ensure that the release rack engages in the cover hole correctly. Tap the cover home and fit the cover securing screws.

35.1a Fit the large thrust washer to input shaft ...

35.1b ... followed by oil pump drive pinion spacer

35.1c Lubricate and fit needle roller bearing ...

35.1d ... then fit pinion with dogs facing outwards

35.2a Check that small rubber peg is in place

35.2b Offer up the clutch drum ...

35.2c ... and insert the spacer assembly

35.2d Grooved face of thrust washer faces inwards

35.3a Inner plain plate is retained by thin wire circlip ...

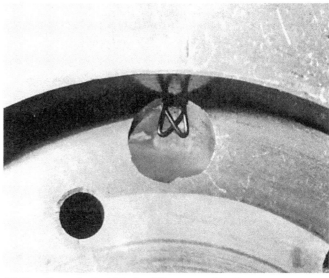

35.3b ... the ends of which should be arranged as shown

35.4a Fit and secure the clutch centre. Note improvised holding tool

35.4b Do not omit to secure the locking tab

35.4c Fit the friction plates and ...

35.4d ... plain plates in alternating sequence

35.5a Assemble the release rack and bearing as shown ...

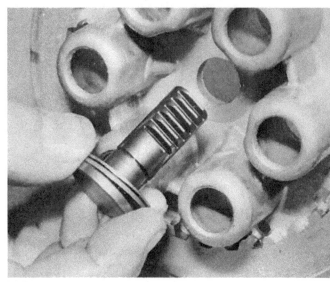
35.5b ... and pass it through the pressure plate

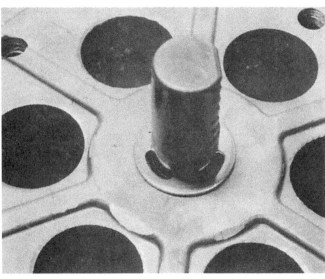
35.5c Secure the rack with its E-clip

35.5d Fit the pressure plate, springs and bolts, tightening them evenly

35.6a Fit a new clutch cover gasket

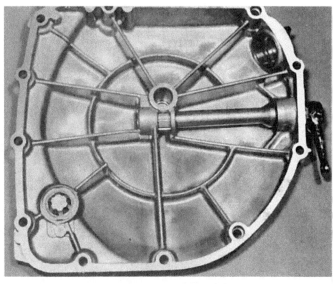
35.6b Check that rack engages correctly when fitting cover

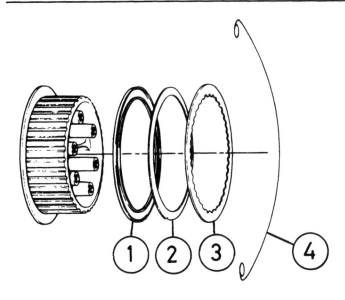

**Fig. 1.34 Clutch anti-snatch spring arrangement**

| | |
|---|---|
| 1   *Spring seat* | 3   *Inner plate* |
| 2   *Diaphragm spring* | 4   *Retaining clip* |

## 36 Engine reassembly: refitting the ignition pickup

1   Clean the crankshaft end and install the automatic timing unit, noting that the slot in the boss must locate over the pin in the recessed end of the crankshaft. Place the crankshaft turning hexagon over the locating lugs, then fit and tighten the retaining bolt to 1.3 – 2.3 kgf m (9.5 – 16.5 lbf ft).

2   Place the timing index plate and the pickup backplate assembly in position, turning it so that the index line is central in relation to the upper screw hole. Hold it in this position and tighten the retaining screws.

3   Apply a film of RTV silicone sealant around the pickup wiring grommet and push it into the casing slot. Route the wiring as shown in the accompanying photograph, bending the clips to hold it in place. Note that the inspection cover should not be fitted until the valve timing has been checked.

36.1a ATU is located by pin in crankshaft end

36.1b Fit the crankshaft turning hexagon noting location

36.1c Fit and tighten the retaining bolt

36.2a Place the timing index plate in position ...

36.2b ... followed by the pickup backplate

36.3 When timing has been checked, fit cover using a new gasket

### 37 Engine reassembly: refitting the starter drive, starter motor and alternator assembly

1   Place the starter idler pinion in its recess, noting that a thrust washer is fitted at each end of its shaft.
2   Check that the starter clutch rollers are located correctly in the clutch body, then fit the starter gear, easing the boss into engagement by turning it. Lubricate and fit the needle roller bearing or bearings to the pinion internal bore. Place the slotted thrust washer over the crankshaft end, then offer up the rotor and clutch assembly. Apply a locking fluid to the rotor nut threads and tighten it to the specified torque setting.

3   Check the condition of the starter motor O-ring and where necessary renew it. Fit the starter motor into the casing recess, ensuring that the motor shaft engages the idler pinion. Once in place, fit the two retaining bolts and tighten them securely.
4   Before refitting the alternator cover and gasket, smear a small quantity of RTV silicone sealant over both the crankcase joint areas; an inch each side of the joint is sufficient. Check that the dowel pin is in place in either the cover or the crankcase, fit a new gasket, and pass the wiring through the casing hole. Offer up the cover and fit and tighten the retaining screws.
5   Route the alternator wiring along the starter motor recess and out through the cutout in the casing. The wiring is held in place by the small rubber block which is wedged between the casing and the motor body.

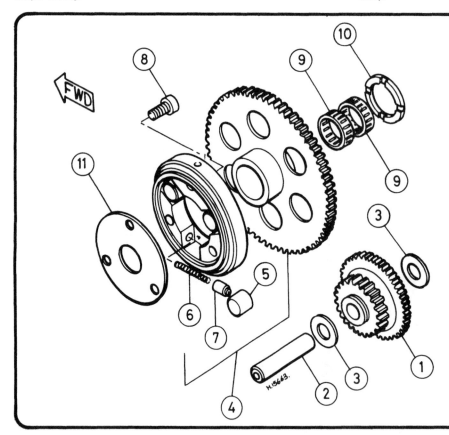

**Fig. 1.35 Starter clutch**

1   Starter idler pinion
2   Pinion shaft
3   Thrust washer – 2 off
4   Starter clutch assembly
5   Roller – 3 off
6   Spring – 3 off
7   Plunger – 3 off
8   Allen bolt – 3 off
9   Needle roller bearing
    – 2 off
10  Slotted thrust washer
11  Backing plate

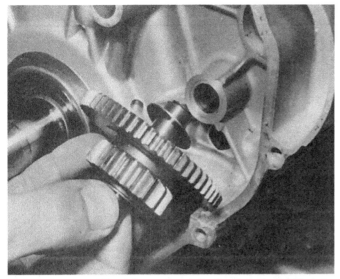

37.1 Fit the starter idler pinion, noting that a thrust washer is fitted at each end

37.2a Check that the starter clutch rollers are located correctly, then install the starter gear

37.2b Place the needle roller bearings in position and lubricate them

37.2c Fit the slotted thrust washer with the chamfered face innermost

37.2d Offer up the rotor assembly, lock the crankshaft and secure the retaining nut

37.3a Fit the starter motor in its recess, ensuring that the teeth mesh as shown ...

37.3b ... then fit and tighten the two retaining bolts

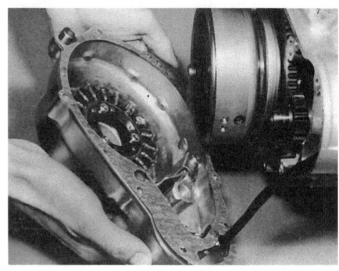

37.4 Fit alternator cover, using a new gasket

37.5 Retain alternator wiring with rubber block (arrowed) then fit the starter motor cable

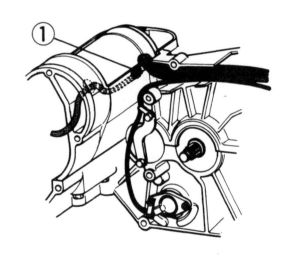

Fig. 1.36 Crankcase wiring layout showing rubber block (1)

### 38  Engine reassembly: refitting the pistons and the cylinder block

1    Pack each crankcase mouth with clean rag to prevent debris or dropped circlips from entering the crankcase. Lay out the pistons in the correct sequence to ensure that they are refitted in the correct relative positions. Note that new circlips should be used even where the old pistons are to be reused; the old clips will have been weakened during removal, and the expense incurred in the event of a failure does not justify making economies.

2    Where the rings have yet to be fitted, position the oil ring spacer, making sure that the ends of the spacer do not overlap. When fitting the 2nd and top rings, note that the 2nd ring is tapered in section and is fitted so that the wider diameter faces downwards. The top ring is of plain section, and like the second ring is marked 'N' on the top face.

3    Lubricate the big-end bearings and the small end eyes, then fit the pistons with the arrow mark on each crown facing forward. If the gudgeon pins are tight, warm the pistons using a rag soaked in near boiling water, taking suitable precautions to avoid scalding. Fit the circlips so that the ends are clear of the removal notch, and ensure that the circlips locate fully in their grooves.

4    Fit the cam chain tensioner blade, using a thread locking

compound on the retaining bolt and tightening it to 0.9 – 1.4 kgf m (6.5 – 10.0 lbf ft). Check that the locating dowels are in position in the underside of the cylinder block. Fit new O-rings to the grooves around each of the cylinder liners. When pushing them into their grooves, work each one in evenly, starting it at three or four points around the groove. This avoids the irritation of having a 'spare' loop of O-ring.

5    When fitting the cylinder block, note that a set of piston ring clamps will prove invaluable. Although each bore has a tapered lead in, it was found that this was so abrupt in the case of the machine shown in the photographs, that it would have been almost impossible to introduce the rings manually. Suzuki ring clamps, Part Number 09916 - 74540, or a proprietary set may be used.

6    Fit a new cylinder base gasket over the holding studs, ensuring that the face marked UP is facing upwards (where applicable). Check that the piston ring ends are staggered, then fit a ring clamp to each, ensuring that all rings are enclosed. Lubricate the cylinder bores with engine oil, then position the cylinder block above the pistons and pass the cam chain loop through the tunnel.

7    Carefully lower the cylinder block over the pistons, keeping the pistons square to the bore as they enter. Once the pistons are well into the bores, the ring clamps may be removed. Check that all the rings are fully engaged, then push the block down onto the gasket.

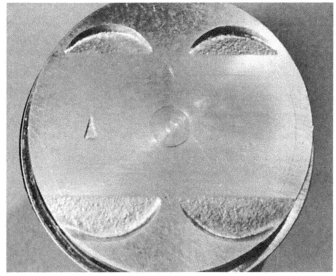

38.3a Pistons should be fitted with arrow mark facing forward

38.3b Ensure that circlips seat fully with ends clear of removal slot

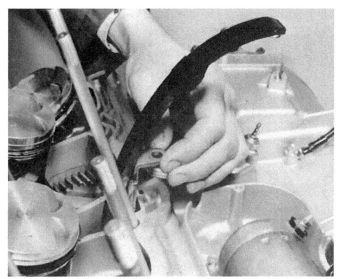

38.4a Refit the cam chain tensioner blade

38.4b Fit new O-ring seals to the groove at the base of each bore

38.6 Position cam chain over tensioner as shown. Fit a new cylinder base gasket

38.7 Lower cylinder block over pistons, using ring clamps to assist in fitting

## 39 Engine reassembly: refitting the cylinder head, camshafts and tensioner

1    Place a new gasket on the cylinder head gasket face, ensuring that the face marked 'TOP' is upwards. Lower the cylinder head into position, feeding the cam chain through the tunnel and securing it by passing a bar or screwdriver through the chain loop.

2    The cylinder head is retained by a diverse selection of fasteners and a mixture of steel and copper washers. The accompanying photograph shows the fasteners in their correct relative positions. In addition, there is a single bolt at each end of the head and a third at the centre of the front of the head.

3    Fit the retaining nuts and washers, tightening them progressively in the sequence indicated by the numbers cast into the head material adjacent to each one. The final torque setting is 3.5 – 4.0 kgf m (25.5 – 29.0 lbf ft). Fit the three 6 mm bolts and tighten them to 0.7 – 1.1 kgf m (5.0 – 8.0 lbf ft). Slide the cam chain guide into the tunnel and position it in its locating pocket.

4    Using a 19 mm socket, turn the crankshaft clockwise to align the '1-4' cylinder 'T' mark with the fixed timing mark. When turning the crankshaft keep the chain taut to prevent it from bunching under the crankshaft sprocket.

5    The inlet camshaft can be identified by the 'IN' marking on its surface, whilst the exhaust camshaft is marked 'EX'. In addition, on some models the exhaust camshaft incorporates the tachometer drive pinion. Note also that the right-hand end of each camshaft has a notch in it.

6    Fit the exhaust camshaft through the chain loop, positioning the sprocket with the notch parallel to the gasket face and the '1' arrow mark pointing forward. The '2' arrow should face upwards. Fit the remaining camshaft, again with the notch lying parallel to the gasket face. The '3' arrow should face upwards and should be 20 pins from that indicated by the '2' arrow on the exhaust camshaft. The arrangement is illustrated in the accompanying line drawing, and should be checked carefully to ensure that the cam timing is correct.

7    Fit the camshaft bearing caps without disturbing the camshafts or the crankshaft. Each cap is marked with an arrow to indicate the front and is numbered to indicate its position. A corresponding number is cast into the cylinder head. Make sure that the locating dowels are in position when fitting the caps, and fit the retaining bolts finger tight only.

8    Tighten the cap bolts by a half turn at a time in a diagonal sequence. It is important that the camshafts are pulled down squarely to avoid damage to the caps or cylinder head. Once the camshaft caps are all in contact with the cylinder head, tighten the bolts, again in a diagonal sequence, to the correct torque setting of 0.8 – 1.2 kgf m (6.0 – 8.5 lbf ft). Suzuki advise that the camshaft cap bolts are made from 'a special material' and are identified by the figure 9 stamped on the heads. On no account should other bolts be used.

9    Check that the tensioner pushrod is fully retracted. Slacken the lock bolt and push the pushrod inwards whilst turning the knurled wheel anti-clockwise against spring pressure. Once the pushrod is fully home, tighten the lock bolt to retain it.

10   Offer up the tensioner using a new gasket, noting that the knurled wheel should face towards the right-hand side. If the tensioner will not seat fully, turn the crankshaft using the 19 mm hexagon on the crankshaft end to obtain chain slack on the tensioner side of the block. Fit and tighten the retaining bolts.

11   Slacken the lock screw to free the pushrod, which should move inwards to apply pressure to the chain, then tighten the locknut whilst holding the lock screw in position. Slowly turn the crankshaft anticlockwise whilst turning the knurled wheel anticlockwise to allow the chain to push the tensioner plunger inwards. Release the knurled wheel, then turn the crankshaft clockwise to check that the mechanism is operating normally. The wheel should revolve as the tensioner adjusts to take up chain slack. Having checked that all is well, do not disturb the knurled wheel further.

39.1a Fit a new cylinder head gasket ...

39.1b ... noting that the "TOP" mark should face upwards

39.1c The cylinder head can now be lowered into position

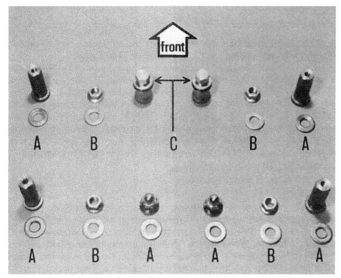

39.2 Note the arrangement of cylinder head fasteners. A: copper washer, B: steel washer, C: O-ring

39.3a Remember to fit the three 6 mm bolts

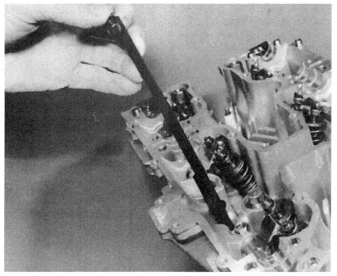

39.3b Fit the cam chain guide into its recess

39.4 Turn the crankshaft to align the "T" mark

39.6 "3" mark should be positioned 20 pins from the "2" mark

39.7 Fit the bearing caps, noting arrow with letter code on each one

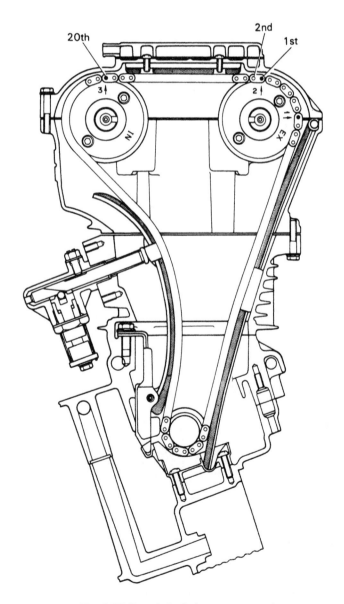

## 40 Engine reassembly: setting the valve clearances

1    If the engine has been overhauled or if the valve train components have been disturbed for any reason, the valve clearances must be checked and reset. The correct clearance for both inlet and exhaust valves is 0.07 – 0.12 mm (0.003 – 0.005 in). Using the 19 mm hexagon on the ignition pickup end of the crankshaft, rotate it until the notches in the camshaft end face outwards and the '1-4 T' mark is aligned. Check and adjust the clearances for the following valves:

Cylinder No.1 Inlet and exhaust
Cylinder No.2 Exhaust only
Cylinder No.3 Inlet only

2    Measure the clearance between each valve and its adjuster, using feeler gauges. If adjustment is required, slacken the locknut with a ring spanner and set the clearance by turning the small square-headed adjuster. When set correctly, the feeler gauge should be a light sliding fit with no free play and no pressure on it from the valve springs. Hold the adjuster in this position and secure the locknut. Note that a Suzuki tool is available to fit the adjuster, Part Number 09917 - 14910. It is useful but not essential.

3    Rotate the crankshaft through 360°, again aligning the '1-4 T' mark. The camshaft notches should now face inwards. Repeat the adjustment sequence on the remaining valves as listed below:

Cylinder No.2 Inlet only
Cylinder No.3 Exhaust only
Cylinder No.4 Inlet and exhaust

39.10 Retract tensioner pushrod and lock it in place prior to installation

40.2a Note position of camshaft notches (see text)

Fig. 1.37 Camshaft timing arrangement

40.2b Check and adjust the valve clearances using feeler gauges

## 41 Engine reassembly: fitting the oil filter and cylinder head cover

**Note**: *If the engine unit has yet to be installed in the frame this operation is best left until it is in position, to give extra clearance during fitting.*

1 Apply grease to the oil filter cover groove to hold the sealing ring in place. Fit a new filter element into its recess, then offer up the cover and spring. Apply thread locking compound to the domed nuts and tighten them evenly and securely.

2 Check that the sump drain plug is secure, then prime the oil pockets around the camlobes with engine oil. Fit new half-moon seals to the ends of the camshafts, sticking them into position using RTV silicone sealant. Coat the gasket face of the head and cover with the same sealant to ensure an oil-tight seal.

3 Fit a new cover gasket and the two locating dowels, then place the cover in position. When fitting the retaining bolts, note that the two longer bolts should be fitted to coincide with the dowels, and that the four outer screws have sealing washers. The bolts should be tightened evenly in a diagonal sequence to avoid warpage.

4 Fit the tachometer drive gear assembly, then fit the camshaft end covers. The retaining screws should be coated with a thread locking compound prior to installation.

41.3a Stick a new cover gasket in place using RTV sealant

## 42 Engine reassembly: fitting the engine unit in the frame

1 Fitting the engine unit will be no easy task without some form of engine hoist, the use of which is advocated where possible. In the absence of this, three persons will be required to lift it into position safely.

2 Lift the unit into position on the right-hand side of the frame and rest it against the frame lower rails. One person should now move to the left-hand side and guide the unit into position. Once it is resting in the frame cradle, the various engine bolts and plates can be fitted loosely. Lever the unit up in the frame using lengths of timber so that the bolt holes can be lined up.

3 When all fasteners are in position, tighten them to the specified torque settings. Note that the three long through bolts are fitted from the left-hand side, with the spacer fitted to the left-hand side of the rear upper bolt.

4 Place the gearbox sprocket in the final drive chain loop and install it. Fit the retaining nut with its recessed face inwards and tighten to 10.0 – 15.0 kgf m (72.5 – 108.5 lbf ft). Remember to reset the final drive chain free play where necessary.

5 Reassemble the exhaust system, using new sealing rings in the ports. These can be held in place with grease during installation and the exhaust pipe collet halves retained by elastic bands or PVC tape. When in position, tighten the exhaust retainer nuts and the clamps below the crankcase, followed by the silencer mounting bolts.

6 Refit the footrests and the gearchange pedal or linkage. Reconnect the spark plug caps and refit the coils. If the horns were removed to gain clearance, these should now be fitted, as should the cylinder head cover and breather cover.

7 Reconnect the alternator, ignition, switch and starter motor wiring. Reconnect the battery and check that the electrical system functions correctly.

8 Reconnect the clutch cable and check its operation and adjustment. Reconnect the throttle and choke cables and manoeuvre the carburettors into position. Check the routing of the fuel and breather hoses. Secure the carburettor retaining clips.

9 Refit the air filter trunking and secure its retaining clip and bolts. Refit and secure the rear brake reservoir. Reconnect the tachometer drive cable. Top up the crankcase with 4.0 litres (8.4/7.0 US/Imp pint) of SAE 10W/40 motor oil, noting that the oil level must be checked after the engine has been run.

41.3b Note sealing washers under heads of the outer screws

42.3 Fit all through bolts from the left-hand side, noting the spacer on rear upper bolt

42.4 Refit the gearbox sprocket, tighten nut and secure the locking tab

42.5a Hold exhaust port seals in place with a dab of grease ...

42.5b ... and split retainers with elastic band or tape

42.5c Tighten exhaust pipe retainers first, then silencer mountings

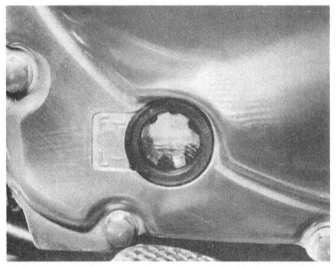

42.9 Top up the crankcase and recheck level after running the engine

## 43 Starting and running the rebuilt engine

1    Attempt to start the engine using the usual procedure adopted for a cold engine. Do not be disillusioned if there is no sign of life initially. A certain amount of perseverance may prove necessary to coax the engine into activity even if new parts have not been fitted. Should the engine persist in not starting, check that the spark plug has not become fouled by the oil used during reassembly. Failing this go through the fault-finding charts and work out what the problem is methodically.

2    When the engine does start, keep it running as slowly as possible to allow the oil to circulate. The oil warning light should go out almost immediately the engine has started, although in certain instances a very short delay can occur whilst the oilways fill and the pressure builds up. If the light does not go out the engine should be stopped before damage can occur, and the cause determined. Open the choke as soon as the engine will run without it. During the initial running, a certain amount of smoke may be in evidence due to the oil used in the reassembly sequence being burnt away. The resulting smoke should gradually subside.

3    Check the engine for blowing gaskets and oil leaks. Before using the machine on the road, check that all the gears select properly, and that the controls function correctly.

4   Once the engine has settled down and is running normally, note that the following adjustments should be checked before the machine is used on the road.

   a)   Rear brake pedal height
   b)   Rear brake lamp switch
   c)   Clutch operation and adjustment
   d)   Final drive chain free play
   e)   Throttle operation and cable adjustment
   f)   Carburettor synchronisation (balance)
   g)   Idle speed adjustment

## 44  Taking the rebuilt engine on the road

1   Any rebuilt machine will need time to settle down, even if parts have been replaced in their original order. For this reason it is highly advisable to treat the machine gently for the first few miles to ensure oil has circulated throughout the lubrication system and that any new parts fitted have begun to bed down.

2   Even greater care is necessary if the engine has been rebored or if a new crankshaft has been fitted. In the case of a rebore, the engine will have to be run-in again, as if the machine were new. This means greater use of the gearbox and a restraining hand on the throttle until at least 500 miles have been covered. There is no point in keeping to any set speed limit; the main requirement is to keep a light loading on the engine and to gradually work up performance until the 500 mile mark is reached. These recommendations can be lessened to an extent when only a new crankshaft is fitted. Experience is the best guide since it is easy to tell when an engine is running freely.

3   If at any time a lubrication failure is suspected, stop the engine immediately, and investigate the cause. If any engine is run without oil, even for a short period, irreparable engine damage is inevitable.

4   When the engine has cooled down completely after the initial run, recheck the various settings, especially the valve clearances. During the run most of the engine components will have settled into their normal working locations. Check the various oil levels, particularly that of the engine as it may have dropped slightly now that the various passages and recesses have filled.

# Chapter 2 Fuel system and lubrication

*Refer to Chapter 7 for information relating to the 1984-on 1135cc-engined GSX1100 and GS1150 models*

## Contents

## Specifications

## GSX 1100 T, ET, LT, EX, EZ, SZ, SD, ESD

### Fuel tank

Overall capacity:

| | |
|---|---|
| T, ET, EX ....................................................................... | 24 lit (6.3/5.3 US/Imp gal) |
| LT .................................................................................... | 15 lit (4.0/3.3 US/Imp gal) |
| EZ, SZ, SD, ESD ........................................................... | 22 lit (5.8/4.8 US/Imp gal) |

Reserve capacity:

| | |
|---|---|
| T, ET, LT ........................................................................ | Not available |
| EX, EZ ............................................................................ | 4.0 lit (8.4/7.0 US/Imp pint) |
| SZ, SD, ESD .................................................................. | 5.0 lit (10.6/8.8 US/Imp pint) |

### Fuel grade ....................................................... Unleaded or low-lead, minimum octane rating 90 RON/RM

### Carburettors

| | |
|---|---|
| Make ............................................................................... | Mikuni |
| Type: | |
| T, ET, LT, EX, EZ, SD, ESD ........................................ | BS34SS |
| SZ ................................................................................... | BS34SS (VM32SS)** |
| Size ................................................................................ | 34 mm (1.34 in) |
| Identity number: | |
| T, ET, LT, EX ................................................................ | 49210 |
| EZ ................................................................................... | 49230 |
| SZ ................................................................................... | 49250 or 49260 (49320)** |
| SD, ESD ......................................................................... | 49340 |
| Idle speed ..................................................................... | 1050 ± 100 rpm |
| Fuel level ....................................................................... | 5.0 ± 0.5 mm (0.20 ± 0.02 in) |
| Float height .................................................................. | 22.4 ± 1.0 mm (0.88 ± 0.04 in) |
| Main jet: | |
| Except SD, ESD ............................................................ | 107.5 (102.5)** |
| SD, ESD ......................................................................... | 112.5 |
| Main air jet .................................................................... | 1.2 (1.5)** |
| Jet needle ...................................................................... | 5D59 (5DL82)** |
| Clip position ................................................................. | 3rd groove from top |
| Needle jet: | |
| T, ET, LT, EX ................................................................ | X-1 |
| EZ, SZ, SD, ESD ........................................................... | X-2 (0-6)** |
| Throttle valve: | |
| T, ET, LT, EX, SZ ......................................................... | Not available |
| EZ, SD, ESD .................................................................. | 135 |
| Pilot jet: | |
| T, ET, LT, EX ................................................................ | 45 |
| EZ, SZ, SD, ESD ........................................................... | 47.5 (20)** |
| Bypass ............................................................................ | 0.8 (0.8)** |
| Pilot outlet: | |
| Except SD, ESD ............................................................ | 0.9 (0.8)** |
| SD, ESD ......................................................................... | 1.0 |

| | |
|---|---|
| Float valve seat ............................................................ | 2.0 (2.3)** |
| Starter jet ..................................................................... | 32.5 (45)** |
| Pilot screw ................................................................... | Preset * |
| Pilot air jet: | |
|    T, ET, LT, SZ ...................................................... | Not available |
|    EX ........................................................................ | 140 |
|    EZ, SD, ESD ...................................................... | 160 |
| Cut-away ...................................................................... | Not available (1.5)** |
| Throttle cable free play .............................................. | 0.5 – 1.0 mm (0.02 – 0.04 in) |
| Choke cable free play: | |
|    Except SZ, SD, ESD ........................................... | Not applicable |
|    SZ, SD, ESD ...................................................... | 0.5 – 1.0 mm (0.02 – 0.04 in) |

*Pilot screw adjustment is preset during assembly, and must not be disturbed
** Items marked apply only to alternative VM32SS fitment on SZ model.

### Engine/transmission oil

| | |
|---|---|
| Capacity: | |
|    Dry ...................................................................... | 4.0 lit (8.4/7.0 US/Imp pint) |
|    Oil and filter change .......................................... | 3.6 lit (7.6/6.4 US/Imp pint) |
|    Oil change only ................................................... | 3.2 lit (6.8/5.6 US/Imp pint) |
| Oil grade ...................................................................... | SAE 10W/40 motor oil SE or SF type |

### Oil pump

| | |
|---|---|
| Type ............................................................................. | Trochoid |
| Pump reduction ratio .................................................. | 1.723 : 1 (87/49 x 33/34 T) |
| Oil pressure: | |
|    Minimum (except SD, ESD) ................................. | 0.1 kg cm$^2$ (1.42 psi) @ 60°C/140°F |
|    SD, ESD ............................................................. | 0.2 kg cm$^2$ (2.84 psi) @ 60°C/140°F |
|    Maximum @ 3000 rpm (except SD, ESD) ............ | 0.5 kg cm$^2$ (7.11 psi) @ 60°C/140°F |
|    SD, ESD ............................................................. | 0.4 kg cm$^2$ (5.68 psi) @ 60°C/140°F |
| Rotor tip clearance ..................................................... | Not available |
| Service limit ................................................................ | 0.20 mm (0.008 in) |
| Outer rotor/body clearance ........................................ | Not available |
| Service limit ................................................................ | 0.25 mm (0.010 in) |
| Rotor end float ........................................................... | Not available |
| Service limit ................................................................ | 0.15 mm (0.006 in) |

## GS 1100 T, ET, LT, EX, EZ, ED, ESD (US)

### Fuel tank

| | |
|---|---|
| Overall capacity: | |
|    T, ET .................................................................. | 19 lit (5.0/4.2 US/Imp gal) |
|    LT, EX ................................................................ | 15 lit (4.0/3.3 US/Imp gal) |
|    EZ, ED, ESD ...................................................... | 22 lit (5.8/4.8 US/Imp gal) |
| Reserve capacity: | |
|    T, ET, LT, EX .................................................... | Not available |
|    EZ, ED, ESD ...................................................... | 4.0 lit (8.4/7.0 US/Imp pint) |

### Carburettors

| | |
|---|---|
| Make ............................................................................ | Mikuni |
| Type ............................................................................. | BS34SS |
| Size .............................................................................. | 34 mm (1.34 in) |
| Identity number: | |
|    T, ET, LT, EX .................................................... | 49200 |
|    EZ ....................................................................... | 49220 |
|    ED, ESD ............................................................. | 49380 |
| Idle speed .................................................................... | 1050 ± 100 rpm |
| Fuel level ..................................................................... | 5.0 ± 0.5 mm (0.20 ± 0.02 in) |
| Float height ................................................................. | 22.4 ± 1.0 mm (0.88 ± 0.04 in) |
| Main jet: | |
|    T, ET, LT, EX .................................................... | 107.5 |
|    EZ ....................................................................... | 110.0 |
|    ED, ESD ............................................................. | 112.5 |
| Main air jet .................................................................. | 1.2 |
| Jet needle .................................................................... | 5D58 |
| Clip position ................................................................ | Not available |
| Needle jet .................................................................... | X-1 |
| Throttle valve: | |
|    T, ET, LT, EX, ED, ESD ..................................... | Not available |
|    EZ ....................................................................... | 135 |
| Pilot jet ....................................................................... | 45 |
| Bypass ......................................................................... | 0.8 |
| Pilot outlet – T, ET, LT, EX, EZ ................................. | 0.9 |
| Pilot outlet – ED, ESD ................................................ | 1.0 |
| Float valve seat ........................................................... | 2.0 |
| Starter jet ..................................................................... | 32.5 |
| Pilot screw ................................................................... | Preset * |

Pilot air jet:
    T, ET, LT, EX ............................................................... Not available
    EZ, ED, ESD ............................................................... 170 – EZ, 180 – ED, ESD
Throttle cable free play ............................................................ 0.5 – 1.0 mm (0.02 – 0.04 in)
Choke cable free play:
    T, ET, LT, EX, ED, ESD ........................................... Not applicable
    EZ ................................................................................... 0.5 – 1.0 mm (0.02 – 0.04 in)

*\* Pilot screw adjustment is preset during assembly, and must not be disturbed.*

## Engine/transmission oil
Capacity:
    Dry ................................................................................. 4.0 lit (8.4/7.0 US/Imp pint)
    Oil and filter change ................................................. 3.6 lit (7.6/6.4 US/Imp pint)
    Oil change only ........................................................... 3.2 lit (6.8/5.6 US/Imp pint)
Oil grade ...................................................................................... SAE 10W/40 motor oil SE or SF type

## Oil pump
Type .............................................................................................. Trochoid
Pump reduction ratio ............................................................... 1.723 : 1 (87/49 x 33/34 T)
Oil pressure:
    Minimum ...................................................................... 0.1 kg cm$^2$ (1.42 psi) @ 60°C/140°F
    Maximum @ 3000 rpm ............................................. 0.5 kg cm$^2$ (7.11 psi) @ 60°C/140°F
Rotor tip clearance ................................................................... Not available
Service limit ................................................................................ 0.20 mm (0.008 in)
Outer rotor/body clearance .................................................... Not available
Service limit ................................................................................ 0.25 mm (0.010 in)
Rotor end float ........................................................................... Not available
Service limit ................................................................................ 0.15 mm (0.006 in)

# GSX 1000 SZ, SD
## Fuel tank
Overall capacity ......................................................................... 22 lit (5.8/4.8 US/Imp gal)
Reserve capacity ........................................................................ 5.0 lit (8.8/10.6 US/Imp pint)

## Carburettors
Make .............................................................................................. Mikuni
Type:
    SZ ................................................................................... VM32SS
    SD ................................................................................... VM34SS
Size:
    SZ ................................................................................... 32 mm (1.26 in)
    SD ................................................................................... 34 mm (1.34 in)
Identity number:
    SZ ................................................................................... 49310
    SD ................................................................................... 49340
Idle speed ................................................................................... 1050 ± 100 rpm
Fuel level ..................................................................................... 5.0 ± 0.5 mm (0.20 ± 0.02 in)
Float height ................................................................................. 22.4 ± 1.0 mm (0.88 ± 0.04 in)
Main jet:
    SZ ................................................................................... 95.0
    SD ................................................................................... 112.5
Main air jet:
    SZ ................................................................................... 1.5
    SD ................................................................................... 1.2
Jet needle:
    SZ ................................................................................... 5DL82
    SD ................................................................................... 5D59
Clip position ............................................................................... 3rd from top
Needle jet:
    SZ ................................................................................... 0-6
    SD ................................................................................... X-2
Throttle valve:
    SZ ................................................................................... Not available
    SD ................................................................................... 135
Pilot jet:
    SZ ................................................................................... 20
    SD ................................................................................... 47.5
Bypass:
    SZ ................................................................................... 0.8
    SD ................................................................................... 0.8
Pilot outlet:
    SZ ................................................................................... 0.8
    SD ................................................................................... 1.0

Float valve seat:
    SZ .................................................................... 2.3
    SD .................................................................... 2.0
Starter jet:
    SZ .................................................................... 45.0
    SD .................................................................... 32.5
Pilot screw ............................................................ Preset*
Throttle cable free play ......................................... 0.5 – 1.0 mm (0.02 – 0.04 in)
Choke cable free play ........................................... 0.5 – 1.0 mm (0.02 – 0.04 in)

*Pilot screw adjustment is preset during assembly, and must not be disturbed.*

## Engine/transmission oil

Capacity:
    Dry ................................................................... 4.0 lit (8.4/7.0 US/Imp pint)
    Oil and filter change ......................................... 3.6 lit (7.6/6.4 US/Imp pint)
    Oil change only ................................................. 3.2 lit (6.8/5.6 US/Imp pint)
Oil grade ............................................................... SAE 10W/40 motor oil SE or SF type

## Oil pump

Type ...................................................................... Trochoid
Pump reduction ratio ............................................. 1.723 : 1 (87/49 x 33/34 T)
Oil pressure @ 60°C, 140°F (service limits):-
    Minimum:
        SZ ............................................................. 0.1 kg cm$^2$ (1.42 psi)
        SD ............................................................. 0.2 kg cm$^2$ (2.84 psi)
    Maximum:
        SZ ............................................................. 0.5 kg cm$^2$ (7.11 psi)
        SD ............................................................. 0.4 kg cm$^2$ (5.68 psi)

# GS 1000 SZ (US)

## Fuel tank

Overall capacity .................................................... 22 lit (5.8/4,8 US/Imp gal)
Reserve capacity .................................................. 5.0 lit (8.8/10.6 US/Imp pint)

## Carburettors

Make ..................................................................... Mikuni
Type ...................................................................... BS34SS
Size ...................................................................... 34 mm (1.34 in)
Identity number .................................................... 49300
Idle speed ............................................................ 1050 ± 100 rpm
Fuel level ............................................................. 5.0 ± 0.5 mm (0.20 ± 0.02 in)
Float height .......................................................... 22.4 ± 1.0 mm (0.88 ± 0.04 in)
Main jet ................................................................ 110.0
Main air jet .......................................................... 1.2
Jet needle ............................................................ 5D58
Clip position ......................................................... Not available
Needle jet ............................................................ X-1
Pilot jet ............................................................... 45
Bypass ................................................................. 0.8
Pilot outlet ........................................................... 0.9
Float valve seat ................................................... 2.0
Starter jet ............................................................ 32.5
Pilot screw .......................................................... Preset *
Throttle cable free play ......................................... 0.5 – 1.0 mm (0.02 – 0.04 in)
Choke cable free play ........................................... 0.5 – 1.0 mm (0.02 – 0.04 in)

*Pilot screw adjustment is preset during assembly, and must not be disturbed*

## Engine/transmission oil

Capacity:
    Dry ................................................................... 4.0 lit (8.4/7.0 US/Imp pint)
    Oil and filter change ......................................... 3.6 lit (7.6/6.4 US/Imp pint)
    Oil change only ................................................. 3.2 lit (6.8/5.6 US/Imp pint)
Oil grade ............................................................... SAE 10W/40 motor oil SE or SF type

## Oil pump

Type ...................................................................... Trochoid
Pump reduction ratio ............................................. 1.723 : 1 (87/49 x 33/34 T)
Oil pressure @ 60°C, 140°F (service limits):
    Minimum ........................................................... 0.1 kg cm$^2$ (1.42 psi)
    Maximum .......................................................... 0.5 kg cm$^2$ (7.11 psi) @ 3000 rpm

## 1  General description

The fuel system comprises a tank from which fuel is fed by gravity to the float chambers of the four carburettors. A single automatic vacuum-operated tap controls the flow of fuel to the carburettors. When the tap is set to the ON or RES positions, fuel can flow only if the engine is running. If the tank is run dry, the PRI position allows the vacuum diaphragm to be bypassed to permit the carburettors to be primed.

All 1100 cc models are equipped with Mikuni BS34SS carburettors, with the exception of some of the UK GSX 1100 SZ Katana models, which may be fitted with the alternative Mikuni VM32SS.

In the case of the UK 1000 cc models, the GSX 1000 SZ is fitted with four Mikuni VM32SS, whilst the GSX 1000 SD uses Mikuni VM34SS instruments

For cold starting, a mixture-enriching circuit, misleadingly called a 'choke', is brought into operation. This is controlled by a handlebar lever or by a rotary control on the side panel, in the case of the Katana machines.

Engine lubrication is by conventional wet sump, the oil reservoir being contained in the bottom of the crankcase. A trochoid pump, driven from the back of the clutch assembly, draws oil from the sump via a strainer and full flow filter, supplying it under pressure to the engine and transmission components.

## 2  Fuel tank: removal and replacement

1   Unlock the seat and lift it away. The fixing bolt(s) at the rear of the tank can now be removed. Lift the rear of the tank slightly, locate the fuel level sender leads and disconnect them.
2   Check that the fuel tap is set to the ON or RES positions. Squeeze together the ends of the fuel pipe retaining clip and slide it up the pipe, clear of the stub. The pipe can now be worked off the stub, using a small screwdriver.
3   It is unnecessary to drain the tank before it can be removed, though a full tank will be somewhat unwieldy. If draining is necessary, take care to avoid any risk of fire. The tank can be drained by turning the fuel tap to the PRI position.
4   The fuel tank is located at the front by two rubber buffers, and can be removed by pulling the tank rearwards. It may help to rock the tank from side to side, but care should be taken to avoid damage to the paintwork as the tank comes free.
5   The tank is installed by reversing the removal sequence. Note that it will be much easier to locate the tank over the rubber buffers if these are lubricated with a little petrol.

## 3  Fuel tap: removal and replacement

1   The fuel tap design differs slightly between models, but is essentially the same in operation. In all but the PRI (prime) position, fuel flow is controlled by engine vacuum by means of a diaphragm assembly connected to the inlet tract. As soon as the engine is started the diaphragm reacts to engine vacuum, opening a plunger valve against spring pressure. When the engine is stopped the valve closes, shutting off the fuel supply. The PRI setting allows fuel to flow even with the engine stopped.
2   It should be noted that replacement parts for the tap are not available, so in the event of failure of the diaphragm or internal seals the whole unit must be renewed. Leakage around the tap mounting flange can be rectified by renewing the O-ring seal.
3   For obvious reasons, the tank should be drained prior to tap removal, and it is preferable to carry out the work with the tank removed and inverted on some soft cloth to protect the paintwork.
4   Remove the two securing bolts and lift the tap away. Examine the O-ring seal on the flange, renewing it if it appears indented or damaged. Examine the gauze strainer, noting any signs of water or dirt. Where necessary, flush out the tank with clean fuel and clean the gauze prior to reassembly. Fit the tap, taking care not to overtighten the retaining bolts.

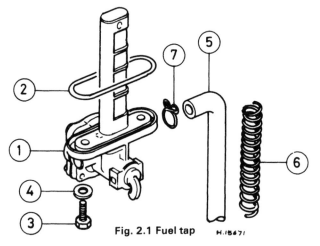

Fig. 2.1 Fuel tap                    H.15471

| | | | |
|---|---|---|---|
| 1 | Fuel tap | 5 | Hose |
| 2 | O-ring | 6 | Spring |
| 3 | Screw | 7 | Clip |
| 4 | Washer | | |

3.4a The fuel tap can be unbolted from the underside of the tank for cleaning or renewal

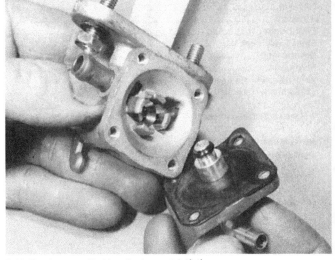

3.4b Tap is controlled by diaphragm and plunger

## 4 Carburettors: removal

1   Remove the seat, fuel tank and side panels to gain access to the carburettor assembly. Note the position of the various drain and overflow pipes, making a sketch to simplify installation. Remove the tank support bracket, then slacken the hose clip and release the two bolts which retain the rear section of the air filter trunking. Pull the trunking back and lift it away.
2   Slacken the hose clips which secure the carburettor assembly to the inlet and air filter stubs. To facilitate removal, remove the right-hand inlet adaptor.
3   Pull the assembly back and twist it to disengage the carburettors from the rubbers. The carburettors are a very close fit between the rubbers, and the help of an assistant will prove advisable. Once free of the rubbers, pull the assembly part way out to the right-hand side.
4   Disconnect the throttle cable by disengaging the outer cable from its anchor point and unhooking the inner cable from the pulley. To free the choke cable, push the retainer against spring pressure and disengage the cable. The carburettors can now be lifted clear.

## 5 Carburettors: dismantling and reassembly

1   The four separate carburettors are joined as an assembly by two steel brackets, linkages for the throttle and choke mechanisms and by fuel feed T-pieces. For almost all practical purposes the carburettors should be dealt with as an assembly, and should not be separated unless the main body, the throttle plate or the cold start plunger assembly requires attention. If separation proves necessary, refer to paragraphs 7 to 13 below.
2   Access to the various jets, the float assembly and the diaphragm assembly can be gained without separating the individual instruments. Always work on one carburettor at a time to ensure that components are not interchanged, or make certain that each part is placed in a marked container as it is removed.
3   Remove the four screws which retain the carburettor top. Lift the top away, taking care not to damage the diaphragm. Invert the assembly and tip out the diaphragm together with the valve and needle. To release the throttle needle, remove the circlip from the inside of the valve, lift out the retainer and displace the needle.
4   Remove the four float bowl screws and remove the float bowl. Displace the float pivot pin to free the float assembly. On some models the float pivot pins are shouldered at the headed end, making them an interference fit in the mounting post. It may therefore prove necessary to tap the pins out of position while fully supporting the mounting posts. The float needle can now be tipped out of its seating. The latter can be removed by releasing the single retaining screw and plate.

5   Prise out the rubber plug which covers the pilot screw and unscrew it. The main jet is screwed into the central pillar and may be removed in a similar manner. The latter also serves to retain the needle jet, which can now be displaced and removed via the throttle valve bore.
6   Reassembly is a straightforward reversal of the above sequence, taking care to avoid overtightening the jets. When fitting the needle jet, ensure that the notch aligns with the locating pin. Do not omit the washer which is fitted below the head of the main jet. When fitting the throttle valve and diaphragm, note that the latter is located by a lug on its outer edge.

### Carburettor separation

7   Remove the throttle cable and choke cable brackets from the carburettor tops. Slacken the four screws which secure the cold start shaft and remove the shaft. The upper and lower mounting brackets are each retained by eight screws, and these should be removed using an impact driver.
8   Carefully pull the carburettors apart. The throttle linkages and the fuel connectors will pull free as the instruments are separated. The throttle stop screw assembly is retained by three screws and may be lifted away once these have been released. The throttle cable lever upon which it acts is held by a single nut. The remaining three carburettors have throttle linkage levers and return springs, these too being retained by a single nut each.
9   If the throttle spindles and plates are to be removed, release the two retaining screws to free each plate and then displace the spindle. The cold start plunger assemblies can be unscrewed from the carburettor bodies.
10  When joining the carburettors, note that the throttle shaft seals have a grooved face which should face outwards. Engage the return spring end over the boss, then tension it by one turn before hooking it over the notch in the lever.
11  Check, and where necessary renew, the fuel connector O-rings prior to installation, and ensure that they and the breather T-pieces engage correctly when the instruments are joined. Note the arrangement of the synchronising screws as shown in the accompanying photograph. A thread locking compound should be used on the throttle stop assembly screws and the mounting bracket screws.
12  Slide the starter shaft into place, aligning the indentations with the screw ends and securing them using a thread locking compound.
13  Whenever the carburettors have been disturbed, they must be synchronised. Preliminary alignment is accomplished by setting the throttle valve plates so that they align with the bypass outlets in the carburettor throats, noting that each instrument is synchronised to the 3rd. The carburettors must be correctly balanced after installation, the procedure being described in Section 9 of this Chapter.

4.4 The choke cable can be freed by pushing the retainer to one side against spring pressure

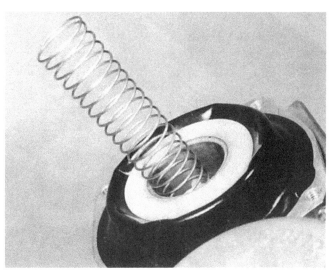

5.3 Release carburettor top and lift away the valve and diaphragm assembly

5.4a Release retaining screws and lift away the float bowl

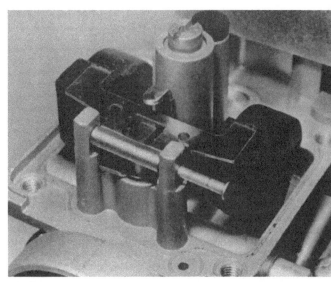

5.4b Displace the float pivot pin and lift away the float assembly

5.4c Float needle can be tipped out of its seating, the latter being retained by a single screw and retainer

5.5a Prise out the rubber plug ...

5.5b ... and unscrew the pilot jet

5.5c Unscrew the main jet from its pillar ...

5.5d ... and remove the washer

5.5e Push the needle jet out ...

5.5f ... and remove it from the carburettor top

5.6 Note locating lug on the throttle diaphragm

5.11a Carburettors are connected by breather T-pieces

5.11b Note the synchronising screw arrangement

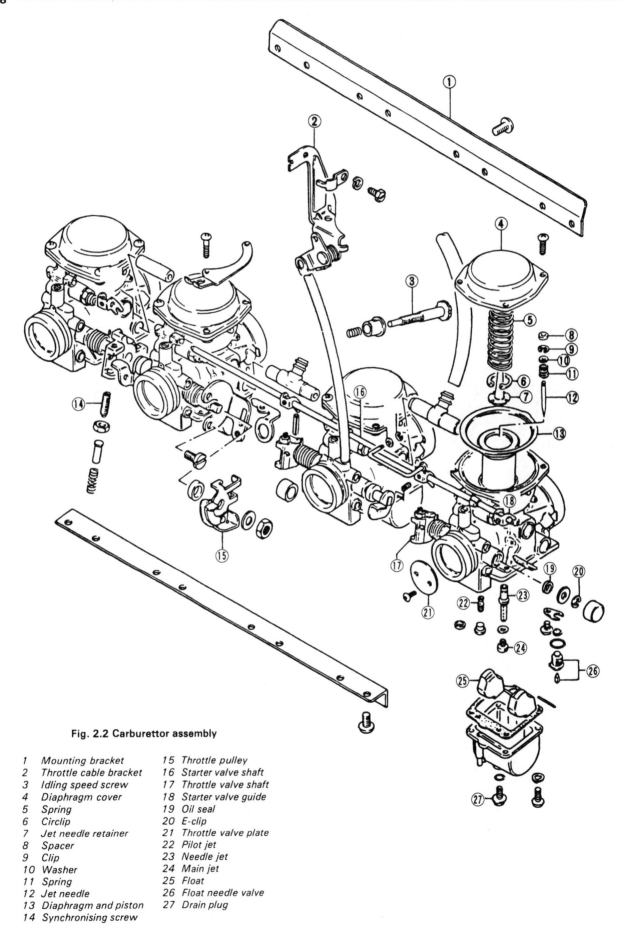

**Fig. 2.2 Carburettor assembly**

| | |
|---|---|
| 1 Mounting bracket | 15 Throttle pulley |
| 2 Throttle cable bracket | 16 Starter valve shaft |
| 3 Idling speed screw | 17 Throttle valve shaft |
| 4 Diaphragm cover | 18 Starter valve guide |
| 5 Spring | 19 Oil seal |
| 6 Circlip | 20 E-clip |
| 7 Jet needle retainer | 21 Throttle valve plate |
| 8 Spacer | 22 Pilot jet |
| 9 Clip | 23 Needle jet |
| 10 Washer | 24 Main jet |
| 11 Spring | 25 Float |
| 12 Jet needle | 26 Float needle valve |
| 13 Diaphragm and piston | 27 Drain plug |
| 14 Synchronising screw | |

## 6 Carburettors: examination and renovation

1 Dismantle each carburettor in turn as described in Section 5, laying out the component parts for examination. Check the float assembly for wear or damage, and renew it if there is any sign of leakage. The float needle and seat should be unworn with no sign of a ridge around the contact faces. If less than perfect, renew the needle and seat as a pair.

2 Inspect the jets, cleaning these and the carburettor passages by blowing them through with compressed air. Do not try to remove obstructions with wire, which will only score or enlarge the drilling and ruin the jet. Avoid the use of rag because lint and fibres can easily block the jets.

3 The diaphragm must be in good condition with no cracks or splits in its surface. If renewal is required, note that the diaphragm is supplied as an assembly with the throttle valve.

4 The needle jet and the needle must be free of visible wear or scoring, and the two items should be renewed as a pair if necessary.

**Note:** Do not disturb or remove the pilot screw. This is considered an emission-related part and is pre-set at the factory. No settings are available, so if removal is unavoidable the screw should first be turned inwards until it seats and the number of turns counted carefully. When refitting the screw, set it to the original position. Note that in some areas, disturbing the pilot screw may violate local emission laws, and is therefore not advised.

6.1 Renew needle valve and seat if grooved or damaged

## 7 Carburettors: settings and adjustment

1 The various component parts of the carburettors are chosen by the manufacturer as providing the best compromise between fuel economy and performance, and take into account the requirements of exhaust emission laws in the various areas in which the machines are sold. It follows that changes to the standard settings should be avoided, particularly where such changes might violate local laws.

2 The pilot mixture is governed by the pilot screw setting, the position of which is determined during manufacture. No nominal screw settings are given by the manufacturer, so they should not be disturbed. Idle speed is controlled by the central throttle stop screw, which can be identified by its large knurled knob. The idle speed should be kept between 950-1150 rpm. The carburettors must be kept synchronised, a procedure which is described in Section 9 of this Chapter.

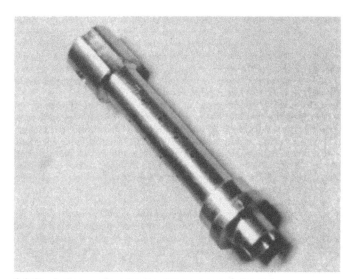
6.2 Check that needle valve bleed holes are unobstructed

## 8 Carburettors: checking the fuel level and float height

1 It is important that the fuel level in each carburettor is kept to the figure given in the specifications, otherwise the mixture will be excessively weak or rich, making normal running impossible and adjustment futile. The level can be measured by connecting a fuel level gauge to each float bowl drain screw hole in turn, and measuring the height of the fuel against the lower edge of the carburettor body.

2 A suitable gauge, consisting of a length of fuel pipe, a clear graduated tube and an adaptor to fit the drain screw thread, can be obtained through Suzuki dealers as Part Number 09913-14540. Alternatively, a similar arrangement can be made up at home without too much difficulty.

3 The fuel level in the gauge should be the specified distance below the lower edge of the carburettor body (see Specifications at the beginning of this Chapter). Note: Hold the gauge securely against the side of the carburettor body so that it is vertical and don't move the gauge whilst the reading is taken.

4 If measurement of the fuel level indicates that adjustment is necessary, remove the carburettors, and release the relevant float bowl. Remove the gasket, then measure the distance between the gasket face and the lower edge of the float, using a vernier caliper. When making the measurement, the carburettor should be inverted and the float valve lightly closed, noting that the small spring-loaded pin in the latter must not be compressed.

5 To adjust the fuel level, carefully bend the small tang which operates the float valve, then recheck the float height as described above.

7.2 Idle speed is controlled by central throttle stop screw

## 9   Carburettors: synchronisation

1   It is of fundamental importance that air and fuel are delivered in identical quantities and proportions to each of the four cylinders. The fuel flow is governed by the jets of each carburettor, but it is essential that the flow of air through each is similar at all engine speeds and throttle openings. It will be appreciated that if this is not the case, one or more cylinders will have to be 'carried' by the remainder, resulting in reduced engine power and fuel efficiency. Poor synchronisation is usually characterised by hesitant throttle response, erratic running and increased engine noise, the latter due to the effects of otherwise normal backlash in the transmission components.

2   If the carburettors have been separated, the throttle plates must be roughly synchronised during assembly as described in Section 5 above. This is a necessary preliminary to full vacuum synchronisation (balancing) which is described below. This procedure should be carried out whenever the engine is running poorly and at the intervals specified in Routine Maintenance. It must be noted that synchronisation cannot be carried out successfully unless all other adjustments are normal, especially the remaining carburettor settings. Unless these are set correctly, any attempt at synchronisation will produce erroneous readings.

3   Note that while carburettor synchronization is basically a simple task it requires the use of an expensive vacuum gauge assembly and of some skill on the owner's part. Furthermore it is complicated by the need to set the outer two carburettors at a higher level than the inner two to compensate for the different lengths of the inlet tracts between the inner and outer carburettors and the air filter element. For this reason it is recommended that the machine be taken to a Suzuki Service Agent for the work to be carried out by an expert using the correct equipment. Those owners who wish to carry out the work themselves should proceed as follows.

4   The most accurate type of vacuum gauge is the mercury manometer, but because these are extremely sensitive they require a great deal of skill to use effectively; since they also require the presence of mercury, which is an extremely toxic and dangerous liquid, their use is not recommended. The best alternative is the Suzuki service tool 09913-13121 which consists of steel balls floating in four calibrated tubes; this is almost as sensitive as the mercury manometer type, but much easier and safer for the average owner to use. The last alternative is to use dial-type vacuum gauges which are widely available and relatively easy to use; **it is essential** however, that the accuracy of such gauges is checked first by connecting each gauge in turn to any one cylinder. In any gauge does not give exactly the same reading as its counterparts, or if any is inconsistent, then the complete set cannot be used. Only the best quality, glycerine-damped, dial type gauges are accurate enough for this task. Once the equipment has been selected proceed according to the relevant paragraphs below.

5   **With both types of balancing equipment** the engine must first be checked thoroughly to ensure that it is in good condition, ie that the air filter element is clean, that there are no leaks in the filter assembly or hoses, that the carburettor fuel levels and (where applicable) the pilot screw settings are correct, that the carburettors are clean and securely fastened and that the ignition timing and spark plugs are correctly adjusted. Check also the valve clearances and engine compression pressures to ensure that the engine is mechanically sound and ensure that the exhaust system is in good condition and is securely fastened, with no leaks. Start the engine and warm it up to normal operating temperature then set the idle speed to the specified amount.

6   Note that if the rear of the tank is raised or if it is removed completely to improve access to the balancing screws, either an alternative fuel supply must be arranged or the fuel and vacuum hoses must be temporarily extended. If the vacuum hose is disconnected it must be plugged securely to allow the engine to run and to ensure that the gauge readings are correct.

7   Remove the blanking Allen screw from each inlet port and substitute the gauge adaptors, then connect the gauge hoses to the adaptors. Check the gauges and readings obtained as described below. If adjustment is necessary a specific sequence must be followed; note that the cylinders are numbered 1, 2, 3 and 4 in sequence from left to right and that all are balanced against No. 3 cylinder. Use the centre screw first to match No. 2 cylinder with no. 3, then the left-hand screw to set No. 1 cylinder. Finally use the right-hand screw to set No. 4 cylinder. Do not press down on the screws when making adjustments and ensure that the setting does not

9.5 Vacuum take-off screw location

alter when the locknuts are tightened. After each adjustment open and close the throttle quickly to settle the linkage, wait for the gauge reading to stabilise and note the effect of the adjustment before proceeding; the idle speed and vacuum reading may vary noticeably.

8   During the course of adjustment do not allow the engine to overheat; stop it and wait for it to cool down if necessary. When the carburettors are correctly synchronised stop the engine, disconnect the gauges and refit the fuel tank and blanking plugs. Check the idle speed and throttle cable free play, resetting each if necessary.

9   **When using the Suzuki service tool,** it must first be calibrated by connecting each gauge in turn to any one cylinder and using the damper screw to ensure that the steel ball is aligned with the tube's centre line. When all four gauges have been calibrated to the same cylinder connect them all to their respective cylinders and start the engine.

10   With the engine idling, the steel balls of the gauges for the two inner cylinders (Nos 2 and 3) should be aligned with their top edges just touching the gauge centre lines, while those for the gauges of the two outer cylinders should be aligned so that they are bisected by the centre line, ie the two inner cylinders should show half of a ball diameter less vacuum than the outer two. If adjustment is necessary, proceed as described in paragraphs 7 and 8 above.

11   **When using dial-type gauges,** first check that they are sufficiently accurate by checking that each produces the same reading on any one cylinder as described above. If the damping is adjustable, set it so that needle flutter is just eliminated, but so that the gauge can record the slightest change in pressure.

12   When the gauges are known to be accurate, connect each to its respective cylinder. Since Suzuki do not give the necessary information the difference in pressure between the inner and outer cylinder can only be approximated; the actual figure will vary greatly depending on the quality of the gauges used.

13   Since this difference in pressure is due to the presence of the air filter element, a reasonably accurate setting can be achieved by temporarily removing the element for the duration of the adjustment and then balancing all four cylinders to exactly the **same** level; if the element is then refitted the difference in pressure can be noted for future reference.

14   With the gauges connected and the element removed, therefore, check that all cylinders give the same reading and make the necessary adjustments as described in paragraphs 7 and 8 above. Do not forget to refit the filter element once adjustment is complete. **Note:** *this operation is described only as an alternative for those owners who have access to a set of dial-type gauges but not to the Suzuki service tool. The results obtained can only approximate the correct setting and will depend on the accuracy of the gauges used and the skill of the owner. If there is any doubt about the results, the machine must be taken to a Suzuki Service Agent for the carburettor synchronization to be checked using the correct tool.*

15   **With both types of gauge,** it should be noted that if aftermarket accessory filters have been substituted for the standard filter assembly

and plenum chamber, some experimentation may be necessary to achieve the correct setting. Since the difference in inlet tract lengths will no longer exist, it is possible that the best results will be obtained with all cylinders set to the same level.

## 10 Air filter: removal and refitting

1    The air filter element is of the dry type, comprising a pleated, resin-impregnated paper element, supported by a metal mesh inner core and

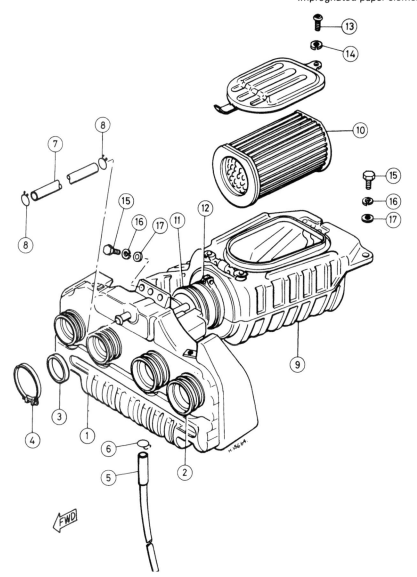

**Fig. 2.3 Air filter assembly**

1    Front section
2    Carburettor hoses
3    Ring
4    Hose clip
5    Drain hose
6    Clip
7    Breather hose
8    Clip
9    Rear section
10   Filter element
11   Connecting hose
12   Clip
13   Screw
14   Spring washer
15   Bolt
16   Spring washer
17   Washer

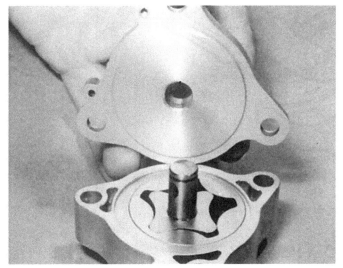

11.3 Lift away the pump cover to gain access to rotors

11.5a Measure rotor tip clearance ...

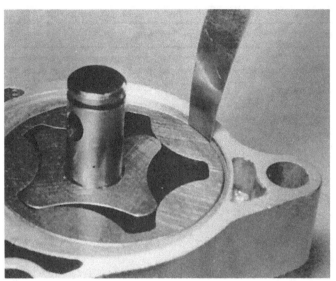

11.5b ... and outer rotor to body clearance

**Fig. 2.4 Oil pump, filter and sump**

1 Sump
2 Sump gasket
3 O-ring
4 Bolt
5 Drain plug
6 Sealing washer
7 Oil filter
8 O-ring
9 Spring
10 Filter cover
11 Stud
12 Domed nut
13 Washer
14 Oil pump
15 O-ring
16 Screw (bolt, later models)
17 Driven gear
18 Washer
19 Drive pin
20 Circlip
21 Strainer
22 Screw

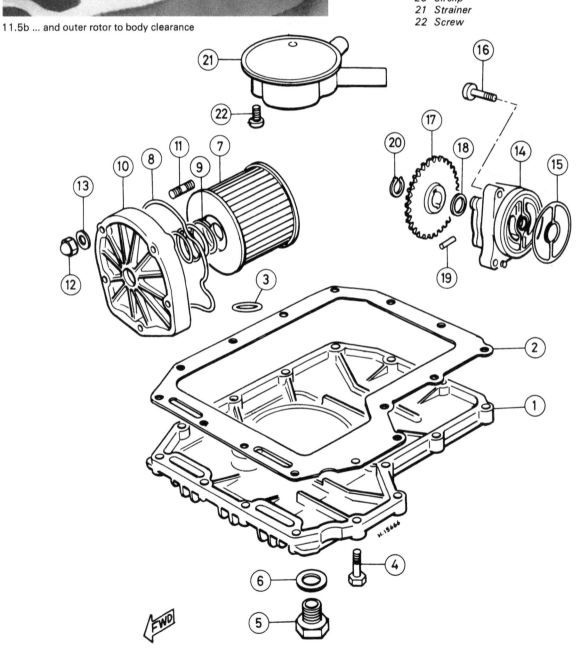

sealed between end caps. The element can be removed after opening the seat and releasing the screw(s) which retains the filter cover. This will necessitate the removal of the seat support bracket on some models. Once this has been removed, release the lid, then grasp the spring retainer and pull the element out of the casing.

2 The element must be cleaned at the intervals given in the Routine Maintenance Section of this Manual. Using compressed air, blow the element clean from the inside. Note that no attempt must be made to clean the element using solvents.

3 If the element has become very dirty, or has already been cleaned 3 or more times, it should be renewed. It is also important to examine the filter surface very carefully, looking for any indications of damage. If torn or holed, renew the element without delay.

4 Note that if the machine is used with a damaged filter element, or if the element is omitted, any airborne dust will be drawn into the engine and will probably result in rapid wear. In addition, the resulting weak mixture will upset the carburation.

## 11 Oil pump: examination and renovation

1 The engine and transmission components are dependent on the reliable supply of oil at the correct pressure. Provided that the oil is changed at the specified intervals and no catastrophic failure of internal components occurs, the oil pump will usually last the life of the engine.

2 The oil pump should only require attention during major engine overhauls, or if indications of low oil pressure or unusual engine noises prompt further investigation. Access to the pump requires the removal of the clutch assembly, and the relevant procedures are described in Sections 10 and 11 of Chapter 1.

3 With the pump removed from the crankcase, the end cover can be removed. On earlier models the cover is retained by three screws, whilst on later machines the cover is held in place by the mounting bolts.

4 Inspect the pump rotors and the working faces of the body and cover for signs of excessive wear or damage. If scoring is found the pump must be renewed and the source of the damage traced and rectified. Note that if debris remains in the oil passages, the new pump will soon be destroyed, and it will be necessary to dismantle the crankcases to ensure that all traces of swarf are removed. It follows that if metal particles are found, an internal component has failed and will require renewal.

5 Using feeler gauges, measure the clearance between the rotor tips and the outer rotor and the body (see accompanying photographs). Check the rotor end float by placing a steel rule or straight edge across the end face of the pump and measuring the gap between it and the rotors with feeler gauges. If the clearances exceed those given in the Specifications, the pump must be renewed.

6 When reassembling the pump, clean the body, cover and rotors with petrol or a cleaning solvent, then lubricate them with engine oil. Make sure that the inner rotor drive pin is correctly located in its slot in the rotor. Where the cover is secured by screws, use a thread locking compound and tighten them firmly. Check the condition of the O-ring seal and renew it if necessary, prior to installation.

## 12 Checking the oil pressure

1 Suzuki recommend that the oil pressure be checked every 3000 miles (5000 km). In addition, the check should be carried out if there

is reason to suspect that there is a lubrication system fault. The test requires the use of the Suzuki pressure gauge, Part Number 09915 - 74510, or an equivalent.

2 Start the engine and run it until it reaches normal operating temperature, then stop it. Remove the plug from the right-hand end of the main gallery, just below the cylinder block. Start the engine again and note the reading on the gauge. If the pump and engine components are sound, a reading above the minimum figure should be obtained. If the reading is below this, dismantle and check the oil pump and the engine to establish the cause.

## 13 Checking the oil pressure switch

1 If the oil pressure lamp comes on and a subsequent check on the oil pressure shows this to be acceptable, it can be assumed that the oil pressure switch is at fault. If necessary, check the switch by substitution.

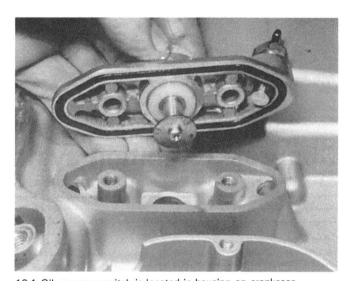

13.1 Oil pressure switch is located in housing on crankcase

## 14 Oil filter: renewing the element

1 It is important that the oil filter element be renewed at the intervals recommended in Routine Maintenance. If renewal is postponed for too long, a bypass valve will open to ensure that the supply of oil is maintained, but this means that the oil now circulating in the engine is unfiltered, and engine wear will be accelerated. The filter renewal procedure is described in detail in Routine Maintenance.

# Chapter 3 Ignition system

*Refer to Chapter 7 for information relating to the 1984-on 1135cc-engined GSX1100 and GS1150 models*

**Contents**

**Specifications**

**Ignition system**

| | |
|---|---|
| Type .................................................................................... | Electronic |
| Minimum spark performance ........................................... | 8 mm (0.3 in) ⊛ 1 atmosphere |
| Firing order ........................................................................ | 1, 2, 4, 3 |

**Ignition timing**

| | |
|---|---|
| Below 1500 ± 150 rpm ...................................................... | 12° BTDC |
| Above 2350 ± 150 rpm ...................................................... | 32° BTDC |

**Spark plug**

| | NGK | ND |
|---|---|---|
| Make ................................................................................... | NGK | ND |
| Type: | | |
| UK models ................................................................. | DR8ES-U | X24ESR-U |
| US models ................................................................. | D8EA | X24ES-U |
| Electrode gap .................................................................. | 0.6 – 0.7 mm (0.024 – 0.028 in) | |

**Ignition coil resistances**

| | |
|---|---|
| Primary windings .............................................................. | 3 – 5 ohms |
| Secondary windings .......................................................... | 31 – 33 k ohms |

**Ignition source coil**

| | |
|---|---|
| Resistance .......................................................................... | Approx 290 – 360 ohms |

**Electrode gap check** - use a wire type gauge for best results

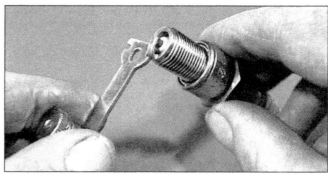

**Electrode gap adjustment** - bend the side electrode using the correct tool

**Normal condition** - A brown, tan or grey firing end indicates that the engine is in good condition and that the plug type is correct

**Ash deposits** - Light brown deposits encrusted on the electrodes and insulator, leading to misfire and hesitation. Caused by excessive amounts of oil in the combustion chamber or poor quality fuel/oil

**Carbon fouling** - Dry, black sooty deposits leading to misfire and weak spark. Caused by an over-rich fuel/air mixture, faulty choke operation or blocked air filter

**Oil fouling** - Wet oily deposits leading to misfire and weak spark. Caused by oil leakage past piston rings or valve guides (4-stroke engine), or excess lubricant (2-stroke engine)

**Overheating** - A blistered white insulator and glazed electrodes. Caused by ignition system fault, incorrect fuel, or cooling system fault

**Worn plug** - Worn electrodes will cause poor starting in damp or cold weather and will also waste fuel

## 1  General description

The machines featured in this manual are equipped with a transistorised ignition system of the type which has become well established in motorcycle applications. An ignition pickup assembly is mounted on the right-hand end of the crankshaft. This comprises a combined rotor and automatic timing unit (ATU), the former inducing a trigger pulse as it sweeps past the two pickup coils. The trigger pulses are fed to a sealed ignition amplifier, the resulting pulse being routed to the appropriate ignition coil.

Two ignition coils are fitted, each coil firing two plugs in a 'spare spark' arrangement. In this system, plugs 1 and 4 spark simultaneously, combustion occurring only in the cylinder in which the fuel/air mixture is under compression, the remaining spark occurring in a cylinder on the exhaust stroke. The same arrangement applies to cylinders 2 and 3.

The ignition system is depicted in the accompanying circuit diagram. It will be seen that the system is very simple in design, and with the reliability expected of transistor circuits, the arrangement will require little attention in normal use.

## 2  Testing the ignition system: general

1    In the event of a fault developing in the ignition system, the spark plugs should always be checked first as described below. Where the fault cannot be attributed to the plugs, plug caps or the high tension leads, further testing of the system will require the use of a simple multimeter. In the absence of the Suzuki 'Pocket Tester', Part Number 09900 – 25002, any similar device will be adequate for most purposes. Note that a multimeter should be considered an essential piece of equipment for ignition and electrical tests. In view of the low cost, it can be considered a good investment.

2    As has been mentioned above, most ignition faults will be traced to the plugs or the high tension side of the system, and will have resulted from normal wear or from corrosion. The second most likely culprit will be damaged, broken or shorted wiring, or a fault in a connector. Failure of the pickup assembly or the ignition unit is rare, but it is worth noting that the nature of the fault will often give a good indication of the source of the problem.

3    If the fault is evident on cylinders 1 and 4, or on cylinders 2 and 3 only, check the relevant pickup coil and ignition coil. In the event of

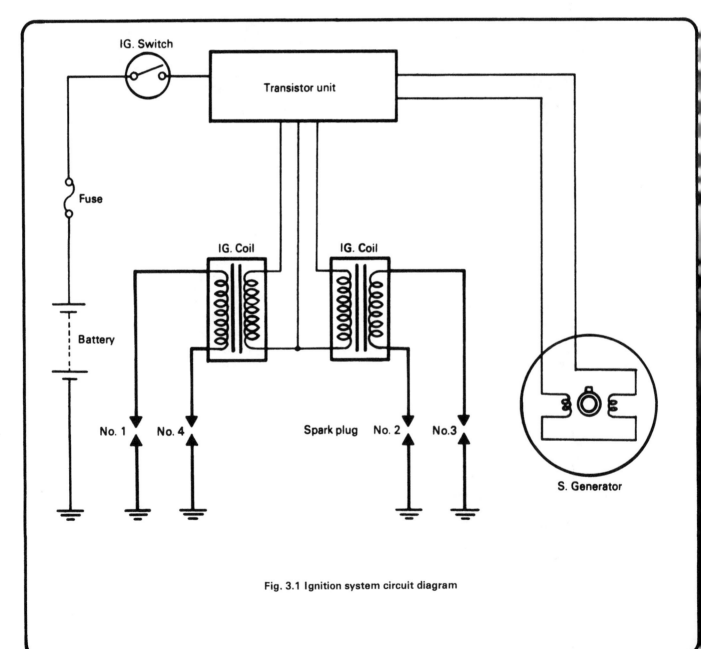

Fig. 3.1 Ignition system circuit diagram

complete ignition failure, the above components are unlikely to be responsible, and attention should be directed towards the supply, switches and the ignition unit.

## 3  Checking the alternator output and battery

The ignition system will not operate normally if the supply to the ignition unit is unstable or non-existent. If other electrical problems are evident, always investigate the alternator output and the condition of the battery before the ignition system is checked. For details, refer to Chapter 6.

## 4  Checking the wiring and connections

Using the wiring diagrams at the end of Chapter 6, trace and examine the ignition wiring, repairing and re-routing any wires which have been damaged or trapped. The associated connectors should be separated and checked for corrosion. This should be removed by scraping it off, and further corrosion prevented by coating the metal parts with silicone grease.

## 5  Spark plugs: checking and resetting the gaps

1   The single most common cause of ignition faults is defects in the spark plugs. The plug condition should be checked regularly at the intervals given in Routine Maintenance, and whenever misfiring or poor starting is evident.
2   All UK models are fitted with NGK DR8ES-U or ND X24ESR-U plugs, whilst the US versions are equipped with NGK D8EA or ND X24ES-U plugs. In all cases the electrode gap is 0.6 – 0.7 mm (0.024 – 0.028 in).
3   Remove the plugs and examine the electrodes and insulator nose, comparing the appearance with the examples shown in the accompanying photographs, as a guide to the general condition of the engine. With experience, a great deal of information may be obtained in this way.
4   Light carbon deposits can be scraped off, taking care not to crack the rather brittle ceramic insulator. If the carbon build-up is heavy, it is preferable to renew the plugs as a set, given the relatively low cost involved.
5   With both new and used plugs, check the electrode gaps prior to installation, using feeler gauges. If adjustment is required, carefully bend the outer earth (ground) electrode only. On no account attempt to bend the inner electrode or the insulator nose will be damaged.
6   Check that the threads are clean and apply a thin film of molybdenum disulphide grease to aid subsequent removal. Fit each plug finger tight only, then use a plug socket or box spanner to tighten it by about a quarter turn, just enough to seat the sealing washer firmly, and no more.
7   If previous overtightening of the plugs has stripped the plug threads in the head casting, they can be reclaimed by fitting Helicoil inserts. This is an inexpensive and convenient service offered by many dealers.
8   It is good practice to carry one or two new spark plugs of the correct type and correctly gapped with the tool kit. This can save a lot of time in the event of a roadside breakdown.

## 6  Plug caps and leads: examination and renovation

1   The plug caps incorporate suppressors to limit radio frequency (RF) interference and form a sealed connection between the plugs and the high tension leads. The most likely problems are failure of the suppressor or tracking and shorting. If a suppressor fails, the increased resistance will usually prevent or reduce sparking at the associated plug.
2   Check first that the plug is operating normally, then check the cap by substitution, swapping it with a sound cap and noting whether this resolves the problem. The seal around the bottom edge of the cap is important because it excludes dirt and moisture at the otherwise vulnerable plug connection. If it is damaged, renew the cap.
3   Another form of shorting, known as tracking, can develop along the HT lead. This is initially due to dirt and moisture on the lead providing an alternative path for the high tension current, and may be visible at night. In time, the repeated discharge will cause a carbon build-up, worsening the problem. If tracking is suspected, clean the leads carefully, using a rag moistened with WD 40 or a similar silicone-based water dispersal spray.
4   Care should be taken to avoid damage to the leads, being especially careful to prevent chafing against the cylinder head cover. Note that the leads are moulded into the coils, and will have to be renewed as a complete assembly if seriously damaged.

## 7  Ignition coils: checking

1   A fault in an ignition coil will almost invariably be indicated by weak or erratic sparking on one pair of plugs only; the chances of both coils failing simultaneously being most unlikely. Always check first that the plugs, plug caps and leads are not at fault (see above) before testing the coils.
2   The spark performance is tested with the coils removed from the machine, using an Electro Tester, a piece of equipment used by most Suzuki dealers, but not worth buying for infrequent home use. An approximate indication of performance can be obtained with the coils in place on the machine.
3   Remove the relevant plugs and lodge them securely against the cylinder head with the plug caps attached. Crank the engine and note the spark at the electrodes. If all is well, a fat blue spark should be produced, whilst a weak coil will show a weak or erratic spark at both plugs, possibly orange in colour. It should be noted that partial failure of the ignition unit could produce similar effects, so check the coil resistances as described below.
4   Remove the fuel tank and separate the low tension connector to the suspect coil, and disconnect the plug caps. Using a multimeter set on the resistance scale, connect the probes to the orange/white and white low tension leads and note the primary winding resistance reading. An approximate value of 3 – 5 ohms should be shown.
5   Connect the meter probes to the two HT leads at the plug caps and note the secondary winding resistance reading. A figure of approximately 31 – 33 ohms should be indicated. Note that the above resistance figures are not intended to be exact, but if the readings obtained differ radically, the coil can be assumed to be defective. A Suzuki dealer will be able to confirm this by performing a spark test.

## 8  Ignition pickup coils: checking

Trace the wiring from the ignition pickup assembly back to the connector near the ignition unit. Separate the connector and measure the resistance between the blue and green leads on the pickup side of the connector. The standard figure is 250 – 360 ohms, a very high or zero reading indicating that the pickup coils are faulty.

## 9  Ignition amplifier unit: checking

Note: The following check of the ignition unit assumes that the coils, leads, plug caps and plugs are not at fault.
1   Remove the spark plugs from the cylinders 3 and 4, refit them in their plug caps and lodge them so that the plug bodies are in firm contact with the cylinder head. Remove the right-hand side panel and disconnect the ignition pickup connector. Set the meter to the ohms x 1 scale, switch on the ignition, and connect the probe leads as follows to the ignition unit side of the connector:

Positive (red) probe to the blue lead
Negative (black) probe to the green lead

2   As the probes touch, a spark should jump across the No.4 plug, whilst a spark should be evident at the No.3 plug as the probes are removed. The test circuit is illustrated in the accompanying diagram.

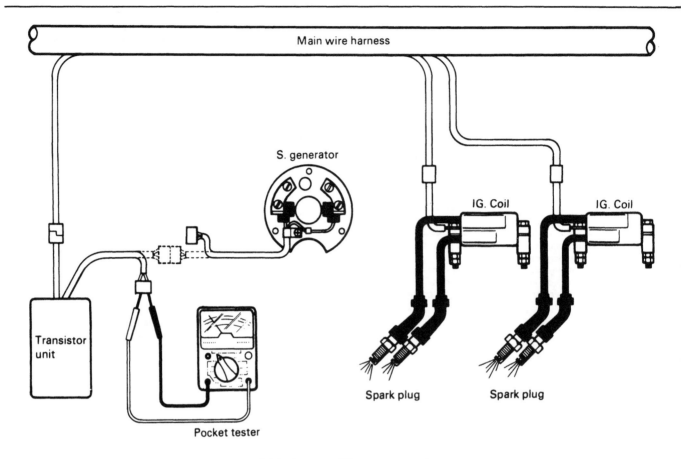

Fig. 3.2 Ignition amplifier test connections

## 10 Ignition timing: checking and adjustment

1   The ignition timing should be checked at the intervals indicated in Routine Maintenance, or when the engine runs or starts erratically. A strobe (stroboscopic timing lamp) will be required, and should preferably be of the more accurate zenon type, rather than one of the cheap neon versions.

2   Remove the ignition pickup inspection cover and connect the strobe according to the manufacturer's instructions to cylinder No. 1 or No.4. Start the engine and direct the lamp at the timing unit via the inspection hole in the pickup backplate. Ensure that the timing marks align as shown in the accompanying line drawings and at the engine speeds indicated.

3   If the marks do not coincide accurately, stop the engine, slacken the backplate screws and adjust the position of the latter. Tighten the screws and repeat the test until the timing is set correctly.

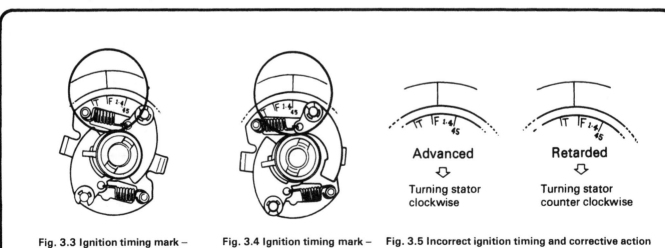

Fig. 3.3 Ignition timing mark –
below 1500 rpm

Fig. 3.4 Ignition timing mark –
above 2300 rpm

Fig. 3.5 Incorrect ignition timing and corrective action

# Chapter 4 Frame and forks

*Refer to Chapter 7 for information relating to the 1984-on 1135cc-engined GSX1100 and GS1150 models*

**Contents**

**Specifications**

## GSX 1100 T, ET, LT, EX, EZ, SZ, SD, ESD

**Frame**

Type ............................................................... Tubular, double cradle

**Front forks**

Stroke:
- T, ET, LT, EX, EZ ............................................. 160 mm (6.3 in)
- SZ, SD, ESD .................................................. 150 mm (5.9 in)

Oil level:
- T ............................................................ 152 mm (6.0 in)
- ET ........................................................... 193 mm (7.6 in)
- LT ........................................................... 260 mm (10.2 in)
- EX ........................................................... 226 mm (8.9 in)
- EZ ........................................................... 195 mm (7.7 in)
- SZ, SD, ESD .................................................. 221 mm (8.7 in)

Oil grade:
- T, ET, EX .................................................... 50% SAE 10W/30, 50% ATF (automatic transmission fluid)
- LT ........................................................... SAE 10W/20
- EZ, SZ ....................................................... 15W fork oil
- SD, ESD ...................................................... SAE 10W/40 SE or SF

Oil quantity (per leg):
- T ............................................................ 302 ml (10.63/10.21 Imp/US fl oz)
- ET ........................................................... 252 ml (8.87/8.52 Imp/US fl oz)
- LT ........................................................... 245 ml (8.63/8.28 Imp/US fl oz)
- EX ........................................................... 232 ml (8.17/7.84 Imp/US fl oz)
- EZ ........................................................... 246 ml (8.66/8.32 Imp/US fl oz)
- SZ, SD, ESD .................................................. 227 ml (7.99/7.67 Imp/US fl oz)

Fork spring free length – service limit:
- T ............................................................ 451 mm (17.8 in)
- ET ........................................................... 514 mm (20.2 in)
- LT ........................................................... 516 mm (20.3 in)
- EX ........................................................... 510 mm (20.1 in)
- EZ ........................................................... 438 mm (17.2 in)
- SZ, SD, ESD .................................................. 463 mm (18.2 in)

Fork air pressure:
- T, ET, EX, EZ ................................................ 0.5 kg cm$^2$ (7.11 psi)
- LT ........................................................... 0.8 kg cm$^2$ (11.38 psi)
- SZ, SD, ESD .................................................. Not applicable

## Rear suspension

| | |
|---|---|
| Type ................................................................................... | Welded alloy swinging arm |
| Travel: | |
|     T, ET, LT, EX, EZ ................................................... | 108 mm (4.25 in) |
|     SZ, SD, ESD ........................................................... | 109 mm (4.29 in) |
| Suspension units ............................................................... | Oil damped, coil spring with adjustable damping |
| Swinging arm pivot shaft runout (service limit) ................... | 0.3 mm (0.012 in) |

## Torque wrench settings

| Component | kgf m | lbf ft |
|---|---|---|
| Front wheel spindle nut ...................................................... | 3.6 – 5.2 | 26.0 – 37.5 |
| Front wheel spindle pinch bolt – T, ET, LT, EX .................. | 1.5 – 2.5 | 11.0 – 18.0 |
| Front wheel spindle clamp nut – EZ, SZ, SD, ESD ............ | 1.5 – 2.5 | 11.0 – 18.0 |
| Front brake caliper mounting bolt ...................................... | 2.5 – 4.0 | 18.0 – 29.0 |
| Fork damper rod bolt: | | |
|     T, ET, EX ................................................................ | 2.0 – 2.5 | 14.5 – 18.0 |
|     LT, SZ ..................................................................... | Not available | |
|     EZ, SD, ESD ........................................................... | 2.0 – 2.6 | 14.5 – 18.8 |
| Fork damper rod nut ........................................................... | 2.5 – 3.5 | 18.0 – 25.5 |
| Air valve: | | |
|     T, ET, LT, EX, EZ ................................................... | 1.0 – 1.3 | 7.0 – 9.5 |
|     SZ, SD, ESD ........................................................... | Not available | |
| Air lock bolt: | | |
|     T, ET, EX ................................................................ | 0.05 – 0.10 | 0.4 – 0.7 |
|     LT, SZ, SD, ESD ..................................................... | Not available | |
| Air union bolt: | | |
|     T, ET, EX ................................................................ | 1.0 – 1.2 | 7.0 – 8.5 |
|     LT, SZ, SD, ESD ..................................................... | Not available | |
| Lower yoke pinch bolt ........................................................ | 1.5 – 2.5 | 11.0 – 18.0 |
| Upper yoke pinch bolt ........................................................ | 2.0 – 3.0 | 14.5 – 21.5 |
| Fork top bolt: | | |
|     T, ET, LT, EX, EZ, SD, ESD ................................... | 1.5 – 3.0 | 11.0 – 21.5 |
|     SZ ........................................................................... | 2.0 – 3.0 | 14.5 – 21.5 |
| Anti-dive air bleed – EZ, SZ, SD, ESD .............................. | 0.6 – 0.9 | 4.5 – 6.5 |
| Anti-dive/brake hose union – EZ, SZ, SD, ESD ................. | 2.0 – 2.5 | 14.5 – 18.0 |
| Anti-dive mounting bolt – EZ, SZ, SD, ESD ...................... | 0.4 – 0.5 | 3.0 – 3.5 |
| Anti-dive valve mounting bolt – EZ, SZ, SD, ESD ............. | 0.6 – 0.8 | 4.5 – 6.0 |
| Steering stem top bolt: | | |
|     EZ, SD, ESD ........................................................... | 2.0 – 3.0 | 14.5 – 21.5 |
|     SZ ........................................................................... | 3.5 – 5.0 | 25.5 – 36.0 |
| Steering stem nut .............................................................. | 4.0 – 5.0 | 29.0 – 36.0 |
| Steering stem pinch bolt .................................................... | 1.5 – 2.5 | 11.0 – 18.0 |
| Steering stem adjuster nut ................................................. | 3.5 – 5.0 | 25.5 – 36.0 |
| Handlebar clamp bolts: | | |
|     T, ET, LT, EX, EZ ................................................... | 1.2 – 2.0 | 8.5 – 14.5 |
|     SZ, SD, ESD ........................................................... | 0.8 – 1.2 | 6.0 – 8.5 |
| Front footrest bolt .............................................................. | 2.7 – 4.3 | 19.5 – 31.0 |
| Swinging arm pivot nut ...................................................... | 5.5 – 8.5 | 40.0 – 61.5 |
| Rear torque arm nut ........................................................... | 2.0 – 3.0 | 14.5 – 21.5 |
| Silencer bracket nut ........................................................... | 1.5 – 2.0 | 11.0 – 14.5 |
| Rear suspension unit nut/bolt ............................................ | 2.0 – 3.0 | 14.5 – 21.5 |
| Rear footrest bolt ............................................................... | 2.7 – 4.3 | 19.5 – 31.0 |
| Rear wheel spindle nut ....................................................... | 8.5 – 11.5 | 61.5 – 83.0 |
| Chain adjuster support bolt: | | |
|     T, ET, LT, EX, EZ ................................................... | 1.5 – 2.0 | 11.0 – 14.5 |
|     SZ, SD, ESD ........................................................... | 1.8 – 2.8 | 13.0 – 20.0 |

# GS 1100 T, ET, LT, EX, EZ, ED, ESD (US)

## Frame

| | |
|---|---|
| Type ................................................................................... | Tubular, double cradle |

## Front forks

| | |
|---|---|
| Stroke ................................................................................. | 160 mm (6.3 in) |
| Oil level: | |
|     T, ET, EX ................................................................ | 216 mm (8.5 in) |
|     LT ........................................................................... | 260 mm (10.2 in) |
|     EZ, ED, ESD ........................................................... | 195 mm (7.7 in) |
| Oil grade ............................................................................ | 15W fork oil |
| Oil quantity (per leg): | |
|     T ............................................................................ | 290 ml (9.80 US fl oz) |
|     ET, EX .................................................................... | 238 ml (8.04 US fl oz) |
|     LT ........................................................................... | 245 ml (8.28 US fl oz) |
|     EZ, ED, ESD ........................................................... | 246 ml (8.31 US fl oz) |

Fork spring free length – service limit:
    T, ET, EX .................................................... 518 mm (20.4 in)
    LT .............................................................. 516 mm (20.3 in)
    EZ, ED ....................................................... 442 mm (17.4 in)
    ESD ............................................................ 454 mm (17.9 in)
Fork air pressure:
    T, ET, EX, EZ, ED, ESD ........................... 0.5 kg cm² (7.11 psi)
    LT .............................................................. 0.8 kg cm² (11.38 psi)

## Rear suspension

Type ................................................................ Welded alloy swinging arm
Travel .............................................................. 108 mm (4.25 in)
Suspension units ............................................. Oil damped, coil spring with adjustable damping
Swinging arm pivot shaft runout (service limit) ................................ 0.3 mm (0.012 in)

## Torque wrench settings (all models except ED, ESD)

| Component | kgf m | lbf ft |
|---|---|---|
| Front wheel spindle nut | 3.6 – 5.2 | 26.0 – 37.5 |
| Front wheel spindle pinch bolt | 1.5 – 2.5 | 11.0 – 18.0 |
| Front wheel spindle clamp nut – EZ | 1.5 – 2.5 | 11.0 – 18.0 |
| Front brake caliper mounting bolt | 2.5 – 4.0 | 18.0 – 29.0 |
| Fork damper rod bolt: | | |
|     T, ET, EX | 2.0 – 2.5 | 14.5 – 18.0 |
|     LT | Not available | |
|     EZ | 2.0 – 2.6 | 14.5 – 18.8 |
| Fork damper rod nut | 2.5 – 3.5 | 18.0 – 25.5 |
| Air valve | 1.0 – 1.3 | 7.0 – 9.5 |
| Air lock bolt: | | |
|     T, ET, EX, EZ | 0.05 – 0.10 | 0.4 – 0.7 |
|     LT | Not available | |
| Air union bolt: | | |
|     T, ET, EX, EZ | 1.0 – 1.2 | 7.0 – 8.5 |
|     LT | Not available | |
| Lower yoke pinch bolt | 1.5 – 2.5 | 11.0 – 18.0 |
| Upper yoke pinch bolt | 2.0 – 3.0 | 14.5 – 21.5 |
| Fork top bolt | 1.5 – 3.0 | 11.0 – 21.5 |
| Anti-dive air bleed – EZ | 0.6 – 0.9 | 4.5 – 6.5 |
| Anti-dive/brake hose union – EZ | 2.0 – 2.5 | 14.5 – 18.0 |
| Anti-dive mounting bolt – EZ | 0.4 – 0.5 | 3.0 – 3.5 |
| Anti-dive valve mounting bolt – EZ | 0.6 – 0.8 | 4.5 – 6.0 |
| Steering stem top bolt – EZ | 2.0 – 3.0 | 14.5 – 21.5 |
| Steering stem nut – EZ | 4.0 – 5.0 | 29.0 – 36.0 |
| Steering stem pinch bolt | 1.5 – 2.5 | 11.0 – 18.0 |
| Steering stem adjuster nut | 3.5 – 5.0 | 25.5 – 36.0 |
| Handlebar clamp bolt | 1.2 – 2.0 | 8.5 – 14.5 |
| Front footrest bolt | 2.7 – 4.3 | 19.5 – 31.0 |
| Swinging arm pivot nut | 5.5 – 8.5 | 40.0 – 61.5 |
| Rear torque arm nut | 2.0 – 3.0 | 14.5 – 21.5 |
| Silencer bracket nut | 1.5 – 2.0 | 11.0 – 14.5 |
| Rear suspension unit nut/bolt | 2.0 – 3.0 | 14.5 – 21.5 |
| Rear footrest bolt | 2.7 – 4.3 | 19.5 – 31.0 |
| Rear wheel spindle nut | 8.5 – 11.5 | 61.5 – 83.0 |
| Chain adjuster support bolt | 1.5 – 2.0 | 11.0 – 14.5 |

# GSX 1000 SZ, SD

## Frame

Type ................................................................ Tubular, double cradle

## Front forks

Stroke ............................................................. 150 mm (5.9 in)
Oil level .......................................................... 221 mm (8.7 in)
Oil grade ......................................................... 15W fork oil
Oil quantity (per leg) ...................................... 227 ml (7.99/7.67 Imp/US fl oz)
Fork spring free length – service limit ........... 463 mm (18.2 in)

## Rear suspension

Type ................................................................ Welded alloy swinging arm
Travel .............................................................. 109 mm (4.29 in)
Suspension units ............................................. Oil damped, coil spring with adjustable damping
Swinging arm pivot shaft runout (service limit) ................................ 0.3 mm (0.012 in)

## Torque wrench settings

| Component | kgf m | lbf ft |
|---|---|---|
| Handlebar clamp bolt | 0.8 – 1.2 | 6.0 – 8.5 |
| Steering stem top bolt: | | |
|     SZ | 3.5 – 5.0 | 25.5 – 36.0 |
|     SD | 2.0 – 3.0 | 14.5 – 21.5 |
| Steering stem clamp bolt | 1.5 – 2.5 | 11.0 – 18.0 |
| Fork top bolt: | | |
|     SZ | 2.0 – 3.0 | 14.5 – 21.5 |
|     SD | 1.5 – 3.0 | 11.0 – 21.5 |
| Upper yoke pinch bolt | 2.0 – 3.0 | 14.5 – 21.5 |
| Lower yoke pinch bolt | 1.5 – 2.5 | 11.0 – 18.0 |
| Anti-dive/brake hose union bolt | 2.0 – 2.5 | 14.5 – 18.0 |
| Air bleed valve | 0.6 – 0.9 | 4.5 – 6.5 |
| Anti-dive valve fitting bolt | 0.6 – 0.8 | 4.5 – 6.0 |
| Anti-dive plunger fitting bolt | 0.4 – 0.5 | 3.0 – 3.5 |
| Front brake caliper mounting bolt | 2.5 – 4.0 | 18.0 – 29.0 |
| Front wheel spindle nut | 3.6 – 5.2 | 26.0 – 37.5 |
| Front wheel spindle clamp nut | 1.5 – 2.5 | 11.0 – 18.0 |
| Front footrest bolt | 2.7 – 4.3 | 19.5 – 31.0 |
| Swinging arm pivot nut | 5.5 – 8.5 | 40.0 – 61.5 |
| Rear footrest bolt | 2.7 – 4.3 | 19.5 – 31.0 |
| Rear suspension unit nut/bolt | 2.0 – 3.0 | 14.5 – 21.5 |
| Rear torque arm nut/bolt | 2.0 – 3.0 | 14.5 – 21.5 |
| Chain adjuster bolt | 1.8 – 2.8 | 13.0 – 20.0 |
| Rear wheel spindle nut | 8.5 – 11.5 | 61.5 – 83.0 |

# GS 1000 SZ (US)

## Frame

| | |
|---|---|
| Type | Tubular, double cradle |

## Front forks

| | |
|---|---|
| Stroke | 150 mm (5.9 in) |
| Oil level | 221 mm (8.7 in) |
| Oil grade | 15W fork oil |
| Oil quantity (per leg) | 227 ml (7.99/7.67 Imp/US fl oz) |
| Fork spring free length – service limit | 467 mm (18.4 in) |

## Rear suspension

| | |
|---|---|
| Type | Welded alloy swinging arm |
| Travel | 109 mm (4.29 in) |
| Suspension units | Oil damped, coil spring with adjustable damping |
| Swinging arm pivot shaft runout (service limit) | 0.3 mm (0.012 in) |

## Torque wrench settings

| Component | kgf m | lbf ft |
|---|---|---|
| Handlebar clamp bolt | 0.8 – 1.2 | 6.0 – 8.5 |
| Steering stem top bolt | 3.5 – 5.0 | 25.5 – 36.0 |
| Steering stem clamp bolt | 1.5 – 2.5 | 11.0 – 18.0 |
| Fork top bolt | 2.0 – 3.0 | 14.5 – 21.5 |
| Upper yoke pinch bolt | 2.0 – 3.0 | 14.5 – 21.5 |
| Lower yoke pinch bolt | 1.5 – 2.5 | 11.0 – 18.0 |
| Anti-dive/brake hose union bolt | 2.0 – 2.5 | 14.5 – 18.0 |
| Air bleed valve | 0.6 – 0.9 | 4.5 – 6.5 |
| Anti-dive valve fitting bolt | 0.6 – 0.8 | 4.5 – 6.0 |
| Anti-dive plunger fitting bolt | 0.4 – 0.5 | 3.0 – 3.5 |
| Front brake caliper mounting bolt | 2.5 – 4.0 | 18.0 – 29.0 |
| Front wheel spindle nut | 3.6 – 5.2 | 26.0 – 37.5 |
| Front wheel spindle clamp nut | 1.5 – 2.5 | 11.0 – 18.0 |
| Front footrest bolt | 2.7 – 4.3 | 19.5 – 31.0 |
| Swinging arm pivot nut | 5.5 – 8.5 | 40.0 – 61.5 |
| Rear footrest bolt | 2.7 – 4.3 | 19.5 – 31.0 |
| Rear suspension unit nut/bolt | 2.0 – 3.0 | 14.5 – 21.5 |
| Rear torque arm nut/bolt | 2.0 – 3.0 | 14.5 – 21.5 |
| Chain adjuster bolt | 1.8 – 2.8 | 13.0 – 20.0 |
| Rear wheel spindle nut | 8.5 – 11.5 | 61.5 – 83.0 |

## 1  General description

The Suzuki GS/GSX 1000 and 1100 models employ a full duplex cradle frame of welded tubular steel. Front suspension is by oil damped, coil sprung telescopic forks. These vary in sophistication and complexity according to the model, and are similar on US and UK versions.

The LT and the spoked-wheel T versions have straightforward forks with air assistance of the coil springs. The two fork stanchion top bolts incorporate air valves to permit pressure adjustment and measurement.

The ET and EX models are similar, with the addition of adjustable damping and fork spring preload. In addition, the fork legs are connected by a pressure hose, allowing pressure adjustments to be made using a single valve. This arrangement automatically equalises pressure between the two legs. The EZ, ESD and all Katana variants have the added sophistication of an anti-dive braking system.

Rear suspension is by a conventional swinging arm, supported by a pair of coil spring suspension units with 4-way damping adjustment, and 5-way spring preload adjustment.

## 2 Front forks: removal and replacement

1 Place the machine on the centre stand and support it securely so that the front wheel is raised clear of the ground. Use a jack or wooden blocks, but take care to ensure stability.
2 Remove the front wheel (see Chapter 5 for details). Release both brake calipers and tie them clear of the forks. In the case of machines fitted with anti-dive units, remove the two Allen screws which retain the plunger units to the anti-dive units on the fork legs. Free the front mudguard by removing the retaining bolts. On Katana models remove the handlebars by withdrawing the retaining screws which secure their mounting clamps.
3 In the case of machines fitted with a single air valve and a connecting hose, remove the air valve cap and release the air pressure by depressing the valve insert. Slacken the bolt on each air valve union to free it.
4 Where separate air valves are fitted, remove the valve caps and depress the inserts to release the air pressure from each leg. Slacken the upper and lower yoke clamp bolts and withdraw the fork legs by pulling them downwards. To facilitate removal, twist each stanchion to and fro.
5 If the fork legs are to be dismantled, leave them partly in position through the lower yoke and temporarily retighten the pinch bolts. Held in this way, it is relatively easy to slacken the fork cap bolts by one or two turns, making subsequent removal much easier.
6 The fork legs are installed by reversing the removal sequence noting that the oil level should be checked prior to installation as described later in this Chapter. Position the stanchions so that the upper edge is in line with the top face of the upper yoke. On Katana models, when refitting the handlebars ensure the fork stanchion is clean and tighten the clamping bolts evenly to the specified torque setting. Check that the handlebars do not touch the fuel tank on full lock. Once the clamp bolts are secured, and before any weight is applied to the forks, check the fork air pressure as described in Section 8 of this Chapter.

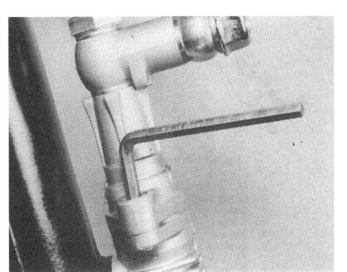
2.3a Remove the two Allen bolts which retain the plunger units ...

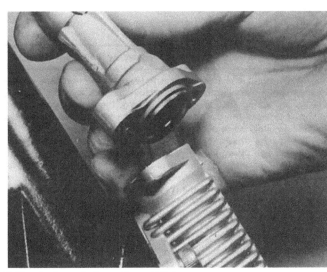
2.3b ... to the anti-dive units and lift them away

2.3c Slacken the air union locating bolts

2.4a Slacken the upper yoke pinch bolt ...

2.4b ... and lower yoke pinch bolts and pull fork leg downwards

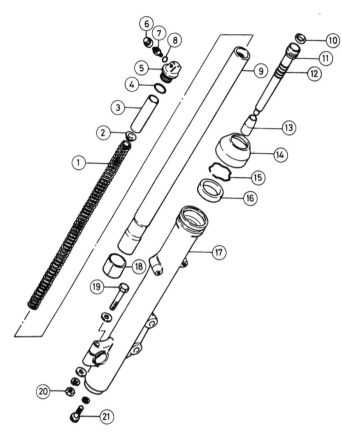

**Fig. 4.1 Front fork – T and LT**

| | | | |
|---|---|---|---|
| 1 | Fork spring | 12 | Rebound spring |
| 2 | Spring seat | 13 | Damper rod seat |
| 3 | Spacer | 14 | Dust seal |
| 4 | O-ring | 15 | Seal retainer clip |
| 5 | Fork cap | 16 | Oil seal |
| 6 | Dust cap | 17 | Lower leg |
| 7 | Air valve | 18 | Bush |
| 8 | O-ring | 19 | Pinch bolt |
| 9 | Stanchion | 20 | Nut |
| 10 | Piston ring | 21 | Damper bolt |
| 11 | Damper rod | | |

## 3   Front forks: dismantling and reassembly – T and LT

1   Dealing with one fork leg at a time to avoid interchanging the internal components, unscrew and remove the fork top bolt. Remove the spacer, spring seat and the fork spring, then invert the leg and 'pump' it to expel the damping oil. Leave the leg inserted for a while to allow residual oil to escape.

2   To separate the stanchion and lower leg it is necessary to release the damper bolt from the underside of the lower leg. This screws into the damper rod which is free to rotate in the stanchion. It will be necessary to hold the rod to prevent it from turning.

3   Suzuki produce a T-handle (Part Number 09940-34520) and an adaptor (Part Number 09940-34580) which locates in the head of the damper rod. In most cases an improvised holding tool can be made by grinding a coarse taper at the end of a length of wooden dowel. This can then be held with a self-locking wrench and pushed hard against the damper rod whilst the bolt is unscrewed.

4   Once the damper rod bolt has been removed, slide the dust seal off the stanchion, separating the stanchion and lower leg by pulling them apart. Invert the stanchion and tip out the damper rod and the rebound spring, then invert the lower leg and tip out the damper rod seat.

5   The split bush at the bottom of the stanchion and the oil seal in the top of the lower leg should not be removed unless examination indicates the need for renewal (see below). The bush can be slid off the stanchion after it has been spread, using a screwdriver to open the split seam. To free the oil seal, prise out the wire retaining clip and work the seal out of its recess by carefully levering it upwards, using a screwdriver, taking care not to damage the top edge of the lower leg.

6   Assemble the leg by reversing the dismantling sequence, having first ensured that all components are clean and dust free. If a new bush is to be fitted, this can be eased over the end of the stanchion and then pushed home by hand. Do not stretch the bush any more than is necessary. Tap the oil seal home using a large socket as a drift and making sure that it seats fully and squarely in the top of the lower leg. Refit the wire retaining clip.

7   The threads of the damper bolt must be clean and dry and should be coated with a thread locking compound. Tighten the bolt to the specified torque figure. Before fitting the fork spring and spacer, compress the fork fully and add the prescribed quantity of fork oil. Check that the oil level is the correct distance below the top of the stanchion (refer to the Specifications for details).

8   When fitting the fork spring, note that the tighter pitched coils must be uppermost. Fit the spring spacer and the top bolt, leaving final tightening until the fork has been refitted in the lower yoke and temporarily clamped in place. After the leg is installed, check and set the air pressure as described in Section 8 of this Chapter.

## 4   Front forks: dismantling and reassembling – ET and EX

1   Dealing with one fork leg at a time to avoid interchanging the internal components, unscrew and remove the fork top bolt. Below the top bolt there is a floating piston with an O-ring seal. This can be displaced by raising very carefully the lower leg until it is pushed free. Remove the fork spring, then invert the leg and 'pump' it to expel the damping oil. Leave the leg inverted for a while to allow residual oil to escape.

2   To separate the stanchion and lower leg it is necessary to release the damper nut on the underside of the lower leg. This screws onto the damper rod which is free to rotate in the stanchion, and it will be necessary to hold the rod to prevent it from turning.

3   Suzuki produce a T-handle (Part Number 09940-34520) and an adaptor (Part Number 09940-34580) which locates in the head of the damper rod. In most cases an improvised holding tool can be made by grinding a coarse taper at the end of the length of wooden dowel. This can then be held with a self-locking wrench and pushed hard against the damper rod whilst the bolt is unscrewed.

4   To gain access to the nut it will first be necessary to remove the damping adjuster by pulling it downwards. Release the detent assembly by removing the small circlip which retains it. Slacken and

remove the damper rod nut, then slide the dust seal off the stanchion, separating the stanchion and lower leg by pulling them apart. Invert the stanchion and tip out the damper rod and the rebound spring, then displace the damper adjuster shaft from inside the damper rod. Invert the lower leg and tip out the damper rod seat and the O-ring.

5    The split bush at the bottom of the stanchion and the oil seal in the top of the lower leg should not be removed unless examination indicates the need for renewal (see below). The bush can be slid off the stanchion after it has been spread, using a screwdriver to open the split seam. To free the oil seal, prise out the wire retaining clip and work the seal out of its recess by carefully levering it upwards, using a screwdriver, taking care not to damage the top edge of the lower leg.

6    Assemble the leg by reversing the dismantling sequence, having first ensured that all components are clean and dust free. If a new bush is to be fitted this can be eased over the end of the stanchion and then pushed home by hand. Do not stretch the bush any more than is necessary. Tap the oil seal home using a large socket as a drift and making sure that it seats fully and squarely in the top of the lower leg. Refit the wire retaining clip.

7    The threads of the damper rod nut must be clean and dry and should be coated with a thread locking compound. When installing the damper rod use a new O-ring below the seat and ensure that the slotted end of the adjuster shaft aligns with the notch in the end of the damper rod. Tighten the nut to the specified torque figure.

8    Check that the detent assembly is arranged as shown in the accompanying line drawing, then fit it over the end of the damper rod. Fit the adjuster, ensuring that it aligns correctly with the end of the adjuster shaft. Without turning the adjuster, set the indicator ring to align the '1' with the raised index mark on the lower leg.

9    Before fitting the fork spring and piston, compress the fork fully and add the prescribed quantity of fork oil. Check that the oil level is the correct distance below the top of the stanchion (refer to the Specifications for details).

10   When fitting the fork spring, note that the tighter pitched coils must be uppermost. Fit the floating piston with its flat face upwards. Back off the spring preload adjuster in the top bolt to the softest setting and fit the top bolt, leaving final tightening until the fork has been refitted in the lower yoke and temporarily clamped in place. After the leg is installed, check and set the air pressure and spring preload as described in Section 8 of this Chapter.

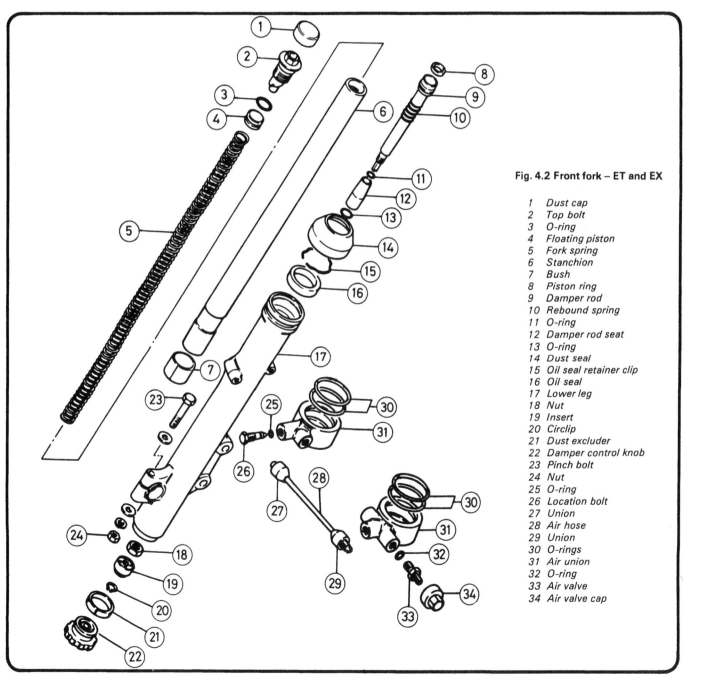

**Fig. 4.2 Front fork – ET and EX**

1    Dust cap
2    Top bolt
3    O-ring
4    Floating piston
5    Fork spring
6    Stanchion
7    Bush
8    Piston ring
9    Damper rod
10   Rebound spring
11   O-ring
12   Damper rod seat
13   O-ring
14   Dust seal
15   Oil seal retainer clip
16   Oil seal
17   Lower leg
18   Nut
19   Insert
20   Circlip
21   Dust excluder
22   Damper control knob
23   Pinch bolt
24   Nut
25   O-ring
26   Location bolt
27   Union
28   Air hose
29   Union
30   O-rings
31   Air union
32   O-ring
33   Air valve
34   Air valve cap

### 5    Front forks: dismantling and reassembly – EZ, ED, ESD and all Katana models

1    Dealing with one fork leg at a time to avoid interchanging the internal components, unscrew and remove the fork top bolt. Below the top bolt there is a floating piston with an O-ring seal. This can be displaced by raising very carefully the lower leg until it is pushed free.

Remove the fork spring, then invert the leg and 'pump' it to expel th damping oil. Leave the leg inverted for a while to allow residual oil t escape.

2    To separate the stanchion and lower leg it is necessary to releas the damper bolt from the underside of the lower leg. This screws int the damper rod which is free to rotate in the stanchion, and it will b necessary to hold the rod to prevent it from turning.

3    Suzuki produce a T-handle (Part Number 09940-34520) and a

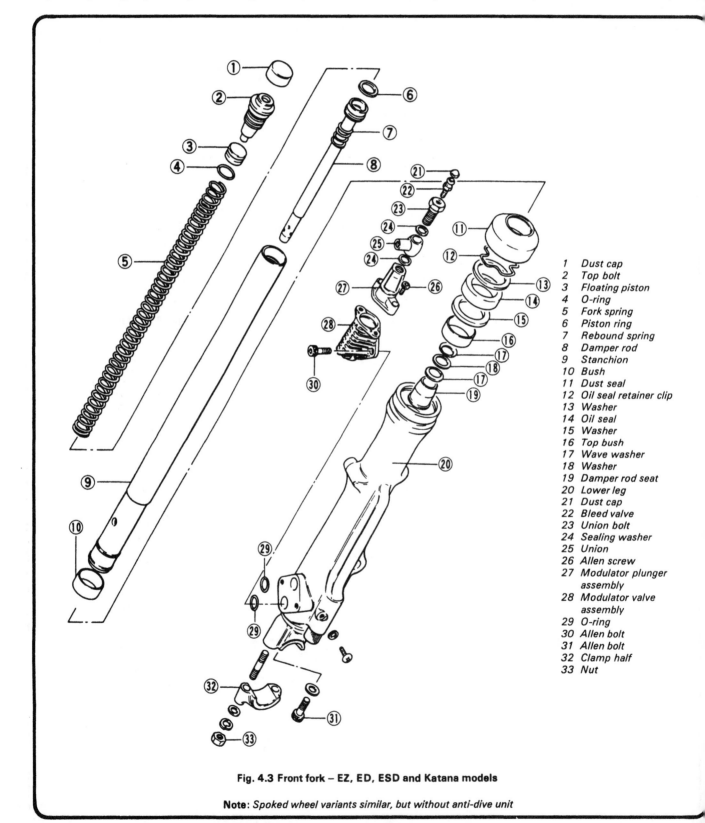

1    Dust cap
2    Top bolt
3    Floating piston
4    O-ring
5    Fork spring
6    Piston ring
7    Rebound spring
8    Damper rod
9    Stanchion
10   Bush
11   Dust seal
12   Oil seal retainer clip
13   Washer
14   Oil seal
15   Washer
16   Top bush
17   Wave washer
18   Washer
19   Damper rod seat
20   Lower leg
21   Dust cap
22   Bleed valve
23   Union bolt
24   Sealing washer
25   Union
26   Allen screw
27   Modulator plunger
     assembly
28   Modulator valve
     assembly
29   O-ring
30   Allen bolt
31   Allen bolt
32   Clamp half
33   Nut

**Fig. 4.3 Front fork – EZ, ED, ESD and Katana models**

**Note:** *Spoked wheel variants similar, but without anti-dive unit*

adapator (Part Number 09940-34561) which locates in the head of the damper rod. In most cases an improvised holding tool can be made by grinding a coarse taper at the end of a length of wooden dowel. This can then be held with a self-locking wrench and pushed hard against the damper rod whilst the bolt is unscrewed.

4    Once the damper rod bolt has been removed, slide the dust seal off the stanchion, and remove the wire circlip which retains the oil seal and top bush. Separate the stanchion and lower leg by pulling them apart sharply until the top bush is displaced together with the oil seal. Invert the stanchion and tip out the damper rod and the rebound spring, then invert the lower leg and tip out the damper rod seat. The anti-dive modulator can be released from the bottom of the lower leg after its retaining screws have been removed.

5    The split bush at the bottom of the stanchion and the oil seal and split bush in the top of the lower leg should be renewed each time the forks are dismantled. The top bush will have been displaced during removal, whilst the bottom bush can be slid off the stanchion after it has been spread, using a screwdriver to open the split seam.

6    Assemble the leg by reversing the dismantling sequence, having

first ensured that all components are clean and dust free. A new bottom bush can be fitted by easing it over the end of the stanchion and then pushing it home by hand. Do not stretch the bush any more than is necessary. Fit the stanchion assembly into the lower leg, then tap the top bush home using a large socket as a drift and making sure that it seats fully and squarely in the top of the lower leg. The oil seal can now be installed by the same method. Refit the wire retaining clip.

7    The threads of the damper bolt must be clean and dry and should be coated with a thread locking compound. Tighten the bolt to the specified torque figure. Before fitting the fork spring and spacer, compress the fork fully and add the prescribed quantity of fork oil. Check that the oil level is the correct distance below the top of the stanchion (refer to the Specifications for details).

8    When fitting the fork spring, note that the tighter pitched coils must be uppermost. Fit the floating piston with its flat face upwards, and fit the top bolt, leaving final tightening until the fork has been refitted in the lower yoke and temporarily clamped in place. After the leg is installed, check and set the air pressure as described later.

5.1a Remove the fork top bolts and compress leg slightly ...

5.1b ... to displace the floating piston

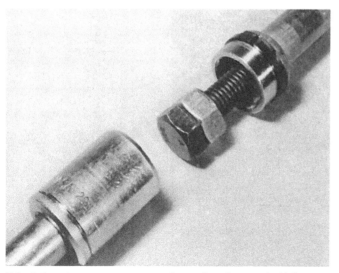

5.2a Bolt and nut arrangement can be used as shown to hold the damper rod head ...

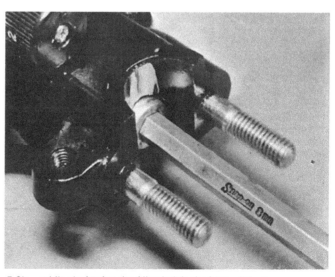

5.2b ... while slackening the Allen bolt in the base of the leg

5.4a Prise out the wire circlip and pull stanchion and lower leg sharply apart ...

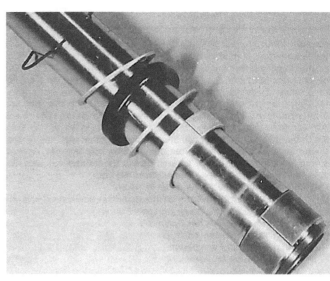

5.4b ... to displace top bush and oil seal

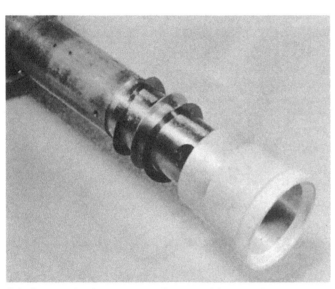

5.4c Remove the damper seat and wave and thrust washers ...

5.4d ... and tip damper rod out of stanchion

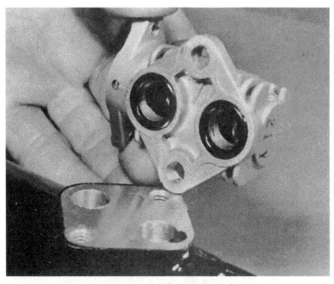

5.4e Remove mounting bolts and lift anti-dive unit away

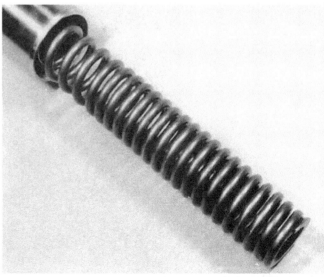

5.8 Tighter pitched coils of fork spring must be uppermost

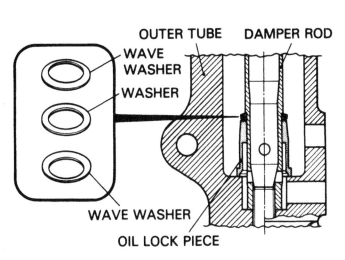

**Fig. 4.4 Arrangement of wave washers on damper rod – anti-dive forks**

## 6 Front forks: examination and renovation

1 Inspect the sliding surfaces of the fork components for signs of wear or damage. Where bushes are used, any wear is easily dealt with by renewing the affected parts, and in the case of models employing two bushes, renewal will be automatic whenever the legs are dismantled. In the case of models using forks with lower bushes only, wear between the stanchion and the top of the lower leg will require the renewal of the latter component, together with the stanchion if this too is worn significantly.

2 Check the stanchion surface for scoring, particularly around the area normally covered by the dust seal. Scoring here will invariably cause oil leakage and rapid seal wear, and if found the stanchion must be renewed.

3 If the forks have been stripped after an accident, check the stanchions for straightness by rolling them on a flat surface. If damage is very slight it may be possible to have them trued by a specialist with press facilities, but it is preferable to renew them to avoid any risk of subsequent stress fracture.

4 The damper rod should be cleaned, ensuring that the oilways are clear. Any residual oil or sediment should be cleaned out of the lower leg and the stanchion prior to reassembly.

### Anti-dive modulator assembly

5 The modulator assemly serves to control the damping effect in the forks in response to pressure in the front brake system, effectively modifying the behaviour of the forks to reduce pitching (diving). The modulator unit can be removed from the fork leg and the O-rings renewed in the event of oil leaks, but if defective it must be renewed as a unit, replacement parts not being available separately.

## 7 Steering head assembly: overhauling and adjustment

1 Start by removing the fork legs from the yokes as described above. Remove the seat and fuel tank to give unobstructed access to the rear of the steering head area. Disconnect the choke cable at the carburettor end, then slacken the choke knob lock ring to allow the choke cable assembly to be withdrawn from the steering stem.

2 Slacken and remove the handlebar clamp bolts, remove the clamp and then lift the handlebar assembly rearwards, resting it across the frame top tubes. Remove the front turn signal lamps from the lower yoke where fitted. In the case of the Katana models, free the separate handlebar sections by removing the retaining screws which secure the mounting clamps.

3 On the Katana models, remove the fairing screen retaining screws and lift it away. Pull off the fairing lowers, then release the fairing

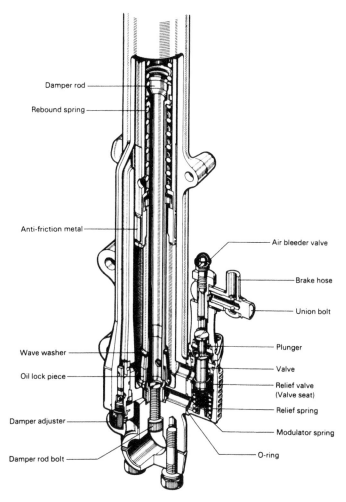

**Fig. 4.5 Sectioned view of fork leg showing anti-dive components**

mounting screws and lift the fairing away. Remove and disconnect the headlamp unit, then trace and disconnect the wiring which passes through the slots in the back of the headlamp shell. Remove the headlamp shell.

4 Disconnect the speedometer and tachometer drive cables at the instrument panel end, then remove the instrument panel mounting bolts and lift it away. Release the ignition switch from the underside of the top yoke. Where the front brake union block is mounted on the lower yoke, it should be detached.

5 Slacken the top yoke clamp bolt and remove the steering stem nut. Using a hide mallet, tap the top yoke upwards and lift it away from the steering stem.

6 Using a deep socket, ring spanner or box spanner, slacken the steering stem adjuster nut. Support the lower yoke, then remove the adjuster nut and lift the yoke clear of the frame. Lift out the upper race and wipe off any grease from the bearings and races.

7 The tapered roller steering head bearings will not normally wear quickly, but should be checked closely for indications of wear or damage. If there is any sign of ridging or indentation of the rollers or races, the bearings must be renewed as a pair. Unless the bearings are to be renewed, they should not be disturbed.

8 The bearing outer races can be driven out using a long drift passed through the steering head, working around the race to keep it square to the recess. Fit the new race using a stepped drift or a drawbolt arrangement, again ensuring that the race is kept square to the sides of the recess.

9 The lower bearing inner race can be drawn off the steering stem

with a bearing extractor. In the absence of the relevant Suzuki tool, many tool hire shops will be able to provide a suitable item.

10 Fit the new outer races using a drawbolt arrangement as shown in the accompanying line drawing. Tap the lower bearing inner race into position using a tubular drift, and grease both bearings prior to installation.

11 Assemble the steering head by reversing the dismantling sequence. Tighten the steering stem adjuster to the specified torque setting. Turn the steering stem from lock to lock several times to seat the bearings, then back off the adjuster by a half turn.

12 On all but the Katana models, position the handbebars so that the punch mark aligns with the flat face of the top yoke, then tighten the clamp bolts evenly to the specified torque figure. The clamp halves must be positioned so that the gap is equal on each side.

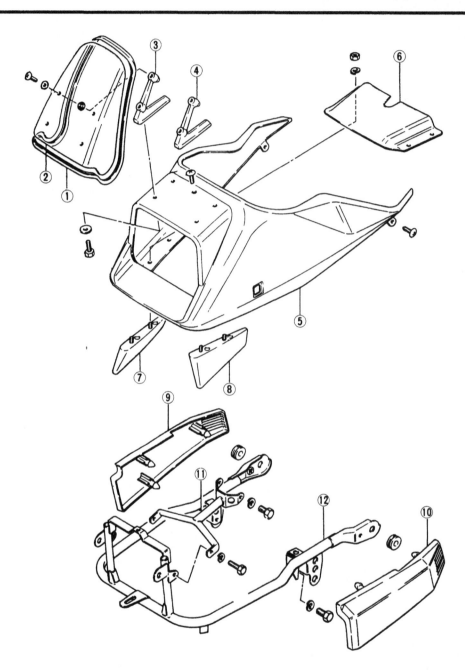

**Fig. 4.6 Fairing components – Katana models**

| | | | |
|---|---|---|---|
| 1 Screen trim | 5 Fairing | 9 Right-hand lower | 11 Fairing mounting |
| 2 Screen | 6 Cover | section | bracket |
| 3 Right-hand screen | 7 Right-hand wing | 10 Left-hand lower | 12 Fairing mounting |
| mounting bracket | 8 Left-hand wing | section | bracket |
| 4 Left-hand screen | | | |
| mounting bracket | | | |

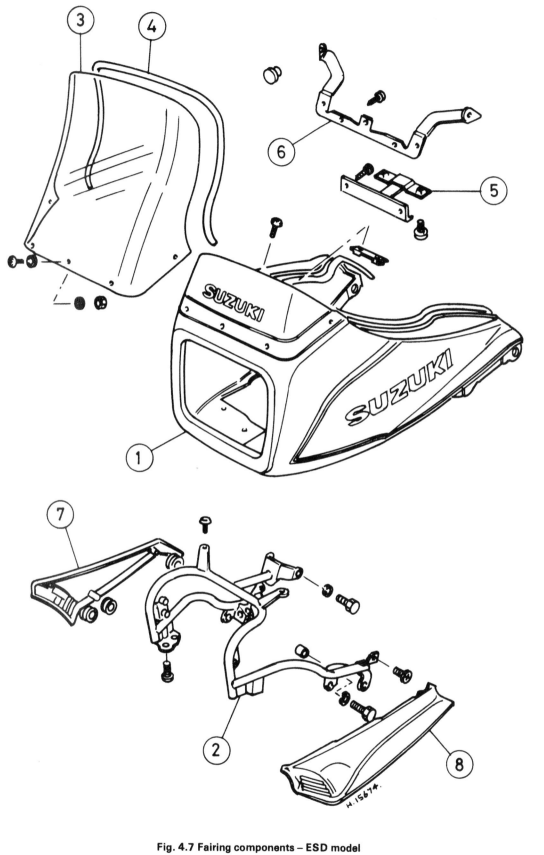

**Fig. 4.7 Fairing components – ESD model**

| | | | | | |
|---|---|---|---|---|---|
| 1 | Fairing | 4 | Screen trim | 7 | Right-hand lower |
| 2 | Fairing mounting | 5 | Mounting bracket | | section |
| | bracket | 6 | Screen mounting | 8 | Left-hand lower |
| 3 | Screen | | bracket | | section |

**132**

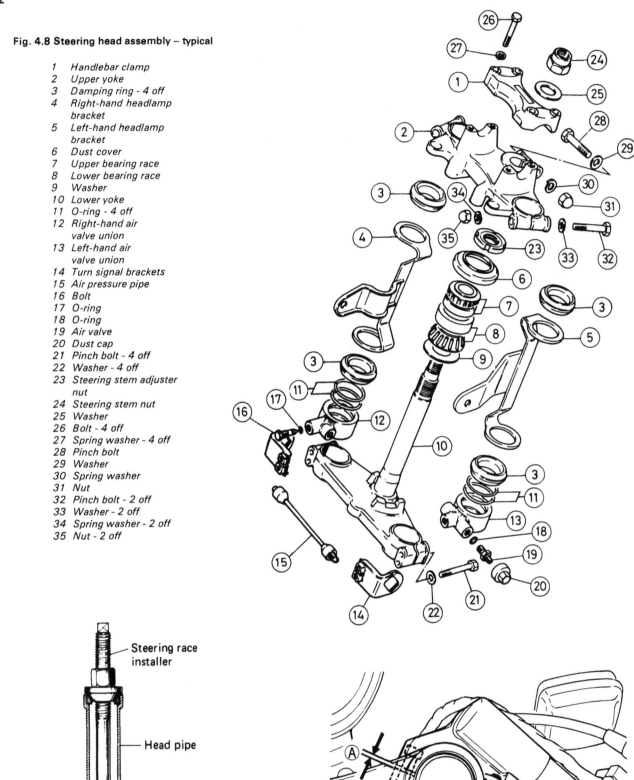

Fig. 4.8 Steering head assembly – typical

1 Handlebar clamp
2 Upper yoke
3 Damping ring - 4 off
4 Right-hand headlamp bracket
5 Left-hand headlamp bracket
6 Dust cover
7 Upper bearing race
8 Lower bearing race
9 Washer
10 Lower yoke
11 O-ring - 4 off
12 Right-hand air valve union
13 Left-hand air valve union
14 Turn signal brackets
15 Air pressure pipe
16 Bolt
17 O-ring
18 O-ring
19 Air valve
20 Dust cap
21 Pinch bolt - 4 off
22 Washer - 4 off
23 Steering stem adjuster nut
24 Steering stem nut
25 Washer
26 Bolt - 4 off
27 Spring washer - 4 off
28 Pinch bolt
29 Washer
30 Spring washer
31 Nut
32 Pinch bolt - 2 off
33 Washer - 2 off
34 Spring washer - 2 off
35 Nut - 2 off

Steering race installer

Head pipe

Fig. 4.9 Using drawbolt arrangement to fit steering head bearings

Fig. 4.10 Fitting the handlebar sections – Katana models

*Note that gaps (A) must be equal*

## 8 Front suspension adjustment

### General

1 The range of front suspension adjustment available on the various models is dependent on the type of fork fitted, but in each case it is important that the general condition of the fork is sound and that the oil level is set correctly. It should be noted that if the oil level is incorrect or unequal in the two legs, any air pressure setting will be correspondingly unbalanced. If the oil has not been changed for some time it is suggested that this is carried out before further adjustment is attempted. To ensure accuracy, remove the fork legs, drain the oil and add oil to the level indicated in the specifications section. For details refer to the preceding Sections.

### Air pressure adjustment

2 As the total volume of each fork leg is small, the accurate setting of the fork air pressure will require the use of a hand pump and gauge designed for this purpose. It is not possible to use an air line without some risk of applying excessive pressure, with an attendant risk of seal damage, and the use of a tyre pressure gauge is likely to waste so much pressure in taking a reading that the resulting fork pressure will not be accurately predictable.

3 Suzuki dealers should be able to order the official Suzuki fork pressure gauge and pump assembly. An alternative available through most large dealers is the S&W equivalent. Both are designed to give an accurate indication of air pressure without losing a significant proportion of the air when the gauge is released.

4 Remove the air valve dust cap and assemble the gauge and pump as indicated in the manufacturer's instructions. Pressurise the forks, taking care not to exceed the maximum safe pressure of 35.0 psi (2.5 kg/cm$^2$) at any time. Gradually release the air until the correct pressure is indicated. Remove the gauge assembly, losing as little pressure as is possible. On models without interconnected forks, repeat the sequence on the remaining fork leg, noting that it is essential that each leg is set at the same pressure. Remember to refit the dust cap(s) after adjustment.

### Fork air pressure (where applicable)

| Model | Std pressure | | Max pressure | |
|---|---|---|---|---|
| | psi | kg/cm$^2$ | psi | kg/cm$^2$ |
| GSX/GS 1100 T, ET, EX, EZ | 7.11 | 0.5 | 35.0 | 2.5 |
| GSX/GS 1100 LT | 11.38 | 0.8 | 35.0 | 2.5 |

### Fork spring preload and damping adjustment

5 The spring preload adjuster allows the forks to be adjusted to suit various load and road surface conditions. The slotted adjuster is

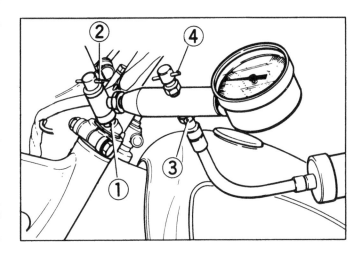

**Fig. 4.11 Using gauge and pump rig to set fork air pressure**

1   Air valve     3   Pump valve
2   Valve handle   4   Pump control handle

located in the top of the fork top bolt, requiring the handlebar assembly to be removed where this obstructs access. The setting chosen is largely a matter of personal preference, but for obvious reasons should be the same on each fork.

6 The damping adjuster controls the overall damping rate and is located on the bottom of the fork leg or the anti-dive modulator unit. In the case of non-Katana models, the recommended standard settings are as follows:

| Model | Spring preload | Damping setting |
|---|---|---|
| GSX 1100 ET, EX | 2 | 3 |
| GSX 1100 EZ | 2 | 2 |
| GS 1100 ET, EX, EZ | 1 | 2 |

7 In the case of the Katana machines, a more detailed range of recommended settings is shown in figure 4.12 below, in conjunction with the rear suspension settings.

| ITEM | FRONT FORK | REAR SHOCK ABSORBER | | REMARKS |
|---|---|---|---|---|
| | | Spring preload | Damping force | |
| SOFTER | 2 | 2 | 2 | —— |
| STANDARD | 3 | 3 *(4) | 2 | *Shows that it is good for circuit running. |
| STIFFER | 4 | 5 | 3 | —— |
| DUAL RIDING | 3 | 4 | 2 | Good for both high speed and ordinary riding. |

Fig. 4.12 Suspension setting chart – Katana models

8.5a Fork top bolt incorporates spring preload adjuster

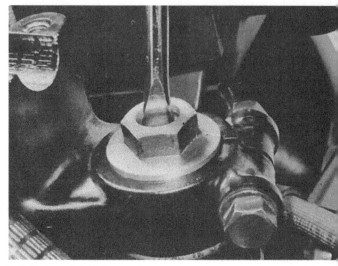

8.5b Release handlebars and adjust using a screwdriver

8.6 Damping adjuster is located at bottom of fork leg or anti-dive unit

## 9   Frame: examination and renovation

1    The frame is unlikely to require attention unless accident damage has occurred. In some cases, renewal of the frame is the only satisfactory remedy if the frame is badly out of alignment. Only a few frame specialists have the jigs and mandrels necessary for resetting the frame to the required standard of accuracy, and even then there is no easy means of assessing to what extent the frame may have been overstressed.

2    After the machine has covered a considerable mileage, it is advisable to examine the frame closely for signs of cracking or splitting at the welded joints. Rust corrosion can also cause weakness at these joints. Minor damage can be repaired by welding or brazing, depending on the extent and nature of the damage.

3    Remember that a frame which is out of alignment will cause handling problems and may even promote 'speed wobbles'. If misalignment is suspected, as a result of an accident, it will be necessary to strip the machine completely so that the frame can be checked, and if necessary, renewed.

## 10   Swinging arm: removal and overhaul

1    Remove the rear wheel (see Chapter 5) then release the single bolt which retains the torque arm. If it is still in position, remove the chainguard mounting screws and lift it away. Remove the rear suspension mounting nuts and lift away the suspension units, supporting the swinging arm as they are pulled clear.

2    It is a good idea to check for bearing wear at this stage. Grasp the ends of the swinging arm and attempt to move it from side to side. If the bearings are in good condition there should be little or no perceptible play, but if there is more than 1-2 mm movement the bearings will probably require renewal.

3    Slacken and remove the swinging arm pivot shaft nut, then use a drift to tap the shaft part way through. Support the swinging arm whilst removing the pivot shaft, then lift it away. Disengage the final drive chain and place the swinging arm on the workbench after thorough cleaning.

4    Lift off the dust covers and then pull out the pivot bearing inner races. Wipe off any grease to allow inspection of the needle roller bearings and races. The outer surface of the inner races should be unmarked, with no sign of corrosion, scoring or indenting. If damage is noted, the bearings must be renewed.

5    Any excessive play noted during removal can be checked by temporarily refitting the inner races and rotating them by hand in the bearings. It is not practicable to assess wear by direct measurement, but if play or unevenness can be felt, the bearings must be renewed.

6    The bearings will probably be destroyed during removal, so do not attempt this unless new ones are to hand. A bearing extractor will be required to draw out the old bearings, though with care and patience it might just be possible to use a long thin drift passed through the swinging arm bore.

7    Fit the new bearings with the marked face outermost, using the Suzuki bearing installer, Part Number 09941-34511, or a drawbolt arrangement. Grease the bearings and the inner races, then slide the latter into place. Examine the condition of the end covers, renewing them if in any doubt as to their condition.

8    The lower ends of the rear suspension units are supported on bonded rubber bushes pressed into eyes at the ends of the swinging arm. If these are worn or damaged, support the swinging arm securely and press or drive out the bush. A large socket can be used as a drift. Where the metal outer bush has become badly corroded it may be necessary to cut out the rubber core and then file or saw through the metal outer section. Fit the new bushes using a press or drawbolt arrangement.

10.1 Remove split pin, castellated nut and shouldered bolt to free the brake caliper torque arm

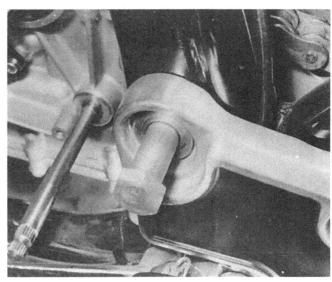

10.3a Remove the swinging arm pivot shaft ...

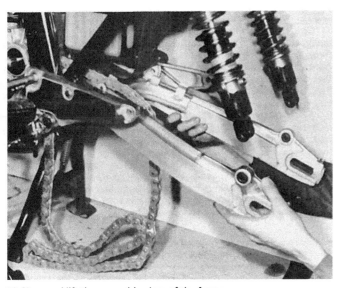

10.3b ... and lift the assembly clear of the frame

10.4a Pull off the dust seals and remove the shim washers

10.4b Remove, clean and examine the bearing inner races

10.4c Renew the needle roller bearings if worn or damaged

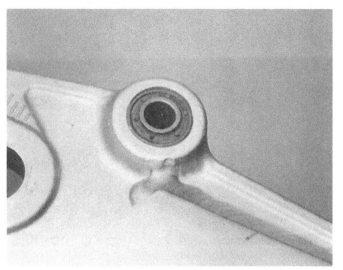

10.8 Rear suspension units are retained on bonded rubber bushes

## 11 Rear suspension units: examination and adjustment

1 The rear suspension units are of the oil-damped coil spring type featuring 4-way adjustable damper settings and 5 spring preload settings. The settings are largely up to the rider, the standard setting for most models being spring preload '3' and damper setting '2'. In the case of the Katana machines a more extensive range of settings will be found in figure 4.12 of this Chapter.

2 Spring preload is altered by turning the stepped collar at the lower end of each unit, using a C-spanner or a round bar in the recesses provided for this purpose. Damping adjustment is effected by turning the knurled wheel set in the top mount. For obvious reasons, ensure that the settings are the same on each unit.

3 The suspension units are of sealed construction, and in the event of failure of the damper insert it will be necessary to renew the units as a pair. Many dealers will be able to offer alternative replacement units from a wide choice of suppliers, and in many cases these can prove a useful way of altering the suspension characteristics. It is important to purchase units matched to the particular model; the dealer should be able to provide advice on choosing a suitable type.

## 12 Stands: examination and maintenance

1 The centre stand is retained by two pivot bolts to lugs on the underside of the frame, each being secured by an R-pin passed through its end. A sleeve is fitted between each bolt and the stand pivots. It is good practice to remove the stand from time to time so that the pivots can be cleaned and greased. Examine the stand for damage or cracking, and have any such damage repaired by welding before it causes the stand to collapse.

2 The side stand (prop stand) is secured by a shouldered pivot bolt to a lug on the left-hand side of the machine. This too should be removed periodically and checked for wear. If the pivot bolt has worn, make sure that the correct hardened bolt is used to replace it.

3 It is important to check the condition of both stands' springs regularly, looking for indications of wear or developing cracks around the hooked ends. Bear in mind that if a spring fails while the machine is being ridden, the stand will drop onto the road. It is all too easy for this type of occurrence to result in a serious accident.

## 13 Footrests: examination and maintenance

1 The footrests are of the folding type, secured to the mounting brackets by clevis pins which are retained in turn by split pins. If the machine is dropped, the footrests will normally fold up, and so are unlikely to become bent. If badly distorted in an accident, or if the rubbers are to be renewed, straighten and remove the split pin then displace the clevis pin and lift away the footrest.

2 It is often easiest to cut off the old rubber where it has become firmly bonded to the steel footrest. To assist in fitting the new rubber use a little petrol as a lubricant. This will soon evaporate after fitting. Check, and where necessary renew, the small spring which holds the footrest in position.

## 14 Brake pedal: examination and maintenance

1 The brake pedal is retained to its splined shaft by a pinch bolt. This should be kept tight to prevent movement of the pedal on the splines and the subsequent wear of these areas. If the pedal is bent in an accident, it may be possible to straighten slight damage by heating the pedal to a cherry red colour with a blowlamp and then hammering it straight. In most cases, however, it will be preferable to renew the pedal to preserve the appearance of the machine.

## 15 Instrument heads: removal and replacement

1 The speedometer and tachometer are fitted in an instrument panel mounted on the top yoke by two rubber-bushed bolts. In the event of either instrument proving faulty, check first that the drive cable is in

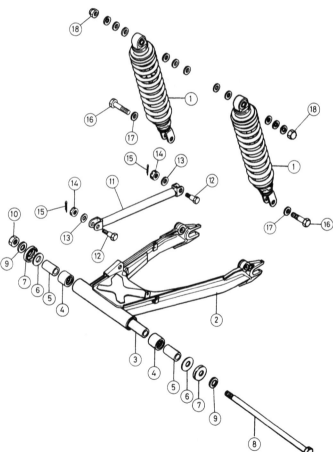

Fig. 4.13 Swinging arm and rear suspension

| | | | |
|---|---|---|---|
| 1 | Rear suspension unit- 2 off | 10 | Nut |
| 2 | Swinging arm | 11 | Torque arm |
| 3 | Spacer | 12 | Bolt - 2 off |
| 4 | Bearing outer race - 2 off | 13 | Washer - 2 off |
| 5 | Bearing inner race - 2 off | 14 | Nut - 2 off |
| 6 | Shim - 2 off | 15 | Split pin - 2 off |
| 7 | Dust cover - 2 off | 16 | Bolt - 2 off |
| 8 | Pivot shaft | 17 | Washer - 2 off |
| 9 | Washer - 2 off | 18 | Nut - 2 off |

good condition as described below. The Katana models are equipped with an electronic tachometer, in which case check the wiring to the instrument.

2   It is not practicable to repair a faulty instrument head, but before ordering a new unit check the local breakers (wreckers) yards for a good secondhand assembly. This is one area where a used part offers a substantial saving with little attendant risk if it fails at a later date.

---

### 16 Instrument drives: examination and renovation

---

1   With the exception of the Katana machines, the speedometer and tachometer are driven mechanically by flexible cables. These should always be checked first in the event of fault, particularly where the instrument needle wavers a lot; this is usually due to a kinked cable.

2   The speedometer drive gearbox is located on the left-hand end of the front wheel spindle and can be inspected once the wheel has been removed. It is good practice to pack the gearbox with grease each time the wheel is detached.

3   The tachometer drive is incorporated in the cylinder head cover and should require little attention. It is well lubricated by the oil feed to the camshafts and is unlikely to wear in use. In the event of damage to either drive unit, the affected parts must be renewed.

15.1 Instrument panel is retained on rubber-bushed mounting

16.2 The speedometer drive gearbox is located on the left-hand end of the front wheel spindle

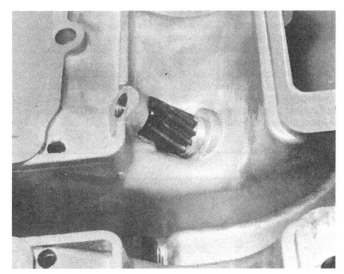

16.3 Tachometer driven gear is incorporated in cylinder head cover

# Chapter 5 Wheels, brakes and tyres

*Refer to Chapter 7 for information relating to the 1984-on 1135cc-engined GSX1100 and GS1150 models*

## Contents

## Specifications

### GSX 1100 T, ET, LT, EX, EZ, SZ, SD, ESD

**Tyre sizes**

Front:
  T, ET, EX, EZ, SZ, SD, ESD .................................................. 3.50V19 4PR
  LT .................................................................................. 100/90-19 57H
Rear:
  T, ET, EX, EZ, SZ, SD, ESD .................................................. 4.50V17 4PR
  LT .................................................................................. 130/90-16 67H

| **Tyre pressures, psi (kg cm²)** | **Front** | **Rear** |
|---|---|---|
| Normal riding - solo: | | |
| T, ET, LT, EX, EZ, SZ, SD, ESD | 25 (1.75) | 28 (2.00) |
| Normal riding - dual: | | |
| T, ET, EX, EZ, SZ, SD, ESD | 28 (2.00) | 36 (2.50) |
| LT | 28 (2.00) | 32 (2.25) |
| Continuous high speed - solo: | | |
| T, ET, EX, EZ, SZ, SD, ESD | 28 (2.00) | 36 (2.50) |
| LT | 28 (2.00) | 32 (2.25) |
| Continuous high speed - dual: | | |
| T, ET, LT, EX, SZ, SD, ESD | 32 (2.25) | 40 (2.80) |
| EZ | 32 (2.25) | 42 (2.90) |

**Minimum recommended tyre tread depth**
  Front .............................................................................. 1.6 mm (0.06 in)
  Rear ............................................................................... 2.0 mm (0.08 in)

**Wheel rim runout**
  Axial (max) ...................................................................... 2.0 mm (0.08 in)
  Radial (max) ..................................................................... 2.0 mm (0.08 in)

**Wheel spindle runout**
  Front (max) ...................................................................... 0.25 mm (0.01 in)
  Rear (max) ....................................................................... 0.25 mm (0.01 in)

## Front brake
| | |
|---|---|
| Type ........................................................................... | Twin hydraulic disc brake |
| Disc thickness ............................................................ | 5.0 ± 0.2 mm (0.20 ± 0.008 in) |
| Service limit .............................................................. | 4.5 mm (0.18 in) |
| Disc runout (max) ...................................................... | 0.30 mm (0.012 in) |
| Master cylinder bore ................................................. | 15.870 – 15.913 mm (0.6248 – 0.6265 in) |
| Master cylinder piston ............................................... | 15.811 – 15.838 mm (0.6225 – 0.6235 in) |
| Caliper bore .............................................................. | 38.180 – 38.219 mm (1.5031 – 1.5047 in) |
| Caliper piston ............................................................ | 38.025 – 38.050 mm (1.4970 – 1.4980 in) |

## Rear brake
| | |
|---|---|
| Type ........................................................................... | Single hydraulic disc |
| Disc thickness ............................................................ | 6.7 ± 0.2 mm (0.26 ± 0.008 in) |
| Service limit .............................................................. | 6.0 mm (0.24 in) |
| Disc runout (max) ...................................................... | 0.30 mm (0.012 in) |
| Master cylinder bore ................................................. | 14.000 – 14.043 mm (0.5512 – 0.5529 in) |
| Master cylinder piston ............................................... | 13.957 – 13.984 mm (0.5495 – 0.5506 in) |
| Caliper bore .............................................................. | 38.180 – 38.256 mm (1.5031 – 1.5061 in) |
| Caliper piston ............................................................ | 38.098 – 38.148 mm (1.4999 – 1.5019 in) |

## Hydraulic fluid
| | |
|---|---|
| Type ........................................................................... | SAE J1703 or DOT 3 specification |

## Torque wrench settings
| Component | kgf m | lbf ft |
|---|---|---|
| Disc mounting bolt ........................................................ | 1.5 – 2.5 | 11.0 – 18.0 |
| Front wheel spindle nut ................................................ | 3.6 – 5.2 | 26.0 – 37.5 |
| Front wheel spindle pinch bolt .................................... | 1.5 – 2.5 | 11.0 – 18.0 |
| Front caliper mounting bolt ......................................... | 2.5 – 4.0 | 18.0 – 29.0 |
| Front caliper spindle bolt: | | |
|     T, ET, LT, EX, EZ ........................................... | 4.0 – 5.5 | 29.0 – 40.0 |
|     SZ, SD, ESD ..................................................... | 1.5 – 2.0 | 11.0 – 14.5 |
| Brake hose union bolt ................................................. | 2.0 – 2.5 | 14.5 – 18.0 |
| Caliper bleed valve ..................................................... | 0.7 – 0.9 | 5.0 – 6.5 |
| Front master cylinder clamp bolt: | | |
|     T, ET, LT, EX, EZ, SD, ESD ............................ | 0.5 – 0.8 | 3.5 – 6.0 |
|     SZ ..................................................................... | 1.0 – 1.6 | 7.0 – 11.5 |
| Rear brake pedal bolt .................................................. | 1.0 – 1.5 | 7.0 – 11.0 |
| Rear master cylinder mounting bolt: | | |
|     T, ET, LT, EX, EZ, SD, ESD ............................ | 1.5 – 2.5 | 11.0 – 18.0 |
|     SZ ..................................................................... | 1.0 – 1.6 | 7.0 – 11.5 |
| Rear brake torque arm ................................................ | 2.0 – 3.0 | 14.5 – 21.5 |
| Rear caliper mounting bolt .......................................... | 2.5 – 4.0 | 18.0 – 29.0 |
| Rear caliper bolt ......................................................... | 2.0 – 3.0 | 14.5 – 21.5 |
| Rear wheel sprocket nut .............................................. | 2.5 – 4.0 | 18.0 – 29.0 |
| Rear wheel spindle nut ................................................ | 8.5 – 11.5 | 61.5 – 83.0 |
| Chain adjuster bolt: | | |
|     T, ET, LT, EX, EZ ........................................... | 1.5 – 2.0 | 11.0 – 14.5 |
|     SZ, SD, ESD ..................................................... | 1.8 – 2.8 | 13.0 – 20.0 |
| Anti-dive air bleed valve ............................................. | 0.6 – 0.9 | 4.5 – 6.5 |
| Anti-dive hose union ................................................... | 2.0 – 2.5 | 14.5 – 18.0 |
| Anti-dive modulator mounting bolt ............................. | 0.4 – 0.5 | 3.0 – 3.5 |
| Anti-dive valve mounting bolt ..................................... | 0.6 – 0.8 | 4.5 – 6.0 |

# GS 1100 T, ET, LT, EX, EZ, ED, ESD (US)

## Tyre sizes
| | |
|---|---|
| Front: | |
|     T, ET, EX, EZ, ED, ESD ................................................. | 3.50V19 4PR |
|     LT ..................................................................................... | 100/90-19 57H |
| Rear: | |
|     T, ET, EX, EZ, ED, ESD ................................................. | 4.50V17 4PR |
|     LT ..................................................................................... | 130/90-16 67H |

## Tyre pressures, psi (kg cm²)
| | Front | Rear |
|---|---|---|
| Normal riding - solo ............................................................ | 25 (1.75), 24 (1.68) – ED, ESD | 28 (2.00) |
| Normal riding - dual: | | |
|     T, ET, EX, EZ, ED, ESD ........................................... | 28 (2.00) | 36 (2.50) |
|     LT ................................................................................. | 28 (2.00) | 32 (2.25) |
| Continuous high speed - solo: | | |
|     T, ET, EX, EZ, ED, ESD ........................................... | 28 (2.00) | 36 (2.50) |
|     LT ................................................................................. | 28 (2.00) | 32 (2.25) |
| Continuous high speed - dual .......................................... | 32 (2.25) | 40 (2.80), 42 (2.94) – ED, ESD |

## Minimum recommended tyre tread depth
Front ............................................................................. 1.6 mm (0.06 in)
Rear .............................................................................. 2.0 mm (0.08 in)

## Wheel rim runout
Axial (max) .................................................................. 2.0 mm (0.08 in)
Radial (max) ................................................................ 2.0 mm (0.08 in)

## Wheel spindle runout
Front (max) .................................................................. 0.25 mm (0.01 in)
Rear (max) ................................................................... 0.25 mm (0.01 in)

## Front brake
Type ............................................................................. Twin hydraulic disc brake
Disc thickness ............................................................. 5.0 ± 0.2 mm (0.20 ± 0.008 in)
Service limit ................................................................ 4.5 mm (0.18 in)
Disc runout (max) ........................................................ 0.30 mm (0.012 in)
Master cylinder bore .................................................... 15.870 – 15.913 mm (0.6248 – 0.6265 in)
Master cylinder piston .................................................. 15.811 – 15.838 mm (0.6225 – 0.6235 in)
Cylinder bore .............................................................. 38.180 – 38.219 mm (1.5031 – 1.5047 in)
Caliper piston .............................................................. 38.025 – 38.050 mm (1.4970 – 1.4980 in)

## Rear brake
Type ............................................................................. Single hydraulic disc
Disc thickness ............................................................. 6.7 ± 0.2 mm (0.26 ± 0.008 in)
Service limit ................................................................ 6.0 mm (0.24 in)
Disc runout (max) ........................................................ 0.30 mm (0.012 in)
Master cylinder bore .................................................... 14.000 – 14.043 mm (0.5512 – 0.5529 in)
Master cylinder piston .................................................. 13.957 – 13.984 mm (0.5495 – 0.5506 in)
Caliper bore ................................................................ 38.180 – 38.256 mm (1.5031 – 1.5061 in)
Caliper piston .............................................................. 38.098 – 38.148 mm (1.4999 – 1.5019 in)

## Hydraulic fluid
Type ............................................................................. SAE J1703 or DOT 3, DOT 4 specification

## Torque wrench settings (all models except ED, ESD)

| Component | kgf m | lbf ft |
|---|---|---|
| Disc mounting bolt | 1.5 – 2.5 | 11.0 – 18.0 |
| Front wheel spindle nut | 3.6 – 5.2 | 26.0 – 37.5 |
| Front wheel spindle pinch bolt | 1.5 – 2.5 | 11.0 – 18.0 |
| Front caliper mounting bolt | 2.5 – 4.0 | 18.0 – 29.0 |
| Front caliper spindle bolt | 4.0 – 5.5 | 29.0 – 40.0 |
| Brake hose union bolt | 2.0 – 2.5 | 14.5 – 18.0 |
| Caliper bleed valve | 0.7 – 0.9 | 5.0 – 6.5 |
| Front master cylinder clamp bolt | 0.5 – 0.8 | 3.5 – 6.0 |
| Rear brake pedal bolt | 1.0 – 1.5 | 7.0 – 11.0 |
| Rear master cylinder mounting bolt | 1.5 – 2.5 | 11.0 – 18.0 |
| Rear brake torque arm | 2.0 – 3.0 | 14.5 – 21.5 |
| Rear caliper mounting bolt | 2.5 – 4.0 | 18.0 – 29.0 |
| Rear caliper bolt | 2.0 – 3.0 | 14.5 – 21.5 |
| Rear wheel sprocket nut | 2.5 – 4.0 | 18.0 – 29.0 |
| Rear wheel spindle nut | 8.5 – 11.5 | 61.5 – 83.0 |
| Chain adjuster bolt | 1.5 – 2.0 | 11.0 – 14.5 |
| Anti-dive air bleed valve | 0.6 – 0.9 | 4.5 – 6.5 |
| Anti-dive hose union | 2.0 – 2.5 | 14.5 – 18.0 |
| Anti-dive modulator mounting bolt | 0.4 – 0.5 | 3.0 – 3.5 |
| Anti-dive valve mounting bolt | 0.6 – 0.8 | 4.5 – 6.0 |

# GSX 1000 SZ, SD

## Tyre sizes
Front ............................................................................. 3.50V19 4PR
Rear .............................................................................. 4.50V17 4PR

## Tyre pressures, psi (kg cm²)

| | Front | Rear |
|---|---|---|
| Normal riding - solo | 25 (1.75) | 28 (2.00) |
| Normal riding - dual | 28 (2.00) | 36 (2.50) |
| Continuous high speed - solo | 28 (2.00) | 36 (2.50) |
| Continuous high speed - dual | 32 (2.25) | 40 (2.80) |

## Minimum recommended tyre tread depth
Front ............................................................................. 1.6 mm (0.06 in)
Rear .............................................................................. 2.0 mm (0.08 in)

## Wheel rim runout
Axial (max) .................................................. 2.0 mm (0.08 in)
Radial (max) ................................................. 2.0 mm (0.08 in)

## Wheel spindle runout
Front (max) .................................................. 0.25 mm (0.01 in)
Rear (max) ................................................... 0.25 mm (0.01 in)

## Front brake
Type .......................................................... Twin hydraulic disc brake
Disc thickness ............................................... 5.0 ± 0.2 mm (0.20 ± 0.008 in)
Service limit ................................................. 4.5 mm (0.18 in)
Disc runout (max) ........................................... 0.30 mm (0.012 in)
Master cylinder bore ........................................ 15.870 – 15.913 mm (0.6248 – 0.6265 in)
Master cylinder piston ...................................... 15.811 – 15.838 mm (0.6225 – 0.6235 in)
Caliper bore ................................................. 38.180 – 38.219 mm (1.5031 – 1.5047 in)
Caliper piston ............................................... 38.025 – 38.050 mm (1.4970 – 1.4980 in)

## Rear brake
Type .......................................................... Single hydraulic disc
Disc thickness ............................................... 6.7 ± 0.2 mm (0.26 ± 0.008 in)
Service limit ................................................. 6.0 mm (0.24 in)
Disc runout (max) ........................................... 0.30 mm (0.012 in)
Master cylinder bore ........................................ 14.000-14.043 mm (0.5512-0.5529 in)
Master cylinder piston ...................................... 13.957 – 13.984 mm (0.5495 – 0.5506 in)
Caliper bore ................................................. 38.180 – 38.256 mm (1.5031 – 1.5061 in)
Caliper piston ............................................... 38.098 – 38.148 mm (1.4999 – 1.5019 in)

## Hydraulic fluid
Type .......................................................... SAE J1703 or DOT 3 specification

## Torque wrench settings

| Component | kgf m | lbf ft |
|---|---|---|
| Front master cylinder clamp bolt: | | |
| SZ | 1.0 – 1.6 | 7.0 – 11.0 |
| SD | 0.5 – 0.8 | 3.5 – 6.0 |
| Front caliper mounting bolt | 2.5 – 4.0 | 18.0 – 29.0 |
| Front caliper spindle bolt | 1.5 – 2.0 | 11.0 – 14.5 |
| Front wheel spindle nut | 3.6 – 5.2 | 26.0 – 37.5 |
| Front wheel spindle clamp bolt | 1.5 – 2.5 | 11.0 – 18.0 |
| Rear master cylinder mounting bolt: | | |
| SZ | 1.0 – 1.6 | 7.0 – 11.0 |
| SD | 1.5 – 2.5 | 11.0 – 18.0 |
| Rear brake torque arm bolt/nut | 2.0 – 3.0 | 14.5 – 21.5 |
| Rear brake caliper mounting bolt | 2.5 – 4.0 | 18.0 – 29.0 |
| Chain adjuster bolt | 1.8 – 2.8 | 13.0 – 20.0 |
| Rear wheel spindle nut | 8.5 – 11.5 | 61.5 – 83.0 |
| Brake hose union bolt | 2.0 – 2.5 | 14.5 – 18.0 |
| Anti-dive hose union bolt | 2.0 – 2.5 | 14.5 – 18.0 |
| Air bleed valve | 0.6 – 0.9 | 4.5 – 6.5 |
| Anti-dive valve fitting bolt | 0.6 – 0.8 | 4.5 – 6.0 |
| Anti-dive modulator fitting bolt | 0.4 – 0.5 | 3.0 – 3.5 |

# GS 1000 SZ (US)

## Tyre sizes
Front ......................................................... 3.50V19 4PR
Rear .......................................................... 4.50V17 4PR

## Tyre pressures, psi (kg cm²)

| | Front | Rear |
|---|---|---|
| Normal riding - solo | 25 (1.75) | 28 (2.00) |
| Normal riding - dual | 28 (2.00) | 36 (2.50) |
| Continuous high speed - solo | 28 (2.00) | 36 (2.50) |
| Continuous high speed - dual | 32 (2.25) | 40 (2.80) |

## Minimum recommended tyre tread depth
Front ......................................................... 1.6 mm (0.06 in)
Rear .......................................................... 2.0 mm (0.08 in)

## Wheel rim runout
Axial (max) .................................................. 2.0 mm (0.08 in)
Radial (max) ................................................. 2.0 mm (0.08 in)

## Wheel spindle runout
Front (max) ........................................................................... 0.25 mm (0.01 in)
Rear (max) ........................................................................... 0.25 mm (0.01 in)

## Front brake
Type .................................................................................... Twin hydraulic disc brake
Disc thickness ...................................................................... 5.0 ± 0.2 mm (0.20 ± 0.008 in)
Service limit ......................................................................... 4.5 mm (0.18 in)
Disc runout (max) ................................................................. 0.30 mm (0.012 in)
Master cylinder bore ............................................................. 15.870 – 15.913 mm (0.6248 – 0.6265 in)
Master cylinder piston .......................................................... 15.811 – 15.838 mm (0.6225 – 0.6235 in))
Caliper bore ......................................................................... 38.180 – 38.219 mm (1.5031 – 1.5047 in)
Caliper piston ...................................................................... 38.025 – 38.050 mm (1.4970 – 1.4980 in)

## Rear brake
Type .................................................................................... Single hydraulic disc
Disc thickness ...................................................................... 6.7 ± 0.2 mm (0.26 ± 0.008 in)
Service limit ......................................................................... 6.0 mm (0.24 in)
Disc runout (max) ................................................................. 0.30 mm (0.012 in)
Master cylinder bore ............................................................. 14.000 – 14.043 mm (0.5512 – 0.5529 in)
Master cylinder piston .......................................................... 13.957 – 13.984 mm (0.5495 – 0.5506 in)
Caliper bore ......................................................................... 38.180 – 38.256 mm (1.5031 – 1.5061 in)
Caliper piston ...................................................................... 38.098 – 38.148 mm (1.4999 – 1.5019 in)

## Hydraulic fluid
Type .................................................................................... SAE J1703 or DOT 3 specification

## Torque wrench settings

| Component | kgf m | lbf ft |
|---|---|---|
| Front master cylinder clamp bolt | 1.0 – 1.6 | 7.0 – 11.0 |
| Front caliper mounting bolt | 2.5 – 4.0 | 18.0 – 29.0 |
| Front caliper spindle bolt | 1.5 – 2.0 | 11.0 – 14.5 |
| Front wheel spindle nut | 3.6 – 5.2 | 26.0 – 37.5 |
| Front wheel spindle clamp bolt | 1.5 – 2.5 | 11.0 – 18.0 |
| Rear master cylinder mounting bolt | 1.0 – 1.6 | 7.0 – 11.0 |
| Rear brake torque arm bolt/nut | 2.0 – 3.0 | 14.5 – 21.5 |
| Rear brake caliper mounting bolt | 2.5 – 4.0 | 18.0 – 29.0 |
| Chain adjuster bolt | 1.8 – 2.8 | 13.0 – 20.0 |
| Rear wheel spindle nut | 8.5 – 11.5 | 61.5 – 83.0 |
| Brake hose union bolt | 2.0 – 2.5 | 14.5 – 18.0 |
| Anti-dive hose union bolt | 2.0 – 2.5 | 14.5 – 18.0 |
| Air bleed valve | 0.6 – 0.9 | 4.5 – 6.5 |
| Anti-dive valve fitting bolt | 0.6 – 0.8 | 4.5 – 6.0 |
| Anti-dive modulator fitting bolt | 0.4 – 0.5 | 3.0 – 3.5 |

## 1  General description

Cast alloy or spoked wheels are fitted to the GS/GSX range, the type used varying according to model. Tyre sizes vary according to the wheel rim diameter and section, the majority of models using a 19 in front and 17 in rear tyre. The LT version, in keeping with its 'custom' image, is equipped with a 19 in front and a fat 16 in rear tyre. The appropriate details of sizes and pressures for the various machines will be found in the Specifications at the front of this Chapter.

Braking on all models is by an hydraulically operated disc. A twin disc arrangement is fitted to the front wheel, a similar single disc assembly being fitted to the rear.

The later models feature an anti-dive arrangement designed to minimise the pitching effect encountered under heavy braking. It does this by utilising pressure in the front brake system to automatically stiffen the front fork damping effect. This in turn allows a more compliant suspension arrangement than would otherwise be possible.

## 2  Wheels: examination and renovation – cast alloy type

1  The wheels require little maintenance apart from regular cleaning. In the event of suspected impact damage, check the wheel for distortion by arranging it so that it is clear of the ground. Using a dial gauge mounted on the front fork or the swinging arm, as appropriate, check the axial (side-to-side) play and radial play (ovality). In each case

this should not exceed 2.0 mm (0.08 in). If distorted beyond this amount the wheel must be renewed. Note that it cannot be repaired or trued.

2  Check for bearing wear by turning the wheel and by rocking it from side to side. Any 'grittiness' or unevenness in its movement, or any discernible free play in the bearing is indicative of the need for renewal. In the case of the rear wheel, take care not to confuse wheel bearing free play with movement in the swinging arm.

3  Give the whole wheel a close visual check for dents or cracks. Any cracking should be viewed with great suspicion, as it can lead to a sudden stress fracture under the high loadings experienced in normal use. If in any doubt, seek professional advice. Small nicks can be dressed out using a fine file or abrasive paper, and the resulting bare metal coated with one of the proprietary alloy wheel lacquers to prevent subsequent corrosion.

## 3  Wheels: examination and renovation – wire spoked type

1  Spoked wheels can go out of true over periods of prolonged use and, like any wheel, as the result of an impact. The condition of the hub, spokes and rim should therefore be checked at regular intervals.

2  An improvised wheel stand is invaluable, but failing this, the wheel can be checked whilst in place on the machine after it has been raised clear of the ground. Make the machine as stable as possible, if necessary using blocks beneath the crankcase as extra support. Spin the wheel and ensure that there is no brake drag. If necessary, remove the disc pads (disc brake models) until the wheel turns freely. In the

case of rear wheels it is advisable to remove the final drive chain.

3 Slowly rotate the wheel and examine the rim for signs of serious corrosion or impact damage. Slight deformities, as might be caused by running the wheel along a curb, can often be corrected by adjusting spoke tension. More serious damage may require a new rim to be fitted, and this is best left to an expert. There is an art to wheel building, and a professional wheel builder will have the facilities and parts required to carry out the work quickly and economically. Badly rusted steel rims should be renewed in the interests of safety as well as appearance. Where light alloy rims are fitted, corrosion is less likely to be a serious problem.

4 Assuming the wheel to be undamaged it will be necessary to check it for runout. Arrange a temporary wire pointer so that it runs close to the rim. The wheel can now be turned and any distortion noted. Check for lateral distortion and for radial distortion, noting that the latter is less likely to be encountered if the wheel was set up correctly from new and has not been subject to impact damage.

5 The rim should be no more than 2.0 mm (0.1 in) out of true in either plane. If a significant amount of distortion is encountered, check that the spokes are of approximately equal tension. Adjustment is effected by turning the square-headed spoke nipples with the appropriate spoke key. This tool is obtainable from most good motorcycle shops or tool retailers.

6 With the spokes evenly tensioned, any remaining distortion can be pulled out by tightening the spokes on one side of the hub and slackening the corresponding spokes from the opposite hub flange. This will allow the rim to be pulled across whilst maintaining spoke tension.

7 If more than slight adjustment is required, the tyre and inner tube should be removed first to give access to the spoke ends. Those which protrude through the nipple after adjustment should be filed flat to avoid the risk of puncturing the tube. It is essential that the rim band is in good condition as an added precaution against chafing. In an emergency, use a strip of duct tape as an alternative; unprotected tubes will soon chafe on the nipples.

8 Should a spoke break, a replacement item can be fitted and retensioned in the normal way. Wheel removal is usually necessary for this operation, although complete removal of the tyre can be avoided if care is taken. A broken spoke should be attended to promptly, because the load normally taken by that spoke is transferred to adjacent spokes which may fail in turn.

9 Remember to check wheel condition regularly. Normal maintenance is confined to keeping the spokes correctly tensioned and will avoid the costly and complicated wheel rebuilds that will inevitably result from neglect. When cleaning the machine do not neglect the wheels. If the rims are kept clean and well polished, many of the corrosion-related maladies will be prevented.

### 4 Front wheel: removal and installation

1 Place the machine on its centre stand, using a jack or an improvised stand to raise the wheel clear of the ground. Detach one of the brake calipers by removing the two retaining bolts. Lift the caliper clear of the disc and place a wooden wedge between the pads to prevent their accidental expulsion. Tie the caliper clear of the forks to avoid placing strain on the hydraulic hose. Free the speedometer drive cable by unscrewing the knurled ring which secures it to the drive gearbox.

2 Straighten and remove the split pin which locks the wheel spindle nut and remove the nut. Make a note of the wheel spacer locations (particularly where shouldered spacers are fitted), so that they may be correctly refitted on reassembly. Slacken the pinch bolt or the clamp nuts then tap the wheel spindle part way out. Grasp the spindle end and withdraw it, supporting the wheel with one hand. The wheel can now be lowered clear of the forks and removed.

3 The wheel is installed by reversing the removal sequence. Refit the wheel spacers in their previously-noted positions, grease the speedometer gearbox and ensure that it locates correctly against the fork lower leg as the wheel is offered up. Fit the wheel spindle and install the spindle nut. Pump the forks up and down several times to align them on the spindle and then tighten the spindle nut to the specified torque setting. When tightening the spindle clamp bolts, ensure that they are tightened evenly so that the gap between the clamp and fork leg casting remains equal. The pinch bolt or clamp half nuts can now be tightened. Refit the brake caliper and check brake operation before using the machine.

4.1 Brake calipers are secured by two bolts (arrowed)

4.2a Slacken the clamp nuts or pinch bolt ...

4.2b ... and remove split pin and nut, then withdraw spindle

4.3 Note headed spacer on right-hand side of wheel

## 5  Front wheel bearings: renewal

1    Remove the wheel as described above. It is recommended that the brake discs are detached to avoid any risk of damage, though this is not essential if care is taken. Support the wheel on the workbench with the right-hand side uppermost. Using a long drift displace the spacer to one side, then drive the left-hand bearing out of the hub. Invert the wheel and remove the spacer, then drive out the remaining bearing.

2    Check the bearing by rotating it by hand, noting any unevenness or free play. Unless it operates smoothly and is free of play, the bearings must be renewed as a pair. The new bearings can be fitted by drawing them into the hub with the Suzuki bearing installer set, Part Number 09924-84510, or an equivalent home-made drawbolt arrangement. Refer to the accompanying line drawing for details, and fit the left-hand bearing first. Position the spacer and then fit the remaining bearing.

3    As an alternative method, a large socket can be used to tap the bearings home (see photograph). Ensure that the drift engages the outer race and that the bearing is kept square to the hub bore as it is tapped into position against the locating shoulder. Do not omit to fit the spacer between the two bearings.

5.1 Knock the spacer to one side and drive out the bearing

5.3a Drop spacer into place in the hub ...

5.3b ... and place the remaining bearing in position ...

5.3c ... driving it home squarely using a socket as a drift

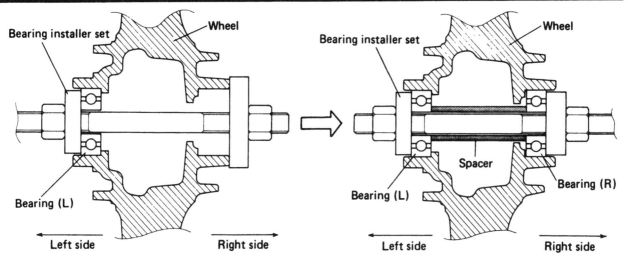

Fig. 5.1 Method of refitting the front wheel bearings

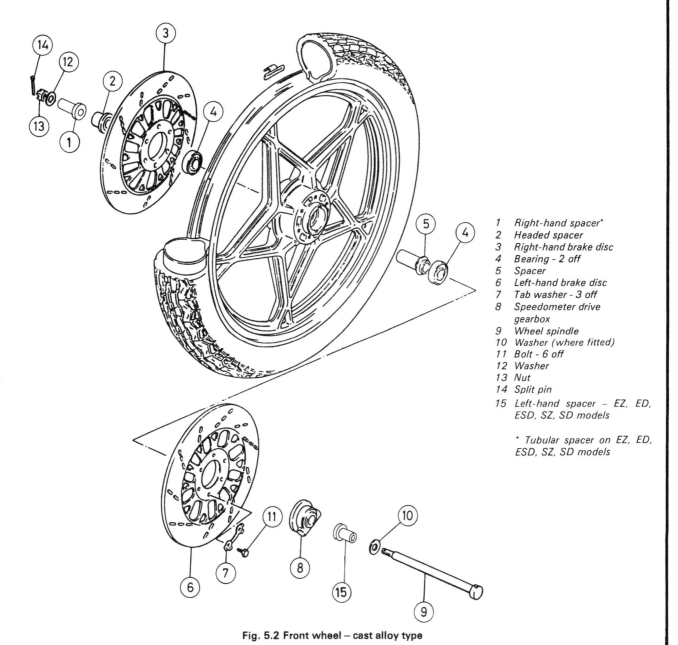

1 Right-hand spacer*
2 Headed spacer
3 Right-hand brake disc
4 Bearing - 2 off
5 Spacer
6 Left-hand brake disc
7 Tab washer - 3 off
8 Speedometer drive gearbox
9 Wheel spindle
10 Washer (where fitted)
11 Bolt - 6 off
12 Washer
13 Nut
14 Split pin
15 Left-hand spacer – EZ, ED, ESD, SZ, SD models

* Tubular spacer on EZ, ED, ESD, SZ, SD models

Fig. 5.2 Front wheel – cast alloy type

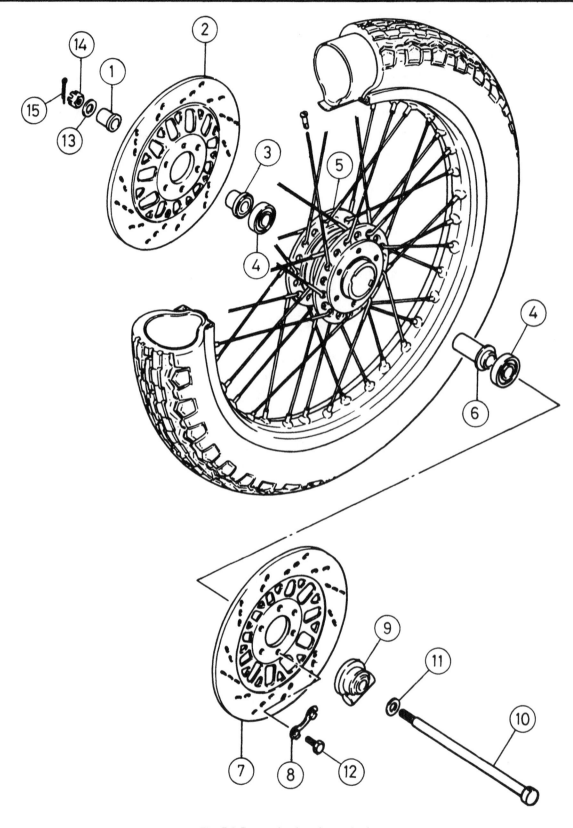

**Fig. 5.3 Front wheel – wire spoked type**

| | | | | | |
|---|---|---|---|---|---|
| 1 | Spacer | 6 | Spacer | 11 | Washer |
| 2 | Right-hand brake disc | 7 | Left-hand brake disc | 12 | Bolt - 6 off |
| 3 | Headed spacer | 8 | Tab washer - 3 off | 13 | Washer |
| 4 | Bearing - 2 off | 9 | Speedometer drive gearbox | 14 | Nut |
| 5 | Wheel hub | 10 | Wheel spindle | 15 | Split pin |

## 6    Rear wheel: removal and installation

1    Place the machine securely on its centre stand. Remove the caliper mounting bolts and the swinging arm end stop retaining bolts, then lift the caliper clear of the disc. Tie the caliper clear of the wheel and swinging arm and place a wooden wedge between the pads to prevent their accidental expulsion. Remove the two bolts which retain the chainguard to the swinging arm and lift it away.

2    Straighten and remove the split pin which locks the wheel spindle nut, then remove the nut. Pull the wheel rearwards and displace the chain adjusters. The wheel can now be pushed forwards and the chain lifted off the sprocket. Holding the chain clear of the sprocket, pull the wheel rearwards and clear of the swinging arm.

3    The wheel can be installed by reversing the above sequence. Note that the drive chain tension should be adjusted before tightening the wheel spindle nut to the specified torque. Use a new split pin to secure the nut and check brake operation before using the machine.

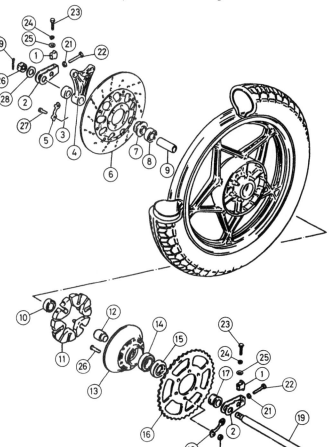

6.3a Note spacer on left-hand side of wheel

6.3b Headed spacer fits on right-hand side of wheel ...

**Fig. 5.4 Rear wheel – cast alloy type**

| | |
|---|---|
| 1   Chain adjuster bracket stop - 2 off | 14   Cush drive hub bearing |
| 2   Chain adjuster bracket - 2 off | 15   Oil seal |
| 3   Spacer | 16   Drive sprocket |
| 4   Brake caliper mounting bracket | 17   Left-hand spacer |
| 5   Tab washer - 3 off | 18   Tab washer - 3 off |
| 6   Brake disc | 19   Wheel spindle |
| 7   Right-hand headed spacer | 20   Nut - 6 off |
| 8   Right-hand bearing | 21   Locknut - 2 off |
| 9   Spacer | 22   Adjusting bolt - 2 off |
| 10   Left-hand bearing | 23   Bolt - 2 off |
| 11   Cush drive rubbers | 24   Spring washer - 2 off |
| 12   Spacer | 25   Washer - 2 off |
| 13   Cush drive hub | 26   Nut |
| | 27   Bolt - 6 off |
| | 28   Washer |
| | 29   Split pin |

6.3c ... followed by caliper bracket and plain spacer

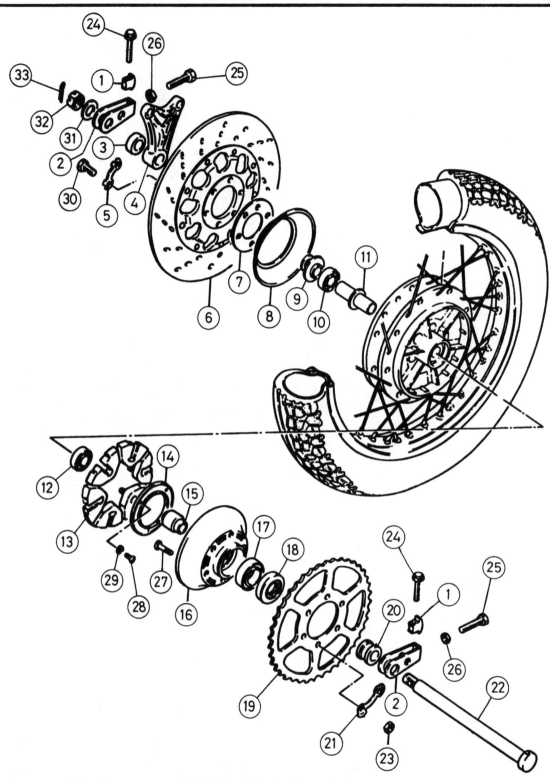

**Fig. 5.5 Rear wheel – wire spoked type**

| | | |
|---|---|---|
| 1 Chain adjuster bracket stop – 2 off | 12 Left-hand bearing | 23 Nut - 6 off |
| 2 Chain adjuster bracket - 2 off | 13 Cush drive rubbers | 24 Bolt - 2 off |
| 3 Spacer | 14 Damping ring | 25 Adjusting bolt - 2 off |
| 4 Brake caliper mounting bracket | 15 Spacer | 26 Locknut - 2 off |
| 5 Tab washer - 3 off | 16 Cush drive hub | 27 Bolt - 6 off |
| 6 Brake disc | 17 Cush drive hub bearing | 28 Screw - 3 off |
| 7 Shim | 18 Oil seal | 29 Spring washer - 3 off |
| 8 Hub flange | 19 Drive sprocket | 30 Bolt - 6 off |
| 9 Right-hand headed spacer | 20 Left-hand spacer | 31 Washer |
| 10 Right-hand bearing | 21 Tab washer - 3 off | 32 Nut |
| 11 Spacer | 22 Wheel spindle | 33 Split pin |

## Tyre changing sequence - tubed tyres

Deflate tyre. After pushing tyre beads away from rim flanges push tyre bead into well of rim at point opposite valve. Insert tyre lever adjacent to valve and work bead over edge of rim.

Use two levels to work bead over edge of rim. Note use of rim protectors

Remove inner tube from tyre

When first bead is clear, remove tyre as shown

When fitting, partially inflate inner tube and insert in tyre

Work first bead over rim and feed valve through hole in rim. Partially screw on retaining nut to hold valve in place.

Check that inner tube is positioned correctly and work second bead over rim using tyre levers. Start at a point opposite valve.

Work final area of bead over rim whilst pushing valve inwards to ensure that inner tube is not trapped

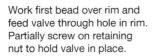

### 7   Rear wheel bearings: renewal

1   Remove the rear wheel as described above and pull off the cush drive hub and sprocket assembly. The bearing arrangement is generally similar to that described above for the front wheel, noting that the right-hand bearing should be removed first. The new bearings should be packed with grease prior to installation; note that the right-hand bearing has a rubber sealed outer face, whilst that of the left-hand bearing has a metal seal.

2   The cush drive/sprocket hub has its own bearing, and this can be driven out, together with the seal, from the inner face using a tubular drift or a suitably sized socket. When fitting the new bearing, grease it thoroughly, then tap it squarely into the recess. Fit a new seal where necessary, pressing it home against the bearing. When refitting the cush drive hub, do not omit the spacer which fits between it and the wheel bearing.

7.2a Bearing and seal can be driven out of the cush drive hub

7.2b Note the spacer which fits from the inner face of the hub

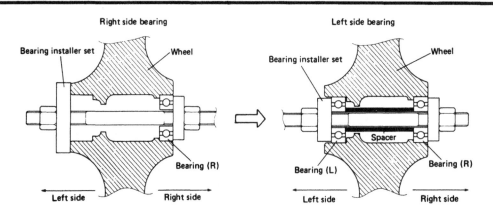

Fig. 5.6 Installing the rear wheel bearings

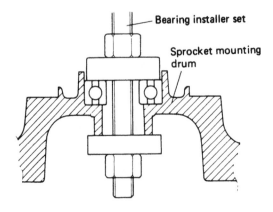

Fig. 5.7 Installing the cush drive hub bearing

## 8 Brake pads: renewal

1    Brake pad wear can be checked with the caliper in position by prising off the plastic inspection cover. If either pad is worn down to the red line which denotes the maximum wear limit, they should be renewed as a pair.

### Front brake

2    Remove the two bolts which retain the caliper body on the mounting bracket, leaving the latter attached to the fork lower leg. Lift away the caliper body, leaving the pads in place on the bracket. Remove the old pads and wipe away residual brake dust using a rag moistened with alcohol or methylated spirit. On no account use an air line to remove the dust and do not inhale it.

3    Push the caliper piston fully into the caliper to provide room for the new pads. Take care that the displaced fluid does not overflow from the reservoir. Check that the pad spring is located correctly and install the new pads on the bracket. Refit the caliper body and tighten the bolts to the specified torque figure.

### Rear brake

4    Prise off the black plastic dust cover to expose the pads. Depress the spring and slide out one of the retaining pins. Once spring pressure has been relieved, remove the remaining pin and the spring. Lift out the pads together with their backing shims.

5    Clean the caliper recess with a rag moistened with alcohol or methylated spirit, taking care not to inhale the dust. Where necessary, gently lever the pistons back into the caliper to provide room for the new pads. Check that the displaced fluid does not overflow from the reservoir.

6    Place the shim on the back of each pad, noting that the arrowhead hole must face forward. Place the pad assemblies into the caliper recess and secure them with the pins and spring. Refit the dust cover.

### Front and rear brake

7    After new pads have been fitted, operate the lever or pedal repeatedly to allow the caliper to adjust to the new pad thickness, then check the level of the hydraulic fluid. For the first 100 miles or so, try to avoid heavy braking so that the pads can bed in properly. Excessively hard braking on new pads may cause glazing of the friction surface and impaired performance.

8.1a Front pad wear can be checked via inspection window

8.1b Rear pads are visible after plastic cover has been removed

8.2a Release the caliper body from the mounting bracket

8.2b Pads can now be lifted clear of the bracket

8.4a Pull out the R-pin ...

8.4b ... and withdraw the pad support pins

8.4c Pads and their backing shims can now be lifted away

8.6 Note that arrow head hole in shim must face forward

## 9 Brake discs: examination and renovation

1 The condition of the brake discs can be checked with the wheel in place. Inspect the disc surface on both sides for scoring. Light scratching of the surface is inevitable and normal, but if excessive the braking performance will be reduced and renewal will be required. The thickness of the disc should be checked using a micrometer and the readings compared with those shown in the Specifications. If worn beyond the service limit or if wear is uneven, renew the discs.

2 Check each disc for runout using a dial gauge mounted so that the probe bears on the outer edge of the disc. If runout is in excess of the service limit, renew the disc. Finally, check carefully for signs of cracking. This is not a common occurrence, but can be very dangerous if the disc breaks up when the machine is being used.

3 To remove the disc(s), remove the relevant wheel as described above. Straighten the locking tabs and remove the mounting bolts, then lift the disc away. When fitting the front discs, note that the left-hand and right-hand components are marked L and R respectively and should not be interchanged. Clean off any dirt from the disc and mounting boss to ensure that the disc locates correctly. Fit new tab washers and tighten the mounting bolts evenly to the prescribed torque setting. Bend over the locking tabs to secure the mounting bolts. Finally, degrease each disc surface thoroughly before use.

## 10 Hydraulic system: general precautions

1 As has been mentioned, the brake systems are of the hydraulic disc type. It is important that the precautions described are observed when working on the system, bearing in mind the consequences of sudden failure on the road.

2 The hydraulic system must be kept free of air at all times. Any air bubbles, however small, will impair braking efficiency, possibly rendering the brake inoperative. If any part of the hydraulic system is disturbed, or if the brake feels 'spongy' in use, bleed the system as described in Section 16 of this Chapter.

3 Use only new hydraulic fluid conforming to SAE J1703 or DOT 3 specifications. The fluid will degrade in time, due to its hygroscopic nature. This tendency to absorb moisture will lower its boiling point, which in extreme cases can create air bubbles due to heat build up under heavy braking. Do not keep old fluid which has been drained from the system.

4 Take great care not to spill fluid on paintwork or plastic surfaces, both of which will be damaged by it. Wash off any splashes promptly, and cover vulnerable areas before starting any dismantling work.

5 Always keep all internal hydraulic components clinically clean. Dust or dirt in the system will invariably get trapped by one of the seals, causing scoring and eventual leakage.

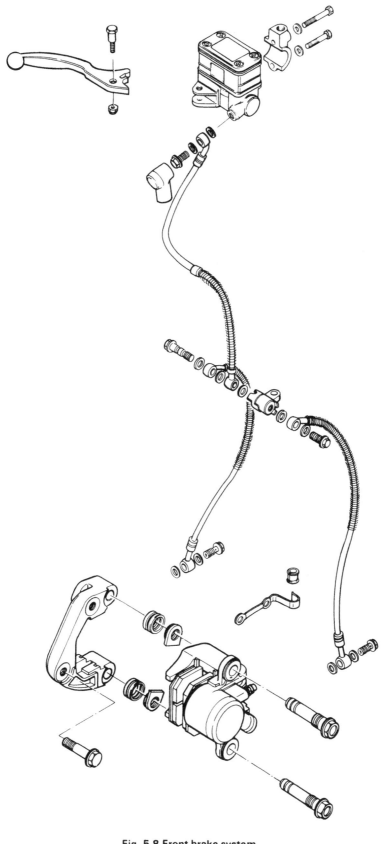

Fig. 5.8 Front brake system

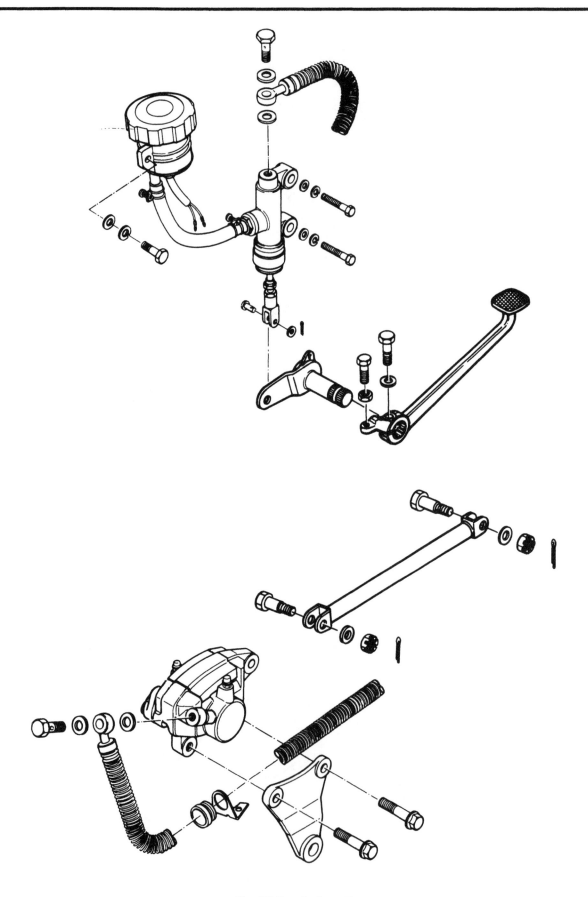

Fig. 5.9 Rear brake system

## 11 Hydraulic hoses: examination

1   The master cylinder and the brake calipers are connected by flexible hydraulic hoses, secured at each end by conventional banjo unions. The hoses must cope with considerable hydraulic pressure and also suspension movement. Although of tough construction, the hoses will deteriorate in time and should be inspected closely, preferably whenever the pads are checked.

2   Look closely for signs of cracking or perishing of the outer casing, and renew the hose if it is in less than perfect condition. It is good practice to renew the hose as a precaution when the rest of the system is to be overhauled.

3   Before a hose can be renewed it will be necessary to drain the system. Remove the dust cap from the caliper bleed valve and fit a length of plastic tubing to it, placing the free end in a jar to catch the fluid. Open the bleed valve by about one turn, then operate the brake lever or pedal to 'pump' the fluid out.

4   Place a rag below the master cylinder union to catch any spilled fluid, then slacken and remove the union bolt. Repeat this sequence on the lower union, release the guide clips and remove the hose. The new hose can be refitted by reversing the removal sequence, noting that new sealing washers should be used. Fill and bleed the system as described in Section 16 and check that the brake works correctly before using the machine.

11.1a Front brake hose is clamped to fork lower leg

11.1b Rear hose is supported by bracket at rear of swinging arm

## 12 Front brake master cylinder: overhaul

1   Before starting any dismantling work, drain the hydraulic system as described in Section 11, paragraph 3. Remove the banjo union bolt to free the hydraulic hose from the master cylinder, taking care to avoid dripping fluid on the painted or plastic parts. Release the front brake switch from the underside of the master cylinder.

2   It is convenient to remove the front brake lever at this stage. Release the locknut, then remove the pivot bolt to free the lever. Unscrew the master cylinder clamp bolts and lift the unit away. Carefully clean the cylinder, then place it on the workbench to await dismantling.

3   Pull off the dust seal from the end of the body to expose the circlip which retains the piston assembly. Release the circlip, then displace the piston using compressed air via the hose union hole. In the absence of a compressed air supply a footpump or a bicycle tyre pump will suffice. To avoid spraying residual fluid around the workshop, wrap some rag around the piston bore. Clean the body and piston assembly, then measure the relevant dimensions using internal and external micrometers, comparing the readings obtained with those shown in the Specifications.

4   In the absence of measuring equipment, any assessment of the condition of the cylinder must be on the basis of careful scrutiny. Examine carefully the condition of the cylinder bore. This must show no sign of scoring or wear, and anything less than a perfect surface

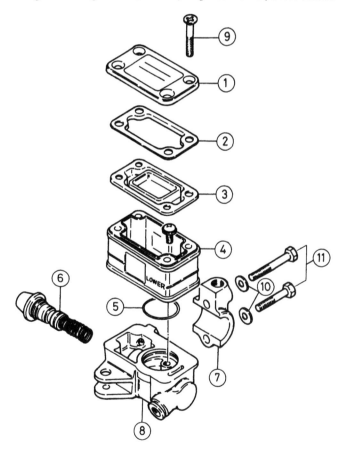

Fig. 5.10 Front master cylinder – E model type

| | | | |
|---|---|---|---|
| 1 | Reservoir cap | 7 | Handlebar clamp |
| 2 | Diaphragm plate | 8 | Master cylinder |
| 3 | Diaphragm | 9 | Screw - 4 off |
| 4 | Reservoir | 10 | Washer - 2 off |
| 5 | O-ring | 11 | Bolt - 2 off |
| 6 | Piston assembly | | |

finish indicates the need for renewal. Note that if the bore is damaged, the cylinder should be renewed complete. If the cylinder is serviceable, check that the small relief port and the main feed port from the reservoir are clear. If necessary the reservoir can be released after the securing screws have been removed.

5    The piston surface will not wear unless the seals are very badly worn and in this case the comparatively soft cylinder bore will invariably have suffered worse damage. The seals are best renewed as a precautionary measure, even if they appear unworn. If the seals are scored, worn or swollen they must be renewed. In an emergency the seals may be reused if in perfect conditon, but remember that they will

have worn to suit the bore surface and may be prone to leakage once disturbed.

6    When assembling the cylinder, soak the piston and seals in hydraulic fluid to provide lubrication. Slide the assembly into place and secure it with the circlip. The dust seal should be renewed if damaged. Complete reassembly by reversing the dismantling sequence, using new sealing washers on the hose union, which should be tightened to the specified torque figure.

7    After assembly, remember that the system must be filled and bled as described in Section 16. Check that there are no leaks and that the brake works correctly before using the machine.

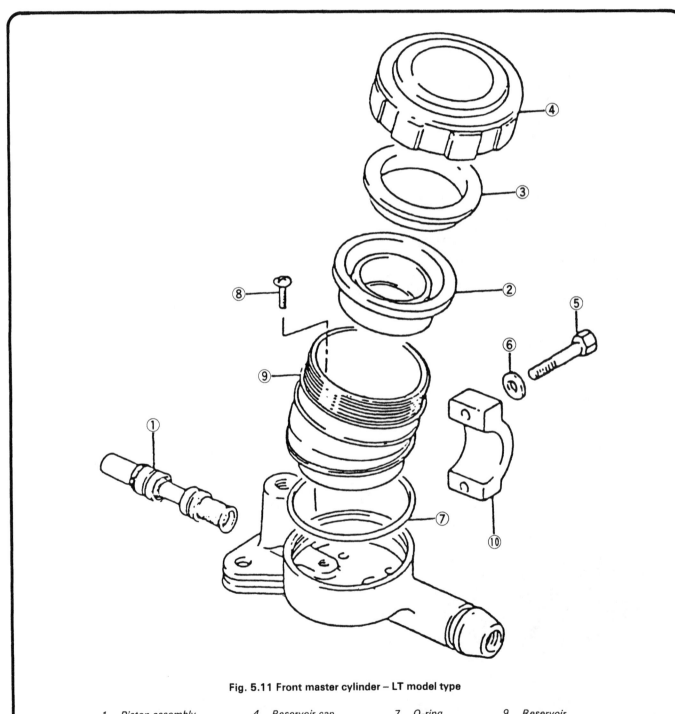

Fig. 5.11 Front master cylinder – LT model type

| 1 | Piston assembly | 4 | Reservoir cap | 7 | O-ring | 9 | Reservoir |
|---|---|---|---|---|---|---|---|
| 2 | Diaphragm | 5 | Bolt - 2 off | 8 | Screw | 10 | Handlebar clamp |
| 3 | Diaphragm plate | 6 | Washer - 2 off | | | | |

### 13 Front brake caliper: overhaul

1    Drain the hydraulic system as described in Section 11, paragraph 3 above, then detach the caliper unit by releasing the two caliper bracket mounting bolts, leaving the caliper bracket and pads attached to the caliper body. Carefully clean the caliper prior to any dismantling work.

2    Slacken and remove the two special bolts on which the caliper body slides in relation to the bracket. Lift away the bracket, together with the pads. Wrap the caliper in clean cloth then displace the piston by applying compressed air via the hose union hole. Take care not to allow fingers to be trapped by the emerging piston. Clean the various components thoroughly using alcohol or methylated spirit only; on no account use petrol or any degreasing solvent because these will attack and damage the seals.

3    The caliper body contains two seals. The outer, convoluted, seal is a dust seal which excludes road and brake dust from the caliper. The importance of this is obvious, and it must be renewed if damaged in any way. The smaller inner seal retains the hydraulic fluid in the caliper. Its second and less obvious function is to control the position of the pads in relation to the disc when the brake is released. This is dependent on the elasticity of the seal, which is why a dragging brake can often be traced to a 'tired' seal. It is recommended that both seals are renewed as a precautionary measure.

4    As in the case of the master cylinder, both the piston and the caliper body must be free from scoring, corrosion and visible wear. Should such damage be found, renewal will be necessary. Note that serious corrosion will often occur at the outer face of the piston where the dust seal has split.

5    The special bolts which secure the caliper body to the caliper holder bracket must slide smoothly in the bores in the caliper, otherwise brake drag will result. This can be prevented by lubricating the pins with PBC (Poly Butyl Cuprysil) grease during assembly. Note that normal general purpose grease must not be used. To prevent later problems, renew the two small dust seals as a precautionary measure.

6    The caliper is assembled by reversing the dismantling sequence, noting that every component must be spotlessly clean. Use hydraulic fluid to lubricate the seals and piston, and ensure that the dust seal locates correctly in the piston and body grooves. Refer to the accompanying line drawing for details of the relative positions of the various components, noting in particular the direction of the pad anti-rattle spring. Use new sealing washers on the hose union. Tighten all fasteners to the recommended torque setting.

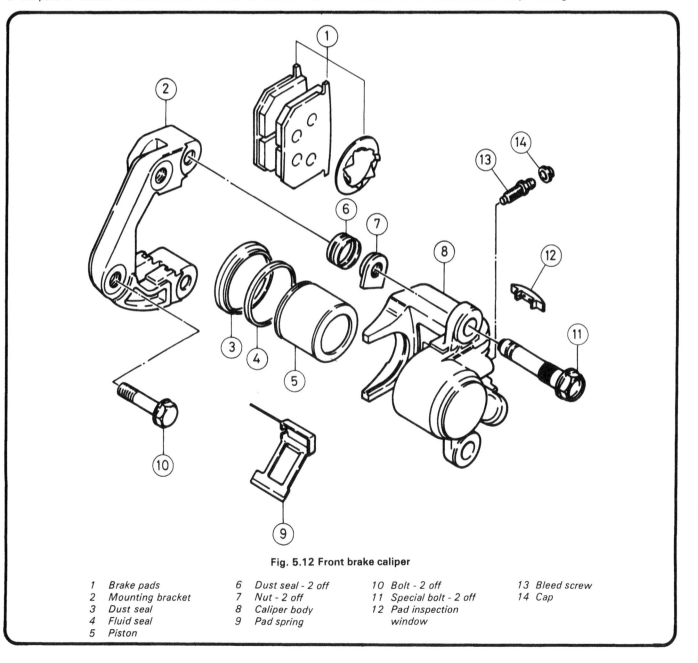

**Fig. 5.12 Front brake caliper**

| | | | |
|---|---|---|---|
| 1 Brake pads | 6 Dust seal - 2 off | 10 Bolt - 2 off | 13 Bleed screw |
| 2 Mounting bracket | 7 Nut - 2 off | 11 Special bolt - 2 off | 14 Cap |
| 3 Dust seal | 8 Caliper body | 12 Pad inspection | |
| 4 Fluid seal | 9 Pad spring | window | |
| 5 Piston | | | |

## 14 Rear brake master cylinder: overhaul

1    Drain the rear brake hydraulic system as described in Section 11. Pull off the right-hand side panel to gain access to the master cylinder and reservoir. Disconnect the master cylinder pushrod from the operating arm inboard of the footrest bracket. The forked end of the pushrod is retained by a clevis pin which is secured in turn by a split pin. Straighten the split pin and remove it, then displace the clevis pin.

2    Remove the hose union bolt and lodge the hose clear of the master cylinder, taking care to avoid fluid spillages on the painted parts. Remove the reservoir mounting bolt and the two master cylinder mounting bolts. Lift away the master cylinder together with the reservoir. Unscrew the reservoir cap and drain any residual hydraulic fluid.

3    Release the clip which secures the reservoir hose to the caliper and separate the two. Pull off the pushrod dust seal to expose the circlip. Remove the circlip and remove the pushrod assembly and the piston, followed by the primary cup and spring.

4    The master cylinder assembly should be examined in the same manner as described in Section 12 for the front master cylinder. It is recommended that the seals are renewed whenever the cylinder is dismantled, soaking them in clean hydraulic fluid prior to installation. When the cylinder has been refitted, refill and bleed the hydraulic system and check that the brake pedal and switch are adjusted correctly.

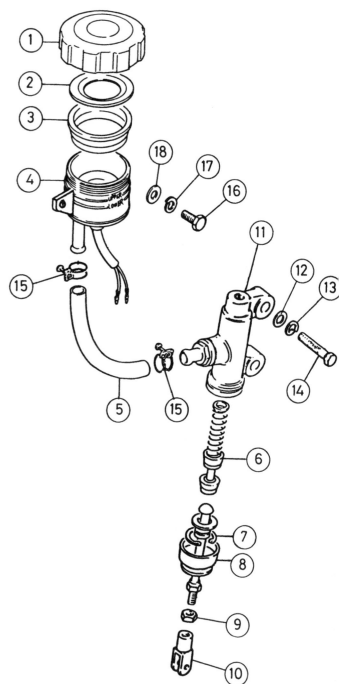

14.1 Remove split pin to free pushrod from brake arm

14.2 Master cylinder is secured by two bolts

Fig. 5.13 Rear master cylinder

| | | | |
|---|---|---|---|
| 1 | Reservoir cap | 10 | Linkage |
| 2 | Diaphragm plate | 11 | Master cylinder |
| 3 | Diaphragm | 12 | Washer - 2 off |
| 4 | Reservoir | 13 | Spring washer - 2 off |
| 5 | Hose | 14 | Bolt - 2 off |
| 6 | Piston assembly | 15 | Hose clip - 2 off |
| 7 | Circlip | 16 | Bolt |
| 8 | Dust seal | 17 | Spring washer |
| 9 | Nut | 18 | Washer |

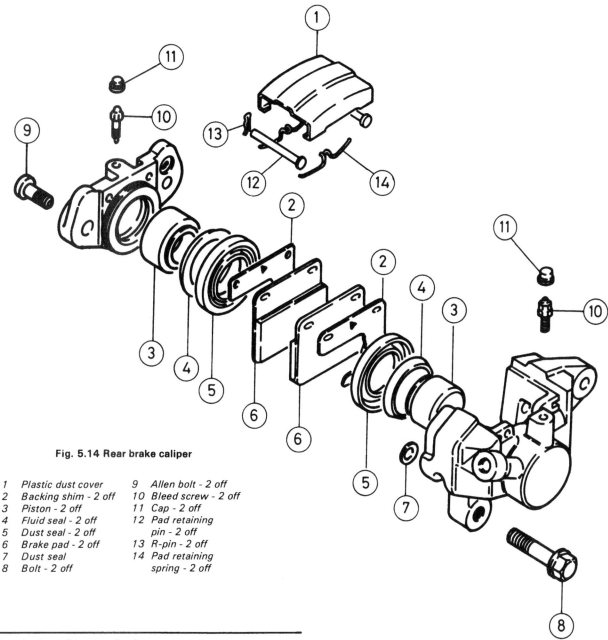

**Fig. 5.14 Rear brake caliper**

| | | | |
|---|---|---|---|
| 1 | Plastic dust cover | 9 | Allen bolt - 2 off |
| 2 | Backing shim - 2 off | 10 | Bleed screw - 2 off |
| 3 | Piston - 2 off | 11 | Cap - 2 off |
| 4 | Fluid seal - 2 off | 12 | Pad retaining |
| 5 | Dust seal - 2 off | | pin - 2 off |
| 6 | Brake pad - 2 off | 13 | R-pin - 2 off |
| 7 | Dust seal | 14 | Pad retaining |
| 8 | Bolt - 2 off | | spring - 2 off |

## 15 Rear brake caliper: overhaul

1   Drain the rear brake hydraulic system as described in Section 11, then remove the pads as described in Section 8. Disconnect the hydraulic hose taking care to avoid dripping residual fluid on the paintwork. Remove the split pin which secures the torque arm nut and remove the latter to free the torque arm. Remove the caliper mounting bolts and lift away the caliper.

2   Remove the two large Allen bolts and separate the caliper halves. It will be noted that the rear caliper is of the twin piston type, unlike the single piston front units. Each half should be dealt with separately, to avoid interchanging the internal components. The general procedure from this point onwards is much the same as that described for the front caliper, and reference should be made to Section 13. It is recommended that the piston seals and the O-ring which seals the passage between the caliper halves are renewed.

## 16 Changing the hydraulic fluid and bleeding the system

1   This Section describes the procedure for changing the hydraulic fluid, an operation which should be carried out at approximately yearly intervals to ensure that the system is kept clean and that the fluid does not become dangerously degraded. Where the system has already been drained during repair or overhaul work, or if it is suspected that air has entered the system, refer to the relevant parts of the procedure.

2   Obtain about two feet of small-bore plastic tubing of the type used for car screen washers or in aquarium aeration systems. The tubing should fit snugly over the caliper bleed valve. Connect the tubing to the valve and place the free end in a glass jar or a similar receptacle. Open the bleed valve by $\frac{1}{2}$ to 1 turn, then operate the brake lever repeatedly until all the fluid has been expelled.

### Front brake

3   Remove the reservoir top and fill it to the upper level mark with new DOT 3 or SAE J1703 hydraulic fluid only. Where the system has been emptied it should first be primed by operating the lever until the fresh fluid starts to emerge through the plastic bleed tube. Take care that the reservoir is kept topped up throughout this stage, and also during the bleeding operation; even a small amount of air drawn in will necessitate starting again, so an assistant to monitor the fluid level would be invaluable. Deal with the left-hand caliper first, then the right-hand unit.

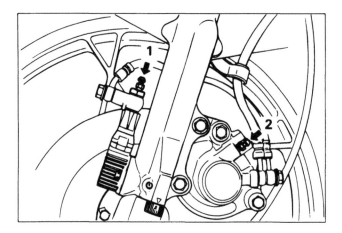

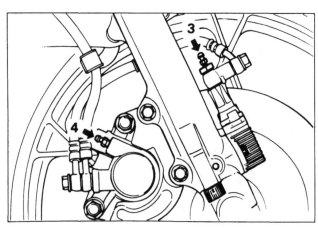

**Fig. 5.15 Front brake air bleeding sequence – anti-dive models**

1    Left-hand anti-dive unit valve
2    Left-hand caliper valve
3    Right-hand anti-dive unit valve
4    Right-hand caliper valve

4    With the bleed valve closed, operate the brake lever until some resistance is felt. Now open the valve by about $\frac{1}{2}$ a turn, whilst maintaining pressure on the lever. As soon as the lever stops moving, close the valve and then release the lever. Check the fluid level and top up as required. This sequence must be repeated until no further air bubbles can be seen emerging from the bleed tube.

5    When bleeding is complete, top up the reservoir and refit the top. Check that the brake feels firm when operated; there must be no sign of 'sponginess'. If the bleeding operation was carried out after poor braking had been noted, check the performance of the brake over the next few days. If any further deterioration is observed, the system must be leaking at some point and should be overhauled. Fluid leakage will be obvious, but note that air can be drawn past worn master cylinder seals without external signs.

**Front brake: anti-dive models**
6    Proceed as described above, noting that it is important that the system is bled in the correct sequence to avoid air pockets being trapped in the various components and connecting hoses. The bleeding sequence is shown in the accompanying line drawing.

**Rear brake**
7    The rear brake system can be dealt with in much the same way, noting that the inner side of the caliper should be bled first, followed by the outer side.

## 17 Final drive chain: maintenance

1    To measure chain wear, remove the chainguard to expose the upper run of the chain. Taking great care to avoid trapped fingers, slowly rotate the rear wheel and check for loose or damaged rollers, pins or side plates. Damage of this type will necessitate chain renewal, so further examination will be pointless. Check both sprockets for damaged or hooked teeth. If serious wear or damage is found, renew the chain and both sprockets as a set. Measure the distance between 21 pins of the chain, taking the measurement in several places. If this exceeds the service limit of 383.0 mm (15.08 in), renew the chain. The chain is of the 'endless' type, and renewal requires the removal of the rear wheel and swinging arm assembly. The correct replacement chain is Daido DID630YL or Takasago RK630GSV.

2    Chain adjustment is correct when there is 20 – 30 mm (0.8 – 1.2 in) free play at the centre of the lower run, measured with the machine on its centre stand. Chains rarely wear evenly, so it is important to rotate the wheel until the tightest point is found.

5    If adjustment is required, slacken the brake torque arm nut, remove the split pin from the wheel spindle nut and slacken it. Slacken the adjuster locknuts, then turn each adjuster by an equal amount to obtain the required tension. Check that the alignment marks on each adjuster are in the same relative position to ensure wheel alignment. Tighten the adjuster locknuts, then secure the wheel spindle nut and torque arm nut to their respective torque settings. Remember to fit a new split pin to secure both nuts.

6    Regular cleaning and lubrication will prolong the life of the chain. Wash off any road dirt with paraffin (kerosene) only, then lubricate the cleaned chain with gear oil. On no account use an aerosol lubricant unless of a type specifically designed for use with O-ring chains. Ordinary types of aerosol lubricant will rot and destroy the O-rings.

17.5 Use marks on swinging arm to check alignment

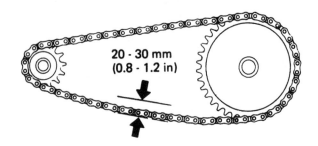

**Fig. 5.16 Measuring drive chain free play**

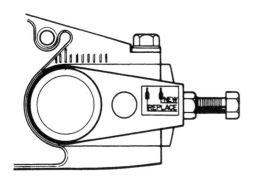

Fig. 5.17 Swinging arm alignment marks and chain wear indicator

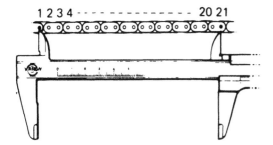

Fig. 5.18 Measuring drive chain wear

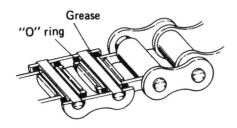

Fig. 5.19 Sectional view of O-ring chain

## 18 Tyres: removal, repair and refitting

1    At some time or other the need will arise to remove and replace the tyres, either as a result of a puncture or because replacements are necessary to offset wear. To the inexperienced, tyre changing represents a formidable task, yet if a few simple rules are observed and the technique learned, the whole operation is surprisingly simple.

2    To remove the tyre from either wheel, first detach the wheel from the machine. Deflate the tyre by removing the valve core, and when the tyre is fully deflated, push the bead away from the wheel rim on both sides so that the bead enters the centre well of the rim. Remove the locking ring and push the tyre valve into the tyre itself.

3    Insert a tyre lever close to the valve and lever the edge of the tyre over the outside of the rim. Very little force should be necessary; if resistance is encountered it is probably due to the fact that the tyre beads have not entered the well of the rim all the way round. if aluminium rims are fitted, damage to the soft alloy by tyre levers can be prevented by the use of plastic rim protectors.

4    Once the tyre has been edged over the wheel rim, it is easy to work round the wheel rim, so that the tyre is completely free from one side. At this stage the inner tube can be removed.

5    Now working from the other side of the wheel, ease the other edge of the tyre over the outside of the wheel rim that is furthest away. Continue to work around the rim until the tyre is completely free from the rim.

6    If a puncture has necessitated the removal of the tyre, reinflate the inner tube and immerse it in a bowl of water to trace the source of the leak. Mark the position of the leak, and deflate the tube. Dry the tube, and clean the area around the puncture with a petrol soaked rag. When the surface has dried, apply rubber solution and allow this to dry before removing the backing from the patch, and applying the patch to the surface.

7    It is best to use a patch of self-vulcanizing type, which will form a permanent repair. Note that it may be necessary to remove a protective covering from the top surface of the patch after it has sealed into position. Inner tubes made from a special synthetic rubber may require a special type of patch and adhesive, if a satisfactory bond is to be achieved.

8    Before replacing the tyre, check the inside to make sure that the article that caused the puncture is not still trapped inside the tyre. Check the outside of the tyre, particularly the tread area to make sure nothing is trapped that may cause a further puncture.

9    If the inner tube has been patched on a number of past occasions, or if there is a tear or large hole, it is preferable to discard it and fit a replacement. Sudden deflation may cause an accident, particularly if it occurs with the rear wheel.

10   To replace the tyre, inflate the inner tube for it just to assume a circular shape but only to that amount, and then push the tube into the tyre so that it is enclosed completely. Lay the tyre on the wheel at an angle, and insert the valve through the rim tape and the hole in the wheel rim. Attach the locking ring on the first few threads, sufficient to hold the valve captive in its correct location.

11   Starting at the point furthest from the valve, push the tyre bead over the edge of the wheel rim until it is located in the central well. Continue to work around the tyre in this fashion until the whole of one side of the tyre is on the rim. It may be necessary to use a tyre lever during the final stages.

12   Make sure there is no pull on the tyre valve and again commencing with the area furthest from the valve, ease the other bead of the tyre over the edge of the rim. Finish with the area close to the valve, pushing the valve up into the tyre until the locking ring touches the rim. This will ensure that the inner tube is not trapped when the last section of bead is edged over the rim with a tyre lever.

13   Check that the inner tube is not trapped at any point. Reinflate the inner tube, and check that the tyre is seating correctly around the wheel rim. There should be a thin rib moulded around the wall of the tyre on both sides, which should be an equal distance from the wheel rim at all points. If the tyre is unevenly located on the rim, try bouncing the wheel when the tyre is at the recommended pressure. It is probable that one of the beads has not pulled clear of the centre well.

14   Always run the tyres at the recommended pressures and never under or over inflate. The correct pressures are given in the Specifications Section of this Chapter.

15   Tyre replacement is aided by dusting the side walls, particularly in the vicinity of the beads, with a liberal coating of french chalk. Washing up liquid can also be used to good effect, but this has the disadvantage, where steel rims are used, of causing the inner surface of the wheel rim to rust.

16   Never replace the inner tube and tyre without the rim tape in position. If this precaution is overlooked there is a good chance of the ends of the spoke nipples chafing the inner tube and causing a crop of punctures.

17   Never fit a tyre that has a damaged tread or sidewalls. Apart from legal aspects, there is a very great risk of a blowout, which can have very serious consequences on a two wheeled vehicle.

18   Tyre valves rarely give trouble, but it always advisable to check whether the valve itself is leaking before removing the tyre. Do not forget to fit the dust cap, which forms an effective extra seal.

## 19 Valve cores and caps

1    Valve cores seldom give trouble, but do not last indefinitely. Dirt under the seating will cause a puzzling 'slow-puncture'. Check that they are not leaking by applying spittle to the end of the valve and watching for air bubbles.

2    A valve cap is a safety device, and should always be fitted. Apart

from keeping dirt out of the valve, it provides a second seal in case of valve failure, and may prevent an accident resulting from sudden deflation.

## 20 Wheel balancing

1    It is customary on all high performance machines to balance the wheels complete with tyre and tube. The out of balance forces which exist are eliminated and the handling of the machine is improved in consequence. A wheel which is badly out of balance produces through the steering a most unpleasant hammering effect at high speeds.

2    Some tyres have a balance mark on the sidewall, usually in the form of a coloured spot. This mark must be in line with the tyre valve, when the tyre is fitted to the inner tube. Even then the wheel may require the addition of balance weights, to offset the weight of the tyre valve itself.

3    If the wheel is raised clear of the ground and is spun, it will probably come to rest with the tyre valve or the heaviest part downward and will always come to rest in the same position. Balance weights must be added to a point diametrically opposite this heavy spot until the wheel will come to rest in ANY position after it is spun.

4    If juddering is noticed, consult a Suzuki dealer for advice on the correct choice of balance weights, noting that the type used on spoked wheels cannot be used on cast alloy wheels, and vice versa. For obvious reasons, ensure that the weights are fitted securely.

# Chapter 6 Electrical system

*Refer to Chapter 7 for information relating to the 1984-on 1135cc-engined GSX1100 and GS1150 models*

**Contents**

**Specifications**

## GSX 1100 T, ET, LT, EX, EZ, SZ, ESD, SD

### Battery
| | |
|---|---|
| Type ............................................................................... | SYB 14L-A2 |
| Voltage ........................................................................... | 12 volts |
| Capacity ......................................................................... | 50.4 kC (14 Ah)/10HR |
| Nominal electrolyte specific gravity (SG) ....................... | 1.2800 @ 20°C (68°F) |

### Alternator
| | |
|---|---|
| Type ............................................................................... | 3-phase |
| No-load voltage .............................................................. | More than 80Vac @ 5000 rpm |
| Regulated voltage .......................................................... | 14.0 – 15.5V @ 5000 rpm |

### Starter motor
Brush length (min):
| | |
|---|---|
| T, ET, LT, EX, EZ, SD, ESD .......................................... | 9.0 mm (0.40 in) |
| SZ ................................................................................... | 6.0 mm (0.24 in) |
| Commutator undercut (min) ........................................... | 0.2 mm (0.008 in) |

### Starter relay
| | |
|---|---|
| Resistance ...................................................................... | Approx 3 – 4 ohms |

**Fuses**

| | |
|---|---|
| Head | 10A |
| Signal | 10A |
| Ignition | 10A |
| Main | 15A |
| Power source | 10A |

**Bulb wattages (all rated at 12V)**

| | |
|---|---|
| Headlamp | 60/55W |
| Parking (city) lamp | 4W |
| Tail/brake lamp | 5/21W |
| Turn signal lamp | 21W |
| Speedometer lamp | 3.4W |
| Tachometer lamp | 3.4W |
| Turn signal warning lamp | 3.4W |
| High beam warning lamp | 3.4W |
| Neutral indicator lamp | 3.4W |
| Oil pressure lamp | 3.4W |
| Fuel meter lamp: | |
|     LT | 3.4W |
|     EZ | 2.0W |
| Check panel lamp – EX | 1.4W |
| Check panel warning lamps (EZ): | |
|     Oil | 3.4W |
|     Battery | 3.4W |
|     Side stand | 3.4W |
|     Turn | 3.4W |
|     Neutral | 3.4W |
|     Headlamp | 3.4W |
|     Tail lamp | 3.4W |
|     Brake lamp | 3.4W |
|     High beam | 3.4W |
| Number (license) plate lamp | 5W |

# GS 1100 T, ET, LT, EX, EZ, ED, ESD (US)

**Battery**

| | |
|---|---|
| Type | YB 14L-A2 or SYB 14L-A2 |
| Voltage | 12 volts |
| Capacity | 50.4 kC (14 Ah)/10HR |
| Nominal electrolyte specific gravity (SG) | 1.2800 @ 20°C (68°F) |

**Alternator**

| | |
|---|---|
| Type | 3-phase |
| No-load voltage | More than 80Vac @ 5000 rpm |
| Regulated voltage | 14.0 – 15.5V @ 5000 rpm |

**Starter motor**

| | |
|---|---|
| Brush length (min) | 9.0 mm (0.40 in) |
| Commutator undercut (min) | 0.2 mm (0.008 in) |

**Starter relay**

| | |
|---|---|
| Resistance | Approx 3 – 4 ohms |

**Fuses**

| | |
|---|---|
| Head | 10A |
| Signal | 10A |
| Ignition | 10A |
| Main | 15A |
| Power source | 10A |

**Bulb wattages (all rated at 12V)**

| | |
|---|---|
| Headlamp | 60/55W |
| Tail/brake lamp | 8/23W (3/32 cp) |
| Turn signal lamp | 23W (32 cp) |
| Front turn/running lamp | 8/23W (3/32 cp) |
| Speedometer lamp | 3.4W |
| Tachometer lamp | 3.4W |
| Turn signal warning lamp | 3.4W |
| High beam warning lamp | 3.4W |
| Neutral indicator lamp | 3.4W |
| Oil pressure lamp | 3.4W |
| Fuel meter lamp – EZ, ED, ESD | 2.0W |

| | |
|---|---|
| Oil temperature lamp – EZ, ED, ESD | 2.0W |
| Check panel lamp – EX | 1.4W |
| Check panel warning lamps (EZ): | |
|     Oil | 3.4W |
|     Battery | 3.4W |
|     Side stand | 3.4W |
|     Turn | 3.4W |
|     Neutral | 3.4W |
|     Headlamp | 3.4W |
|     Tail lamp | 3.4W |
|     Brake lamp | 3.4W |
|     High beam | 3.4W |
| License plate lamp | 8W (4 cp) |

# GSX 1000 SZ, SD

### Battery
| | |
|---|---|
| Type | SYB 14L-A2 or YB 14L-A2 |
| Voltage | 12 volts |
| Capacity | 50.4 kC (14 Ah)/10HR |
| Nominal electrolyte specific gravity (SG) | 1.2800 @ 20°C (68°F) |

### Alternator
| | |
|---|---|
| Type | 3-phase |
| No-load voltage | More than 80Vac @ 5000 rpm |
| Regulated voltage | 14.0 – 15.5V @ 5000 rpm |

### Starter motor
| | |
|---|---|
| Brush length (min): | |
|     SZ | 6.0 mm (0.24 in) |
|     SD | 9.0 mm (0.40 in) |
| Commutator undercut (min) | 0.2 mm (0.008 in) |

### Starter relay
| | |
|---|---|
| Resistance | Approx 3 – 4 ohms |

### Fuses
| | |
|---|---|
| Head | 10A |
| Signal | 10A |
| Ignition | 10A |
| Main | 15A |
| Power source | 10A |

### Bulb wattages (all rated at 12V)
| | |
|---|---|
| Headlamp | 60/55W |
| Parking (city) lamp | 4W |
| Tail/brake lamp | 5/21W |
| Turn signal lamp | 21W |
| Meter lamp | 3.4W |
| Turn signal warning lamp | 3.4W |
| High beam warning lamp | 3.4W |
| Neutral indicator lamp | 3.4W |
| Oil pressure lamp | 3.4W |
| Number (license) plate lamp | 5W |

# GS 1000 SD (US)

### Battery
| | |
|---|---|
| Type | YB 14L-A2 or SYB 14L-A2 |
| Voltage | 12 volts |
| Capacity | 50.4 kC (14 Ah)/10HR |
| Nominal electrolyte specific gravity (SG) | 1.2800 @ 20°C (68°F) |

### Alternator
| | |
|---|---|
| Type | 3-phase |
| No-load voltage | More than 80Vac @ 5000 rpm |
| Regulated voltage | 14.0 – 15.5V @ 5000 rpm |

### Starter motor
| | |
|---|---|
| Brush length (min) | 9.0 mm (0.40 in) |
| Commutator undercut (min) | 0.2 mm (0.008 in) |

**Starter relay**

Resistance ................................................................................ Approx 3 – 4 ohms

**Fuses**

Head ........................................................................................ 10A
Signal ...................................................................................... 10A
Ignition ................................................................................... 10A
Main ........................................................................................ 15A
Power source .......................................................................... 10A

**Bulb wattages (all rated at 12V)**

Headlamp ................................................................................ 60/55W
Tail/brake lamp ...................................................................... 8/23W (3/32 cp)
Turn signal lamp .................................................................... 23W (32 cp)
Meter lamp ............................................................................. 3.4W
Turn signal warning lamp ..................................................... 3.4W
High beam warning lamp ....................................................... 3.4W
Neutral indicator lamp .......................................................... 3.4W
Oil pressure lamp .................................................................. 3.4W
License plate lamp ................................................................ 8W (4 cp)

## 1  General description

The electrical system is supplied by a crankshaft-mounted alternator, the rotor being retained on the tapered end of the crankshaft and the stator windings being housed inside the left-hand outer cover. The alternating current (ac) from the alternator is fed to a combined electronic regulator/rectifier unit, where it is converted to direct current (dc) and the system voltage limited to a nominal 12 volts before being passed to the battery and the electrical circuit of the machine.

A 12 volt lead-acid battery provides a store of electrical power for starting and to ensure a stable supply of power to the system, irrespective of engine speed. The various models are equipped with a diverse array of electrical devices and circuits and these are described in detail under the relevant Section headings.

## 2  Testing the electrical system: general

1   Many of the tests described in this Chapter necessitate checking a circuit or component for continuity. In many instances a simple dry battery and bulb can be connected as shown in the accompanying line drawing to force a cheap and effective tester.
2   Where it is necessary to measure a specific voltage or resistance,

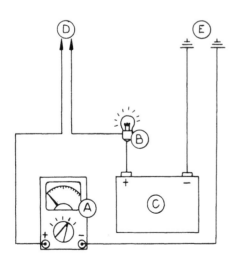

**Fig. 6.1 Simple testing arrangement for checking continuity**

A   *Multimeter*          D   *Positive probe*
B   *Bulb*                E   *Negative probe*
C   *Battery*

a multimeter will be required. An inexpensive instrument can be obtained from many electrical or electronic suppliers as well as from most good motorcycle dealers. Suzuki dealers can provide a suitable meter as Part Number 09900-25002.
3   Care must be taken with all electrical tests, but particularly where the rectifier is involved. On no account must this be shorted or subjected to high currents or it will be destroyed in a fraction of a second. For this reason **do not** use a 'megger' or any high current test instrument.
4   Where test equipment is not available, or if the owner does not feel confident about a particular test, it is strongly recommended that the work be entrusted to a Suzuki dealer or an electrical expert. Although the tests are not unduly complex, a single error could prove costly.
5   In the event of an electrical fault it is generally best to remove the seat, fuel tank and side panels prior to any testing. In the case of the headlamp/instrument panel area, removal of the fairing, where fitted will improve access considerably.
6   Finally, always check the obvious first. Most faults can be traced to loose or broken leads or connectors, or switch faults. Checking these first may avoid a lot of unnecessary work.

## 3  Wiring and connectors: examination

1   The wiring used throughout the machine is colour-coded and will be found to correspond with the colour wiring diagrams at the end of this Chapter. Where multi-pin connectors are used these are handed and thus cannot be connected wrongly.
2   Inspect the wiring for signs of breakage or damage to the outer covering which might indicate an open or short circuit. On rare occasions a wire may be trapped or crushed, leaving the outer cover intact but the inner core partially or completely broken. This can cause mysterious intermittent faults, as can poor connections or water in connector blocks. A wiring run is easily checked using the meter set on the resistance scale as a continuity tester. Note that the machine's battery should be disconnected to avoid damage to the meter.

## 4  Testing the charging output

1   Before making this test, it is important that the battery is sound and fully charged. A faulty or discharged battery will give misleading results.
2   Remove the left-hand side panel to gain access to the starter relay terminals. Set the multimeter to the 0-20 volts dc range (or a higher range where necessary) and connect the positive probe to the positive (red) starter relay terminal. Start the engine and set the lights to high beam.
3   With the engine running at 5000 rpm, touch the negative probe to earth (ground) and note the reading. If a reading of less than 14 volts or more than 15.5 volts is indicated, the regulator/rectifier unit is faulty and must be renewed.

## 5 Testing the alternator no-load performance

1 Trace the alternator wiring back to the three bullet connectors and separate them. Set the meter to a range capable of handling at least 80 volts ac (note that the meter will be damaged if a dc range is selected). Start the engine and measure the alternator output between successive pairs of output leads, a total of three tests. The relevant connections are shown in the accompanying line drawing. Note that the polarity of the probe is unimportant when measuring ac output.

2 In each case, the above test should show a reading of at least 80 volts ac at 5000 rpm. If one or more pairs of leads shows a lower or zero reading, check the stator winding resistances as described in Section 6.

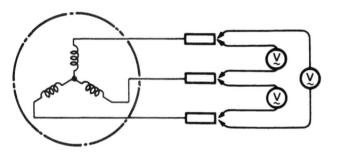

Fig. 6.2 The alternator no-load output test connections

## 6 Testing the alternator stator winding resistances

1 If the no-load test has indicated a fault in the stator, stop the engine and set the meter to the resistance range. Using the same pattern of tests described in Section 5, check for continuity between successive pairs of leads. The exact figure is unimportant, but should be approximately equal in each case. A zero reading or a reading of infinite resistance indicates a short or open circuit respectively and will require the renewal of the stator assembly.

2 Check that the insulation between each set of windings and the stator core is intact. If continuity exists between any lead and the core, a short circuit has occurred and renewal will be necessary.

6.1 If renewal is required, stator is retained by three screws to inside of casing

## 7 Testing the regulator/rectifier unit

1 The regulator/rectifier unit is mounted beneath the left-hand side panel and can be identified by its finned alloy casing. If checks have indicated that the alternator is operating correctly and charging faults persist, check the unit as described below.

2 Trace the leads back from the regulator/rectifier unit and separate them at their connectors. Set the multimeter to the ohms x1 scale, then measure the resistance between the various pairs of leads as indicated in the accompanying table. If the unit fails to produce the appropriate values, it should be renewed.

7.1 Regulator/rectifier is mounted behind the left-hand side panel

**Tester range: X1Ω**

Unit: Ω

| ⊕ probe of tester to: | | | | | |
|---|---|---|---|---|---|
| ⊖ probe of tester to: | R | W/Bl | W/R | Y | B/W |
| R | | ∞ | ∞ | ∞ | ∞ |
| W/Bl | 5 – 6 | | ∞ | ∞ | ∞ |
| W/R | 5 – 6 | ∞ | | ∞ | ∞ |
| Y | 5 – 6 | ∞ | ∞ | | ∞ |
| B/W | 35 – 45 | 5 – 6 | 5 – 6 | 5 – 6 | |

Fig. 6.3 Regulator/rectifier unit testing table

### Wire colour code

| | | | |
|---|---|---|---|
| R | Red | W/R | White and red |
| Y | Yellow | W/Bl | White and blue |
| B/W | Black and white | | |

## 8 Battery: examination and maintenance

1 The battery is housed in a tray below the seat and air filter housing. Access is very restricted, and it must be removed for inspection and maintenance. Start by opening the seat and removing the bracket which spans the air filter housing. Free the rear brake

master cylinder after releasing its single retaining bolt. Release the housing bolts and the hose clip at the front of the housing and manoeuvre it clear of the frame. Disconnect the battery leads, battery level sensor lead and the vent hose and lift the battery out of its recess.

2    The transparent plastic case of the battery permits the upper and lower levels of the electrolyte to be observed without disturbing the battery by removing the side cover. Maintenance is normally limited to keeping the electrolyte level between the prescribed upper and lower limits and making sure that the vent tube is not blocked. The lead plates and their separators are also visible through the transparent case, a further guide to the general condition of the battery. If electrolyte level drops rapidly, suspect over-charging and check the system.

3    Unless acid is spilt, as may occur if the machine falls over, the electrolyte should always be topped up with distilled water to restore the correct level. If acid is spilt onto any part of the machine, it should be neutralised with an alkali such as washing soda or baking powder and washed away with plenty of water, otherwise serious corrosion will occur. Top up with sulphuric acid of the correct specific gravity (1.260 to 1.280) only when spillage has occurred. Check that the vent pipe is well clear of the frame or any of the other cycle parts.

4    It is seldom practicable to repair a cracked battery case because the acid present in the joint will prevent the formation of an effective seal. It is always best to renew a cracked battery, especially in view of the corrosion which will be caused if the acid continues to leak.

5    If the machine is not used for a period of time, it is advisable to remove the battery and give it a 'refresher' charge every six weeks or so from a battery charger. The battery will require recharging when the specific gravity falls below 1.260 (at 20°C – 68°F). The hydrometer reading should be taken at the top of the meniscus with the hydrometer vertical. If the battery is left discharged for too long, the plates will sulphate. This is a grey deposit which will appear on the surface of the plates, and will inhibit recharging. If there is sediment on the bottom of the battery case, which touches the plates, the battery needs to be renewed. Prior to charging the battery, refer to the following Section for correct charging rate and procedure. If charging from an external source with the battery on the machine, disconnect the leads, or the rectifier will be damaged.

6    Note that when moving or charging the battery, it is essential that the following basic safety precautions are taken:

(a)    Before charging check that the battery vent is clear or, where no vent is fitted, remove the combined vent/filler caps. If this precaution is not taken the gas pressure generated during charging may be sufficient to burst the battery case, with disastrous consequences.

(b)    Never expose a battery on charge to naked flames or sparks. The gas given off by the battery is highly explosive.

(c)    If charging the battery in an enclosed area, ensure that the area is well ventilated.

(d)    Always take great care to protect yourself against accidental spillage of the sulphuric acid contained within the battery. Eyeshields should be worn at all times. If the eyes become contaminated with acid they must be flushed with fresh water immediately and examined by a doctor as soon as possible. Similar attention should be given to a spillage of acid on the skin.

Note also that although, should an emergency arise, it is possible to charge the battery at a more rapid rate than that stated in the following Section, this will shorten the life of the battery and should therefore be avoided if at all possible.

7    Occasionally, check the condition of the battery terminals to ensure that corrosion is not taking place, and that the electrical connections are tight. If corrosion has occurred, it should be cleaned away by scraping with a knife and then using emery cloth to remove the final traces. Remake the electrical connections whilst the joint is still clean, then smear the assembly with petroleum jelly (NOT grease) to prevent recurrence of the corrosion. Badly corroded connections can have a high electrical resistance and may give the impression of complete battery failure.

## 9    Starter circuit and relay: preliminary checks

1    In the event of a starter malfunction always check first that the battery is fully charged. A partly discharged battery may provide

enough power for the lighting circuit, but not the very heavy current required to crank the engine. This can be checked by switching the lights on and operating the starter button. If the starter relay clicks and the lights dim or extinguish, a discharged battery is indicated.

2    If the relay has failed, it will not emit the characteristic click as the contacts close, unless the battery is completely discharged. To check its operation, remove the right-hand side panel. The relay is easily identified by its heavy duty starter motor lead, which should be disconnected from the relay terminal. Connect a 12 volt bulb between the starter lead terminal and a convenient earth (ground) point. Switch on the ignition and operate the starter button. If the bulb lights, the fault lies with the starter motor, whilst if it fails to light, the relay is faulty and should be renewed.

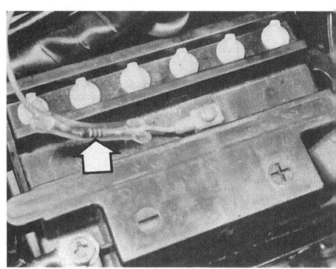

8.1 When removing sensor lead, take care not to damage the resistor (arrowed)

## 10    Starter motor: removal and overhaul

1    Remove the starter motor cover and disconnect the starter cable, having first checked that the ignition is switched off. Remove the two bolts which retain the starter motor, pull the motor to the right to disengage it, then lift it clear of its recess.

2    Release the two long screws which retain the motor end covers and lift away the brush assembly. Measure the length of the brushes, renewing them if they have worn below the minimum length.

3    Inspect the commutator segments on which the brushes bear. If badly burnt or scored, the only alternative to renewal is to have the commutator skimmed and the segments re-cut by a vehicle electrical specialist. The minimum undercut is 0.2 mm (0.008 in). A dirty commutator can be restored by carefully cleaning it with fine (400 grit) emery paper.

4    The armature windings may be checked by testing between each commutator segment and the armature core with a multimeter set on the resistance scale. If any segment is shorted to the core, the armature must be renewed. Check that there is conductivity between each pair of segments, any open circuit indicating the need for renewal.

5    When assembling the motor, ensure that the brushes are fitted in their respective holders. Fit the brush plate over the commutator, and check that it aligns properly when the end cover is fitted. The brush plate and the end cover have locating notches to ensure alignment. Apply a thread locking compound to the two retaining screws and tighten them securely.

10.2a Remove the two retaining screws ...

10.2b ... and lift away the cover and brush plate

10.2c Unhook spring and withdraw the brush ...

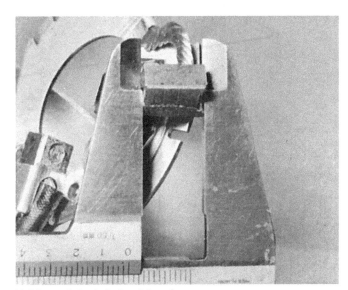

10.2d ... and check the brush length

10.3 Check the commutator segments for wear or burning

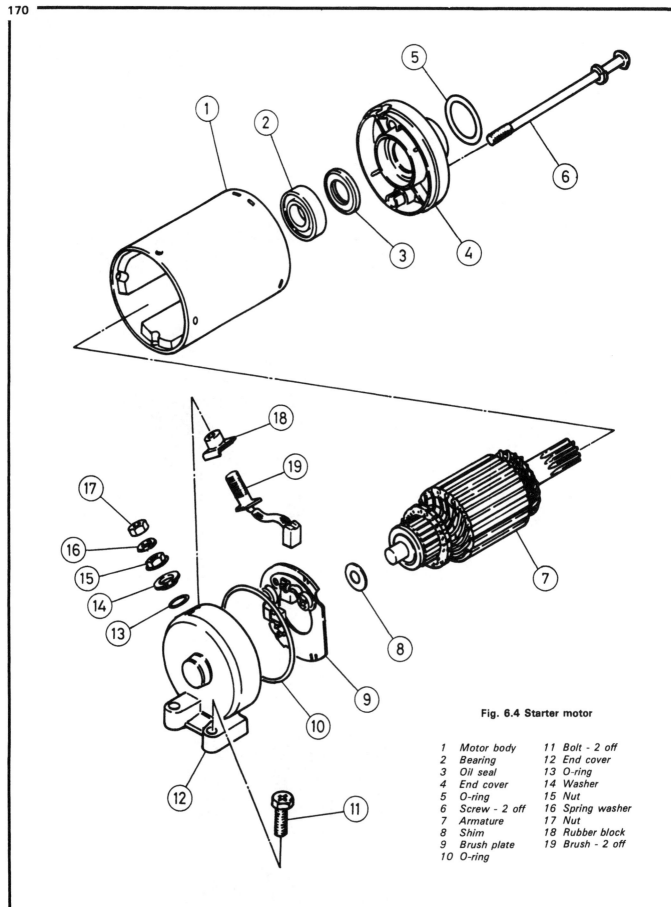

**Fig. 6.4 Starter motor**

| | |
|---|---|
| 1 | Motor body |
| 2 | Bearing |
| 3 | Oil seal |
| 4 | End cover |
| 5 | O-ring |
| 6 | Screw - 2 off |
| 7 | Armature |
| 8 | Shim |
| 9 | Brush plate |
| 10 | O-ring |
| 11 | Bolt - 2 off |
| 12 | End cover |
| 13 | O-ring |
| 14 | Washer |
| 15 | Nut |
| 16 | Spring washer |
| 17 | Nut |
| 18 | Rubber block |
| 19 | Brush - 2 off |

## 11 Headlamp: adjustment

### T, ET and EX models

1 The rectangular headlamp unit is housed in a plastic moulding which is secured to the steering head and which forms a backing for the instrument panel. The housing is fixed in relation to the forks, and to provide horizontal and vertical alignment two separate adjustment screws are located in the edge of the rim: the lower screw controlling vertical alignment and the upper screw controlling horizontal alignment.

### LT and EZ models

2 A round headlamp unit is fitted to brackets on each fork leg. Vertical alignment is adjusted by slackening the two mounting bolts and turning the assembly to the desired position. Horizontal alignment is controlled by way of a screw in the rim edge.

### ESD models

3 A rectangular unit is housed inside the fairing. Adjustment is controlled by two knobs which can be accessed from the inside of the fairing.

### Katana models

4 The Katana models employ a rectangular unit housed in the fairing. Vertical adjustment can be set after the two mounting bolts have been slackened. A knob located inside the fairing controls horizontal alignment, permitting adjustment whilst riding the machine.

### All models

5 The horizontal alignment should be set so that the beam shines straight ahead on main beam. The correct position is best found on an experimental basis during a night-time ride. Vertical alignment must comply with local laws, and these may vary from state to state in the US. In the UK, lighting regulations require that the headlamp is adjusted so that the light will not dazzle a person standing at a distance of not less than 25 feet from the machine and on the same plane, and whose eye level is not less than 3 ft 6 inches from the ground. Adjustment should be made with the rider seated normally, plus any regular passenger or luggage.

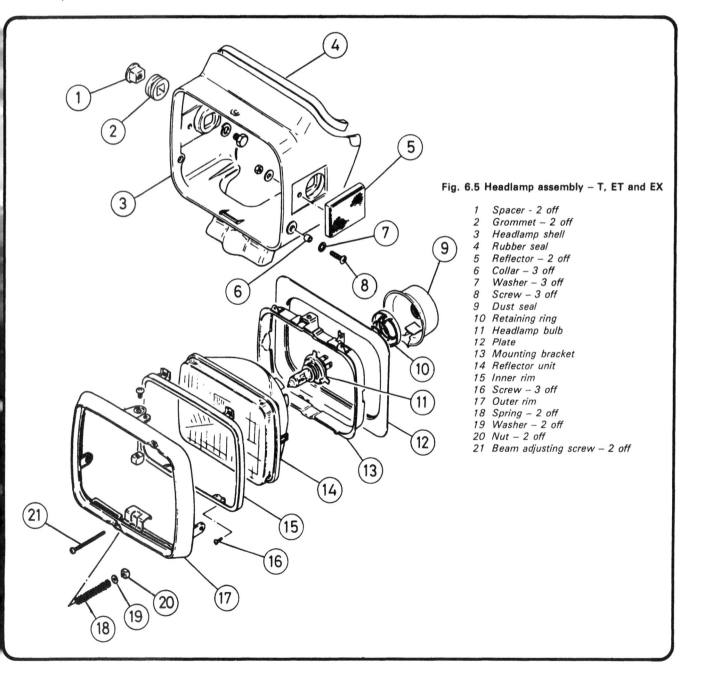

**Fig. 6.5 Headlamp assembly – T, ET and EX**

1 Spacer - 2 off
2 Grommet – 2 off
3 Headlamp shell
4 Rubber seal
5 Reflector – 2 off
6 Collar – 3 off
7 Washer – 3 off
8 Screw – 3 off
9 Dust seal
10 Retaining ring
11 Headlamp bulb
12 Plate
13 Mounting bracket
14 Reflector unit
15 Inner rim
16 Screw – 3 off
17 Outer rim
18 Spring – 2 off
19 Washer – 2 off
20 Nut – 2 off
21 Beam adjusting screw – 2 off

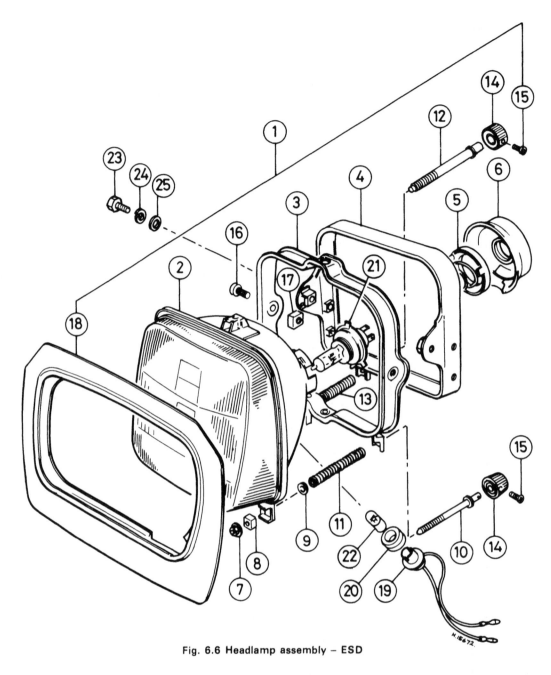

**Fig. 6.6 Headlamp assembly – ESD**

| | | | |
|---|---|---|---|
| 1 | Headlamp assembly | 8 | Nut |
| 2 | Reflector unit | 9 | Washer |
| 3 | Mounting bracket | 10 | Horizontal adjustment screw |
| 4 | Housing | 11 | Spring |
| 5 | Retaining ring | 12 | Vertical adjustment screw |
| 6 | Dust cover | 13 | Spring |
| 7 | Nut | | |

| | | | |
|---|---|---|---|
| 14 | Adjusting knob – 2 off | 20 | Grommet |
| 15 | Screw – 2 off | 21 | Headlamp bulb |
| 16 | Screw – 2 off | 22 | Parking lamp bulb |
| 17 | Nut – 2 off | 23 | Bolt – 2 off |
| 18 | Rim | 24 | Spring washer – 2 off |
| 19 | Parking lamp bulbholder | 25 | Washer – 2 off |

## 12 Headlamp: bulb renewal

1   In the case of the unfaired machines, remove the screws which pass through the front edge of the headlamp housing and lift the headlamp assembly clear of its recess. Pull off the rubber dust seal from the back of the reflector unit and unplug the wiring connector. On machines equipped with a front parking (city) lamp, pull the bulb holder out of its rubber collar.

2   On the machines equipped with a fairing it will be necessary to release the two bolts which retain the headlamp support frame in the fairing recess. Once the assembly has been disengaged, the wiring connector and parking lamp can be disconnected as described above.

3   The headlamp bulb is retained by a ring which can be released by twisting it anti-clockwise. Lift away the ring and spring, then lift out the bulb, holding it by the metal base only. On some models, a modified retainer consisting of a wire spring is used. This can be freed by depressing the spring end and disengaging it from the collar. On no account should the bulb envelope be touched. If this is done accidentally, wipe off any finger marks carefully, using a paper tissue moistened with methylated spirit or alcohol. Where a parking lamp is fitted, its bulb can be removed from the bulbholder by turning it anti-clockwise.

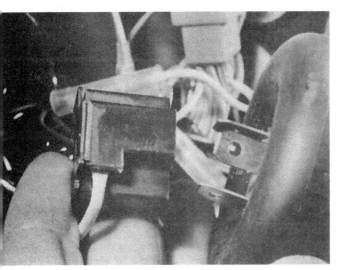

12.1 Pull off the wiring connector and rubber dust seal

12.3a Where spring retainer is fitted, depress the end and disengage it from the collar

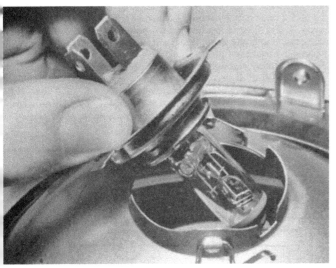

12.3b The bulb should be handled by the metal cap only

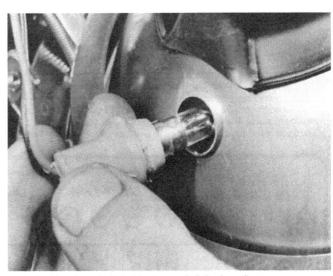

12.3c Parking lamp bulb holder is a push fit in the reflector

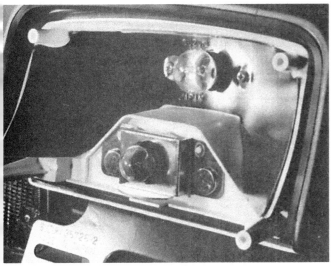

13.1 Rear lamp contains separate stop/tail and licence plate lamp

## 13 Tail/brake lamp: bulb renewal

1   The design of the tail lamp assembly varies according to the model, but in each case access to bulbs is gained after removing the lens, this being retained by two or more screws. The lamp contains a twin filament stop/tail bulb and a separate number (license) plate bulb. In each case the bulbs are of the bayonet fitting type and are removed by pushing them inwards slightly and twisting anti-clockwise. The twin filament types have offset pins to ensure that they are fitted correctly. The metal parts of the bulbholder should be checked for corrosion. It is important to replace bulbs with those of the same wattage.

## 14 Turn signal lamps: bulb renewal

1   Release the screws which retain the lens and lift it away. Remove the bulb by pushing it inwards slightly and twisting anti-clockwise. In certain areas the front turn signal lamps double as running lamps, in which case twin filament bulbs are fitted. These have offset pins to ensure that they are fitted correctly. Check that the rubber seal is sound before refitting the lens.

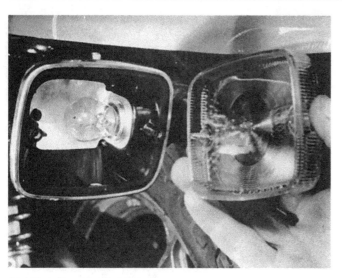

14.1 Turn signal bulbs are a bayonet fitting

lamp bulb. If a fault occurs, the appropriate warning lamp is switched on in the panel. When the ignition switch is turned on the warning lamps in the check panel are illuminated. When the engine is started the warning lamps should extinguish as soon as the oil pressure lamp goes out. Any remaining warning lamps indicate a fault in the relevant circuit. The check light display circuit is shown in the accompanying diagram.

2  The check panel is dependent on its supply from the oil pressure switch. If this fails, the check panel will not operate normally. If this condition is suspected, remove the oil pressure switch lead and connect it to earth (ground) with the ignition switch on. If the check panel now comes on, the oil pressure switch should be tested by temporarily disconnecting its lead and connecting it to earth (ground). If when the ignition switch is turned on the check panel operates normally, the oil pressure switch is faulty and should be renewed.

3  If the check panel is still malfunctioning, check the associated wiring for continuity, using a multimeter set on the resistance (ohms) scale. Refer to the circuit diagram and the main wiring diagram for details. If the wiring, fuses and warning lamp bulbs are in good condition, check that the headlamp and stop/tail lamp bulbs are working. Note that if the fuses protecting the check panel are renewed use only Suzuki genuine parts to ensure full protection of the check panel circuit. If this fails to resolve the problem, check the rear brake fluid sensor and battery electrolyte sensor as described below.

4  Disconnect the sensor leads from the rear brake reservoir and connect to them the probe leads of the multimeter. Set the meter on its resistance range and note the reading. If the sensor is working correctly, no continuity should exist when the fluid is at the upper level mark. If the float is pushed down, continuity should be shown to correspond with the lower level mark. If the sensor is faulty, renew the reservoir assembly.

5  The battery electrolyte sensor can be checked by measuring the voltage between it and the battery positive terminal. If the battery is fully charged, a serviceable sensor will indicate 1.5 – 2.0 volts. If this is not the case, renew the sensor. If the above tests fail to locate a fault, the check panel must be considered faulty and should be renewed.

## 15  Check panel assembly: description and testing – T, ET, EX

1  A comprehensive check light display is incorporated in the instrument panel. It monitors the status of the brake fluid level, battery electrolyte level, headlamp bulb high and low beam and the tail/brake

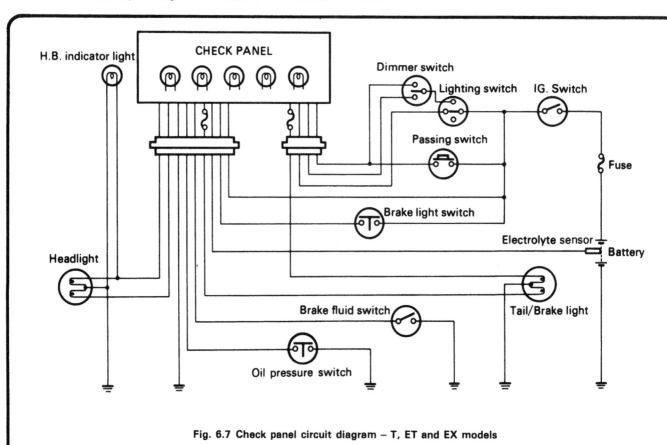

Fig. 6.7 Check panel circuit diagram – T, ET and EX models

## 16 Instrument panel: examination – T, ET and EX models

1 The instrument panel assembly houses the speedometer, tachometer, fuel gauge, check panel and the various warning and instrument illumination lamps. Access to the panel can be gained after disconnecting the headlamp unit, releasing the instrument panel wiring connectors and instrument drive cables and removing the two mounting nuts. Once removed, withdraw the bottom cover, then release the screws which secure the top of the panel to the base. Note that to change bulbs, other than those of the check panel, it will be necessary to remove the bottom cover only.

2 The accompanying circuit diagram shows the instrument panel wiring connections in detail, and should be used in conjunction with the main wiring diagram to locate any fault in the panel. A multimeter can be used as a continuity tester to trace a suspected break in the wiring or a blown bulb, without having to remove the instrument panel from the machine. Access to the connectors is via the headlamp housing.

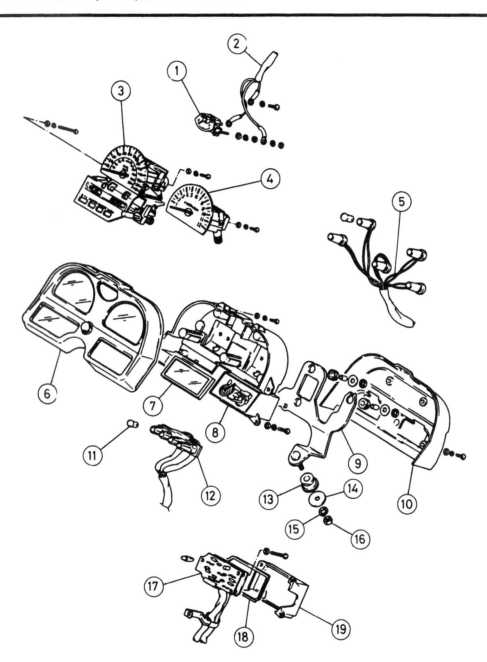

Fig. 6.8 Instrument panel assembly – T, ET and EX

| | | | |
|---|---|---|---|
| 1 Fuel gauge | 6 Top cover | 11 Bulb – 4 off | 16 Nut – 2 off |
| 2 Fuel gauge wiring | 7 Check panel lens | 12 Bulbholder wiring | 17 Check panel |
| 3 Speedometer | 8 Instrument panel housing | 13 Grommet – 2 off | 18 Sealing ring |
| 4 Tachometer | 9 Mounting bracket | 14 Washer – 2 off | 19 Check panel housing |
| 5 Instrument panel wiring | 10 Bottom cover | 15 Spring washer – 2 off | |

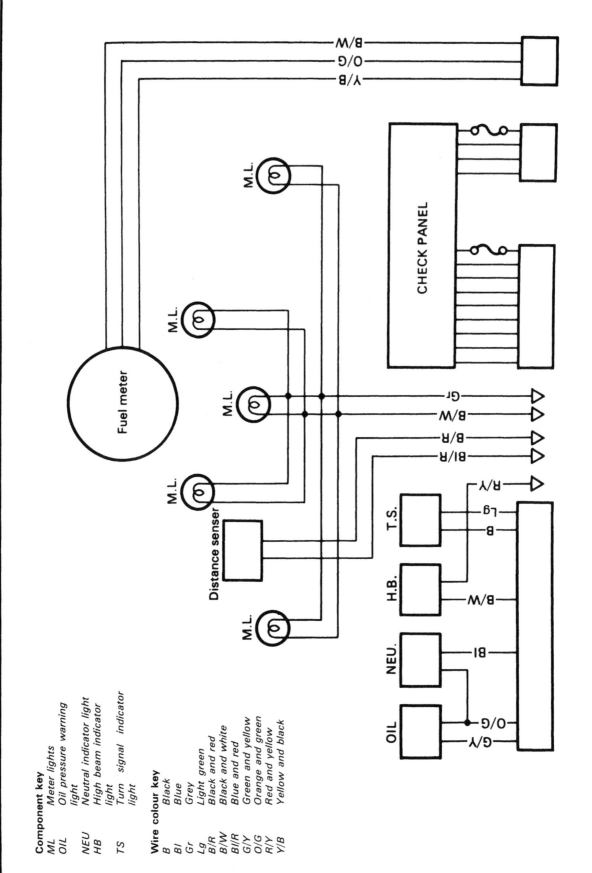

**Component key**

| | |
|---|---|
| ML | Meter lights |
| OIL | Oil pressure warning light |
| NEU | Neutral indicator light |
| HB | High beam indicator light |
| TS | Turn signal indicator light |

**Wire colour key**

| | |
|---|---|
| B | Black |
| Bl | Blue |
| Gr | Grey |
| Lg | Light green |
| B/R | Black and red |
| B/W | Black and white |
| Bl/R | Blue and red |
| G/Y | Green and yellow |
| O/G | Orange and green |
| R/Y | Red and yellow |
| Y/B | Yellow and black |

Fig. 6.9 Instrument panel circuit diagram – T, ET and EX models

## 17 Instrument panel: examination – LT model

1   The instrument panel assembly comprises separate speedometer and tachometer heads together with a central panel housing the fuel gauge, gear position indicator and the various warning lamps. In the event of a suspected electrical fault, the affected circuit can be checked using a multimeter set on its resistance range as a continuity tester. The instrument panel wiring connectors can be accessed via the headlamp shell. Refer to the accompanying circuit diagram for details, using it in conjunction with the main wiring diagram.

2   To renew the illumination bulbs in the speedometer and tachometer heads, release the domed nuts which secure the instrument to its mounting bracket, pulling it clear to reveal the push-in bulb holder. Access to the fuel gauge meter, gear position indicator bulbs and the warning lamp bulbs requires the removal of the back of the central panel, this being secured by four screws.

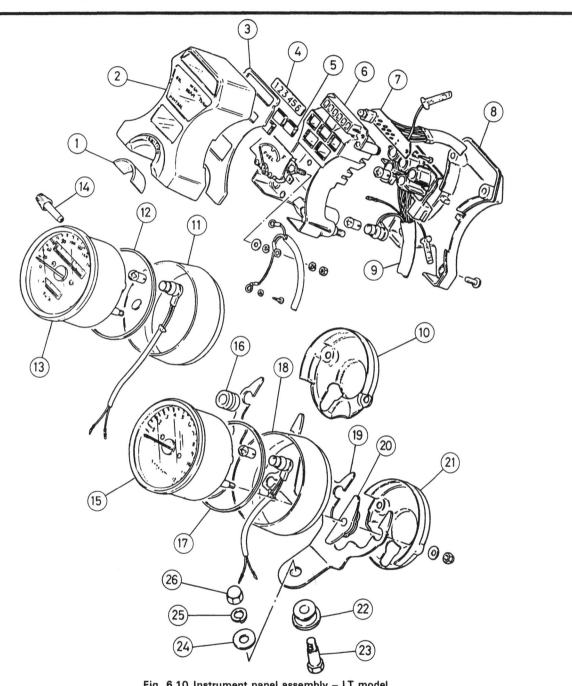

Fig. 6.10 Instrument panel assembly – LT model

| | | | |
|---|---|---|---|
| 1 Damping rubber | 8 Bottom cover | 15 Tachometer | 21 Tachometer lower cover |
| 2 Top cover | 9 Instrument panel wiring | 16 Grommet | 22 Grommet |
| 3 Gasket | 10 Speedometer lower cover | 17 Sealing ring | 23 Bolt – 2 off |
| 4 Gear position indicator | 11 Speedometer housing | 18 Tachometer housing | 24 Washer – 2 off |
| 5 Warning lamps | 12 Sealing ring | 19 Mounting bracket | 25 Spring washer – 2 off |
| 6 Instrument panel housing | 13 Speedometer | 20 Damping rubber | 26 Nut – 2 off |
| 7 Gear position indicator wiring | 14 Trip meter knob | | |

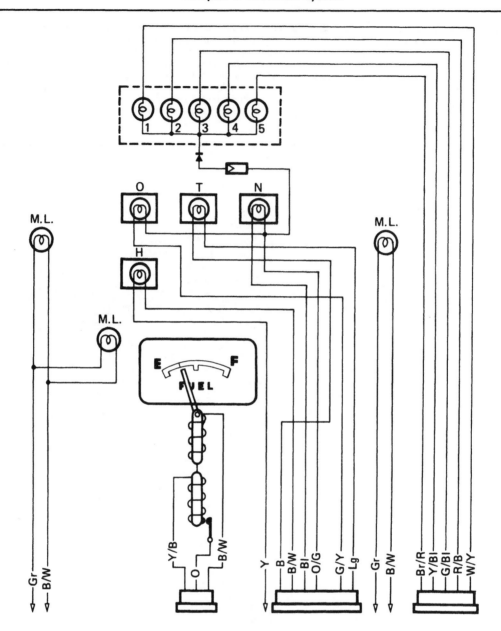

**Fig. 6.11 Instrument panel circuit diagram – LT model**

## Component key

| | | | |
|---|---|---|---|
| 1-5 | Gear position indicator | N | Neutral indicator light |
| O | Oil pressure warning light | H | High beam indicator light |
| T | Turn signal indicator light | ML | Meter lights |

## Wire colour key

| | | | | | | | |
|---|---|---|---|---|---|---|---|
| B | Black | Lg | Light green | O/G | Orange and green | G/Bl | Green and blue |
| Y | Yellow | Gr | Grey | Br/R | Brown and red | R/B | Red and black |
| Bl | Blue | B/W | Black and white | Y/Bl | Yellow and blue | W/Y | White and yellow |
| O | Orange | Y/B | Yellow and black | | | | |

## 18 Instrument panel: examination – EZ and ED models

1   The instrument panel comprises a moulded plastic housing in which are mounted the speedometer and tachometer heads. Flanking these are the smaller fuel and oil temperature gauges, the warning lamps and the LCD gear indicator filling the space between them.

2   As with the arrangements described in earlier Sections, the panel wiring and bulbs can be checked for continuity at the connectors inside the headlamp shell, using a multimeter set on the resistance range and referring to the wiring diagram for details of the wiring colours and connections.

3   If bulb renewal is necessary, disconnect the instrument driv cables then remove the screws which secure the back cover of the panel assembly. The moulded rubber bulb holders are a push-fit in the panel. If further dismantling is required, the entire panel can be removed after disconnecting the wiring and releasing the pane mounting nuts. The relative positions of the panel components is shown in the accompanying line drawing.

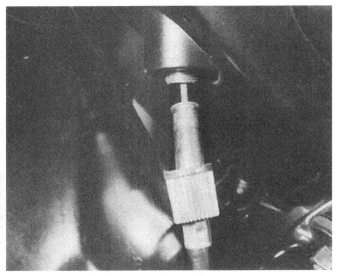

18.3a Disconnect the instrument drive cables ...

18.3b ... and remove the back of the panel to gain access to the bulb holders

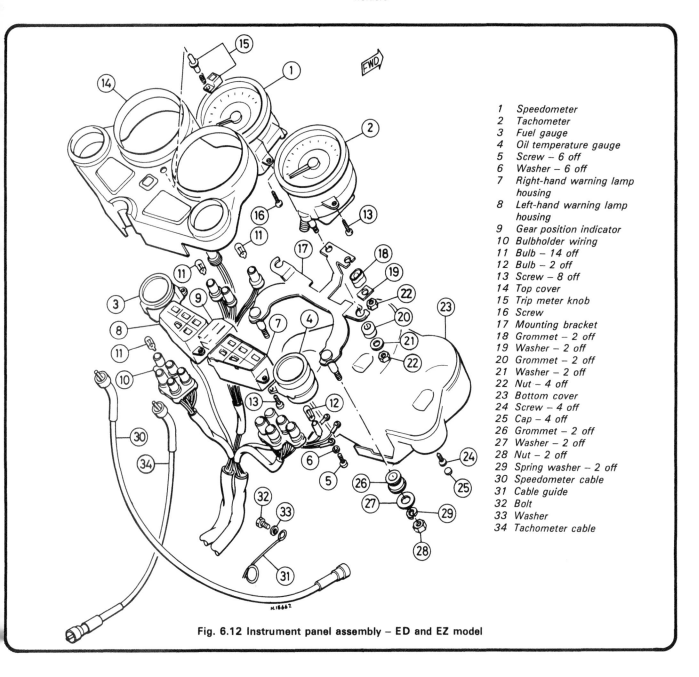

1 Speedometer
2 Tachometer
3 Fuel gauge
4 Oil temperature gauge
5 Screw – 6 off
6 Washer – 6 off
7 Right-hand warning lamp housing
8 Left-hand warning lamp housing
9 Gear position indicator
10 Bulbholder wiring
11 Bulb – 14 off
12 Bulb – 2 off
13 Screw – 8 off
14 Top cover
15 Trip meter knob
16 Screw
17 Mounting bracket
18 Grommet – 2 off
19 Washer – 2 off
20 Grommet – 2 off
21 Washer – 2 off
22 Nut – 4 off
23 Bottom cover
24 Screw – 4 off
25 Cap – 4 off
26 Grommet – 2 off
27 Washer – 2 off
28 Nut – 2 off
29 Spring washer – 2 off
30 Speedometer cable
31 Cable guide
32 Bolt
33 Washer
34 Tachometer cable

Fig. 6.12 Instrument panel assembly – ED and EZ model

### 19 Instrument panel: examination – ESD model

1 The ESD model employs yet another type of instrument panel assembly in which the speedometer and tachometer heads are combined into a single unit, to form the main face of the panel. The separate fuel and oil temperature gauges are let into the main panel as are the warning lamps. The entire assembly is housed in a fla enclosure and shares a common lens. The accompanying line drawing shows the relative positions of the panel components.

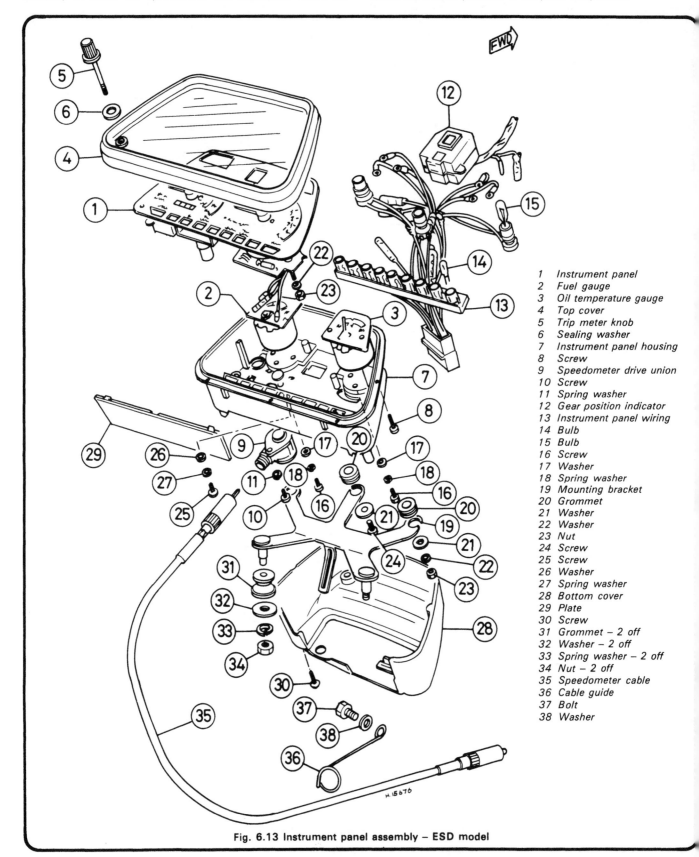

1 Instrument panel
2 Fuel gauge
3 Oil temperature gauge
4 Top cover
5 Trip meter knob
6 Sealing washer
7 Instrument panel housing
8 Screw
9 Speedometer drive union
10 Screw
11 Spring washer
12 Gear position indicator
13 Instrument panel wiring
14 Bulb
15 Bulb
16 Screw
17 Washer
18 Spring washer
19 Mounting bracket
20 Grommet
21 Washer
22 Washer
23 Nut
24 Screw
25 Screw
26 Washer
27 Spring washer
28 Bottom cover
29 Plate
30 Screw
31 Grommet – 2 off
32 Washer – 2 off
33 Spring washer – 2 off
34 Nut – 2 off
35 Speedometer cable
36 Cable guide
37 Bolt
38 Washer

Fig. 6.13 Instrument panel assembly – ESD model

## 20 Instrument panel: examination – 1000 and 1100 Katana models

1    The Katana range shares a common instrument panel assembly in which the speedometer and tachometer have been combined in a single compact housing together with the warning lamps. Access to the bulbs can be gained after removing the rear of the panel, this being retained by three screws. The tachometer on these models is electronic, using the ignition pulses as a trigger reference. Like the other models, the panel is retained by a pressed steel mounting bracket secured by rubber-bushed bolts, and may be detached after these and the speedometer cable have been released.

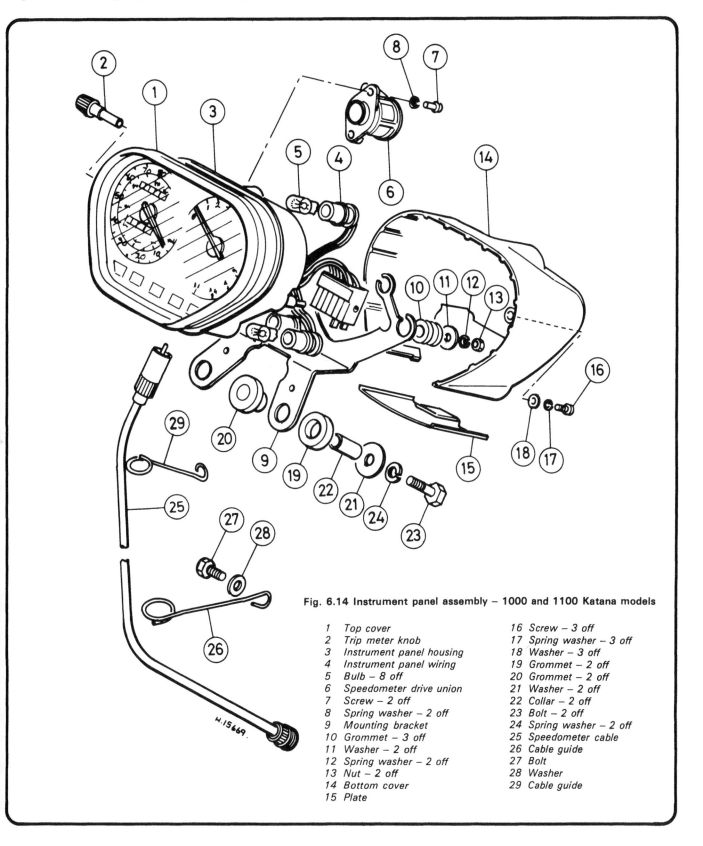

**Fig. 6.14 Instrument panel assembly – 1000 and 1100 Katana models**

| | | | |
|---|---|---|---|
| 1 | Top cover | 16 | Screw – 3 off |
| 2 | Trip meter knob | 17 | Spring washer – 3 off |
| 3 | Instrument panel housing | 18 | Washer – 3 off |
| 4 | Instrument panel wiring | 19 | Grommet – 2 off |
| 5 | Bulb – 8 off | 20 | Grommet – 2 off |
| 6 | Speedometer drive union | 21 | Washer – 2 off |
| 7 | Screw – 2 off | 22 | Collar – 2 off |
| 8 | Spring washer – 2 off | 23 | Bolt – 2 off |
| 9 | Mounting bracket | 24 | Spring washer – 2 off |
| 10 | Grommet – 3 off | 25 | Speedometer cable |
| 11 | Washer – 2 off | 26 | Cable guide |
| 12 | Spring washer – 2 off | 27 | Bolt |
| 13 | Nut – 2 off | 28 | Washer |
| 14 | Bottom cover | 29 | Cable guide |
| 15 | Plate | | |

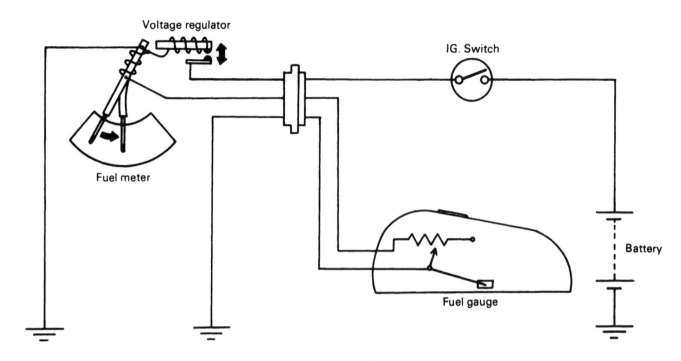

Fig. 6.15 Fuel gauge circuit – T, ET and EX models

### 21 Fuel gauge circuit: testing – all models, where applicable

1   If the fuel gauge appears to be faulty, the source of the fault is easily traced. Open the seat and locate the fuel gauge sender leads. Separate the connectors, then join the harness side of the two leads with a length of wire. This eliminates the sender unit from the circuit, and if the ignition is now switched on, the fuel gauge should indicate full, indicating that the fault must lie in the sender unit. If the gauge does not respond, the fault lies in the instrument itself or its supply from the ignition switch. This can be checked using the accompanying circuit diagram in conjunction with the main wiring diagram.

2   Note that on later models the fuel gauge meter has a heavily-damped movement designed to leave the meter needle at its read-off value when the ignition is switched off. When checking operation, allow a few minutes for the gauge to respond.

3   If the fuel gauge sender is suspect, remove the fuel tank as described in Chapter 2. With the tank inverted on soft rag to protect the paintwork, remove the sender holding bolts and manoeuvre it out of the tank, taking care not to damage or bend the float. Check the sender resistances at the full, half and empty positions. In the case of the T, ET and EX models, the appropriate float positions are indicated for the 19 and 24 litre tanks in the accompanying line drawing. In the case of the remaining models, no equivalent details are available, but it is still possible to check the full and empty positions. If the resistances are not as shown below, the sender unit must be renewed.

**Fuel gauge sender unit resistances – ohms (approx)**

| Model | Full | Half | Empty |
|---|---|---|---|
| T, ET, EX | 1-5 | 25-40 | 100-120 |
| LT | 7 | 32.5 | 95 |
| EZ | 2-7 | 25-40 | 90-110 |
| ED, ESD | | No information | |
| All Katana models | | Not applicable | |

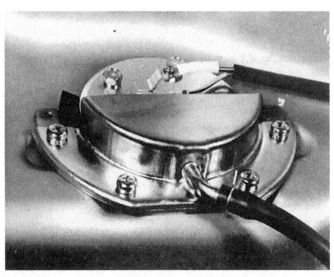

21.3a Remove the drip shield by removing the four screws ...

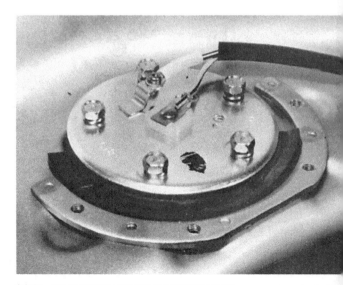

21.3b ... then remove the sender retaining bolts ...

21.3c ... and manoeuvre the sender out of the tank

The relation between the needle of fuel meter and fuel amount is as follows:

### 19L tank

| | | |
|---|---|---|
| F | Approx.: | 16L (16.9/14.1 US/Imp qt) |
| H | Approx.: | 9.5L (10.0/8.4 US/Imp qt) |
| E | Approx.: | 3L (3.2/2.6 US/Imp qt) |

### 24L tank

| | | |
|---|---|---|
| F | Approx.: | 21L (22.2/18.5 US/Imp qt) |
| H | Approx.: | 13L (13.7/11.4 US/Imp qt) |
| E | Approx.: | 3L (3.2/2.6 US/Imp qt) |

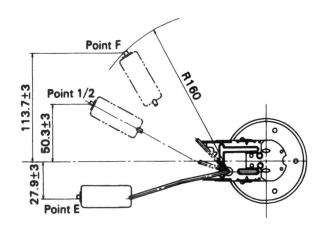

**19 LITER TANK**

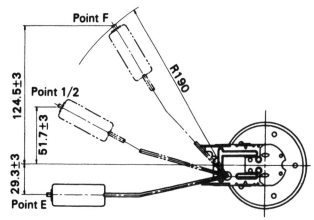

**24 LITER TANK**

Fig. 6.16 Fuel gauge sender float positions – T, ET and EX models

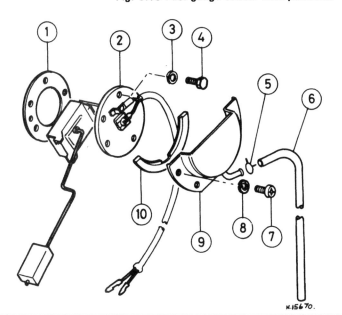

Fig. 6.17 Fuel gauge sender unit

1  Seal
2  Sender unit
3  Spring washer
4  Bolt
5  Hose clip
6  Overflow pipe
7  Screw
8  Spring washer
9  Shield
10  Seal

K.15670.

## 22 Turn signal self-cancelling circuit: general description and checks

1    The turn signals are equipped with a self-cancelling circuit based on a timer arrangement. The timer measures the degree of discharge of a capacitor and operates only when the machine is travelling above 15 kph (24 mph). It follows that if the machine is travelling at a constant speed above this minimum, the system will reset after about 10 seconds. If the machine slows to 12 kph (19.5 mph) or stops at a junction, the timing function is suspended until it again exceeds the minimum. The operation of the self-cancelling circuit is shown in the accompanying flow chart. Details of the circuit are shown in the accompanying line drawing. Note that these relate to the T, ET and EX models. In the absence of detailed information on the remaining models, it has been assumed that the same circuit is used throughout the range.

2    In the event of a fault in the turn signal circuit, ensure that the battery is fully charged, the bulbs are working and of the correct wattage and that wiring and switch contacts are sound. If this fails to resolve the fault, disconnect the control unit. If the turn signals now operate, the control unit is defective and should be renewed. If the fault persists, check the turn signal relay as described later in this Chapter.

22.2 Turn signal control unit location

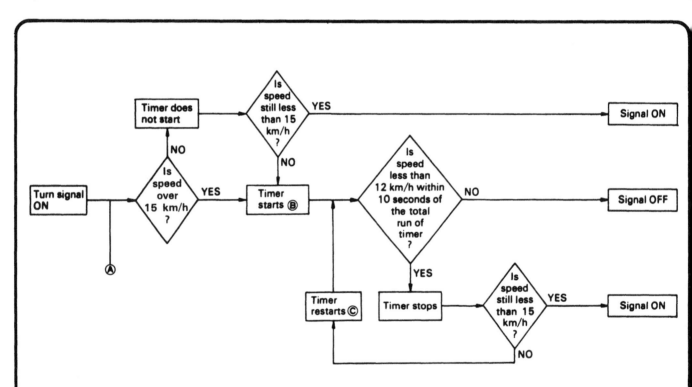

**Fig. 6.18 Turn signal self-cancelling circuit fault diagnosis flow chart**

**A**    *If after the turn signals have been switched on the switch is reset manually, the timing circuit is reset. If turned back on, the timing sequence will start from this point.*

**B**    *At this point in the timing sequence the capacitor is fully charged.*

**C**    *The capacitor is partly discharged at this point. If the timing sequence stops whilst the machine has slowed or stopped, it will resume without recharging the capacitor unless reset manually.*

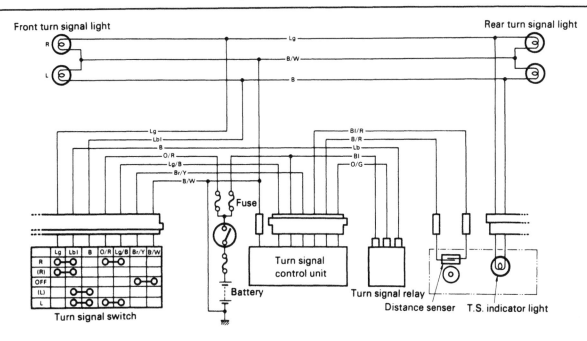

**Fig. 6.19 Turn signal self cancelling circuit diagram**

### Wire colour key

| | | | |
|---|---|---|---|
| Lg | Light green | Lbl | Light blue |
| B | Black | O/R | Orange and red |
| B/W | Black and white | Lg/B | Light green and black |

| | | | |
|---|---|---|---|
| Br/Y | Brown and yellow | Bl | Blue |
| Bl/R | Blue and red | O/G | Orange and green |
| B/R | Black and red | | |

## 23 Turn signal relay: examination and renewal

1 The turn signal relay is retained in a resilient mounting between the frame top tubes, below the fuel tank. If the unit is functioning normally, a series of audible clicks will be heard as the lamps flash. It should be noted that the most likely cause of problems with the turn signal system is a failed bulb or a break in the wiring. In such cases, there is usually one click, after which the remaining lamp on that side of the machine will flash rapidly and weakly. It follows that the bulbs and wiring should be checked before turning attention to the relay.

2 If the system is inoperative on both sides, check the supply to the relay, and check that the self-cancelling unit is working as described in Section 22. If a faulty relay is indicated, it is best checked by substitution. Note that if only one side of the system operates, the relay is serviceable and the fault must lie in the lamps, switch and wiring on the faulty side.

23.1 Turn signal relay location

## 24 Warning lamp switches: testing

1 Each machine is fitted with various switches which control the warning and indicator lamps in the instrument panel. In the event of a failure not attributable to the bulb or wiring, check the relevant switch using a multimeter or a continuity tester. If the switch fails to show an open circuit in one direction and a closed circuit in the other, it should be renewed. Further details on the various switches are given below.

### Neutral switch – T, ET and EX models
2 Remove the engine sprocket cover to gain access to the switch. Trace and disconnect the switch lead and check for continuity between it and the crankcase in the various gear positions. If continuity is not shown in neutral, remove the switch and check the contacts. If cleaning does not resolve the problem, renew the switch.

### Gear position indicator switch
3 Trace the wiring back from the gear position switch and separate the connector. Refer to the wiring diagram and check each lead in turn whilst selecting the relevant gear. In each case, the lead should be earthed when its gear is selected and isolated in all other positions. If the switch is operating normally, check and renew any blown bulbs. If the entire system is inoperative, check that power is being supplied to the display panel, noting that this is taken from the neutral lamp by way of a diode which prevents back feeding.

### Oil pressure switch
4 Disconnect the oil pressure switch lead and connect the meter probes between its terminal and earth (ground). With the engine off, continuity should be indicated. If the engine is now started, the meter should indicate isolation as the switch responds to rising pressure. If the switch does not operate as described, it must be renewed. If the switch operates normally, check the warning lamp bulb and the associated wiring. Note that in the case of machines fitted with a check panel, its operation will be affected by a fault in the oil pressure circuit.

**Side stand switch**

5    Trace and disconnect the leads from the side stand switch and test for continuity when the stand is down. If the stand is now retracted, the meter should indicate isolation. If required, the switch can be adjusted by altering its position, using the adjuster nuts on the body.

## 25 Handlebar switches: maintenance and testing

1    Generally speaking, the switch clusters will give little trouble despite their necessarily complicated construction. The switches take the form of two halves which are clamped around the handlebar ends, the right-hand unit doubling as the housing for the throttle twistgrip. The assemblies are in a somewhat exposed position and occasionally the ingress of water may cause short circuits. This can be resolved by separating the switch halves and spraying the switches with a water dispersal fluid such as WD 40. To prevent recurrent problems, pack the switch halves with silicone grease.

2    In the event of more serious malfunctions, trace the switch wiring back to their connectors and check for continuity, using a multimeter set on the resistance range. The switch contacts and wiring colours are shown in the wiring diagrams at the end of this Chapter. It is difficult to repair a faulty switch, but worth trying before ordering a new assembly. Note that it is well worth checking the local breakers (wreckers) for a used switch cluster before resorting to a new unit.

## 26 Ignition switch: removal and replacement

1    The combined ignition and lighting master switch is bolted to the underside of the upper yoke and may be removed after the instrument panel has been detached to gain access to the retaining bolts. Trace back the switch wiring to the connector inside the headlamp shell and separate it before lifting the switch away.

2    It is not possible to repair or dismantle a defective switch because of its sealed construction; a new unit must be fitted.

## 27 Brake lamp switches: location and maintenance

1    The front brake lamp switch takes the form of a small sealed unit retained by two screws to the underside of the right-hand handlebar switch assembly In the event of a fault, try soaking the switch with WD 40 or similar. If this fails to resolve the problem, the switch must

be renewed. A certain amount of adjustment is possible if the retaining screws are slackened and the switch body moved to the required position.

2    The rear brake switch is located next to the rear master cylinder. The switch is of the plunger type and is operated by a spring from the brake arm. The switch is adjusted by moving the body in relation to the mounting bracket. Slacken the switch body locknut and turn the nuts in the required direction until the desired setting is achieved, holding the switch body to prevent it from turning.

3    Both switches should be adjusted so that the brake lamp comes on just before the brake begins to operate. If either switch malfunctions, it must be renewed.

## 28 Fuses: location

1    The fuses are housed in a moulded plastic box behind the left-hand side panel, the function and rating of each fuse being clearly marked. On the earlier models, conventional cylindrical fuses are used, whilst later machines are equipped with blade type fuses. In each case, spare fuses are provided in the fuse box, and these should always be replaced if used.

2    Before renewing a blown fuse, check for any obvious fault or short circuit in the associated wiring, particularly if any one fuse has failed more than once.

3    The fuse box incorporates an accessory terminal. Note that the terminal has a 10 amp fuse, and any accessory connected to it must be within this rating. In the case of the Katana models, two switches are incorporated in the left-hand side panel for the control of any electrical accessory. These can be wired to the accessory terminal as shown in the accompanying diagram. Note that the accessory load must not exceed 60W with the headlamp off or 12W with it on.

## 29 Horn(s): location and examination

1    Depending on the model, a single or twin horn arrangement is fitted and is retained on a flexible steel bracket to the front of the frame. If the horn fails or operates feebly, it can be adjusted using the small screw provided after the locknut has been slackened. The screw should be set to give the best consistent note. If adjustment fails to restore the horn, it must be renewed; it is not possible to dismantle it for repair.

24.5 Side stand switch location

26.1 The ignition switch can be removed after instrument panel has been displaced

28.1 The later models employ blade-type fuses

28.3 An accessory terminal is fitted at the bottom of the fuse box

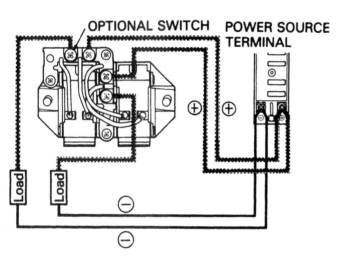

Fig. 6.20 Accessory circuit switch diagram – Katana models

OPTIONAL SWITCH    POWER SOURCE TERMINAL

Load

Load

29.1 Horn can be adjusted by way of small screw (arrowed)

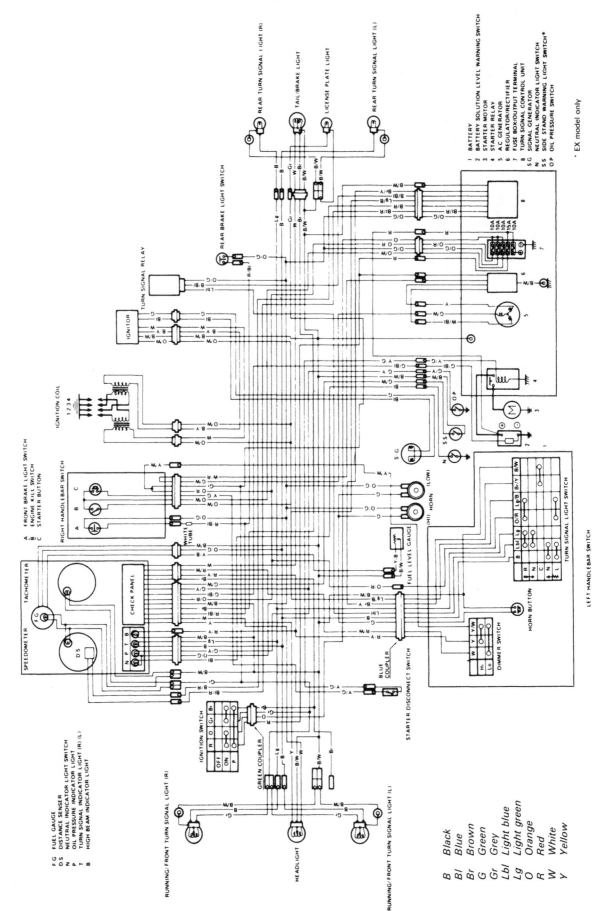

**Wiring diagram – GS1100 ET and EX US models**

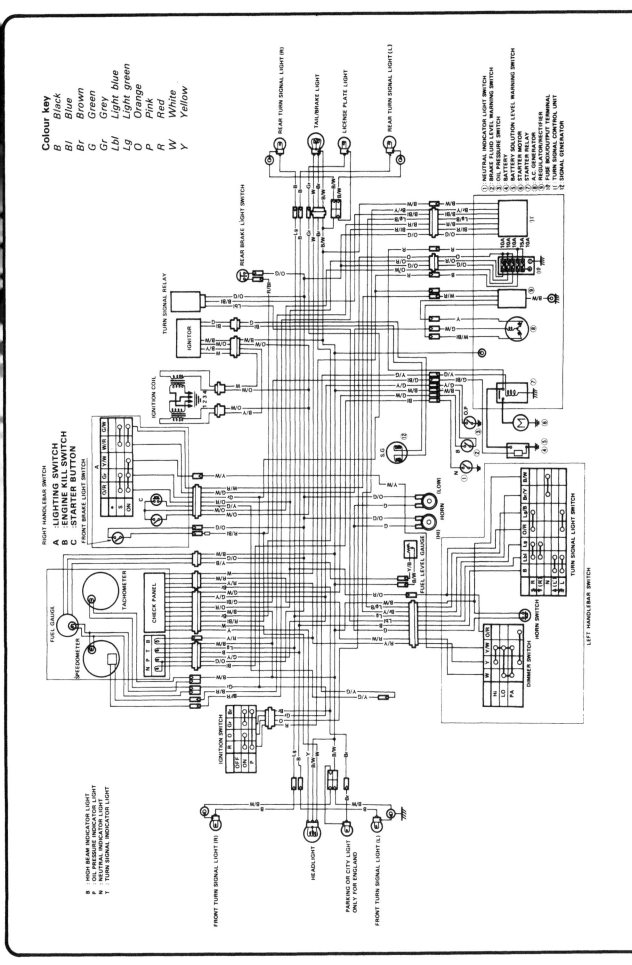

**Wiring diagram – GSX1100 T and ET UK models**

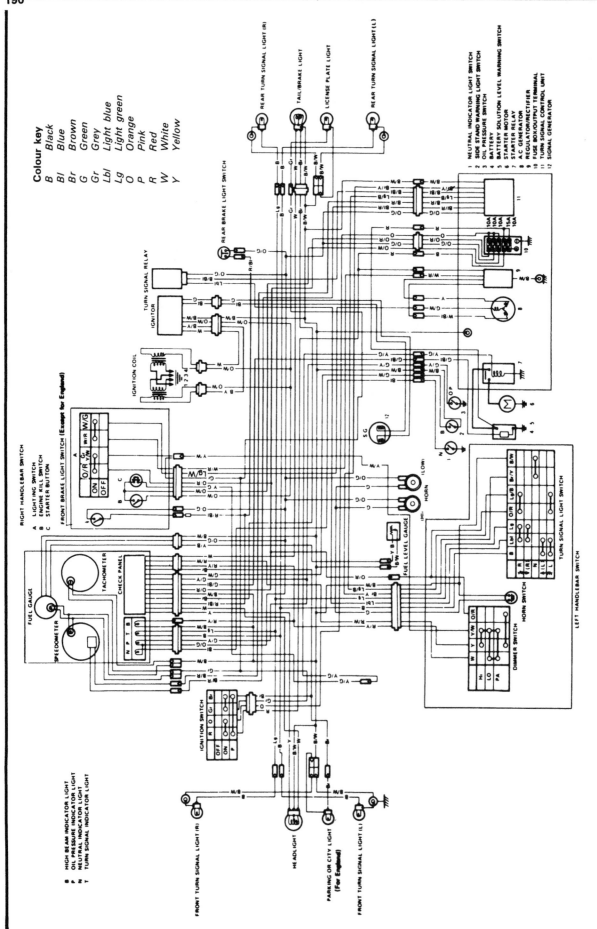

**Colour key**

B — Black
Bl — Blue
Br — Brown
G — Green
Gr — Grey
Lbl — Light blue
Lg — Light green
O — Orange
P — Pink
R — Red
W — White
Y — Yellow

Wiring diagram – GSX1100 EX UK model

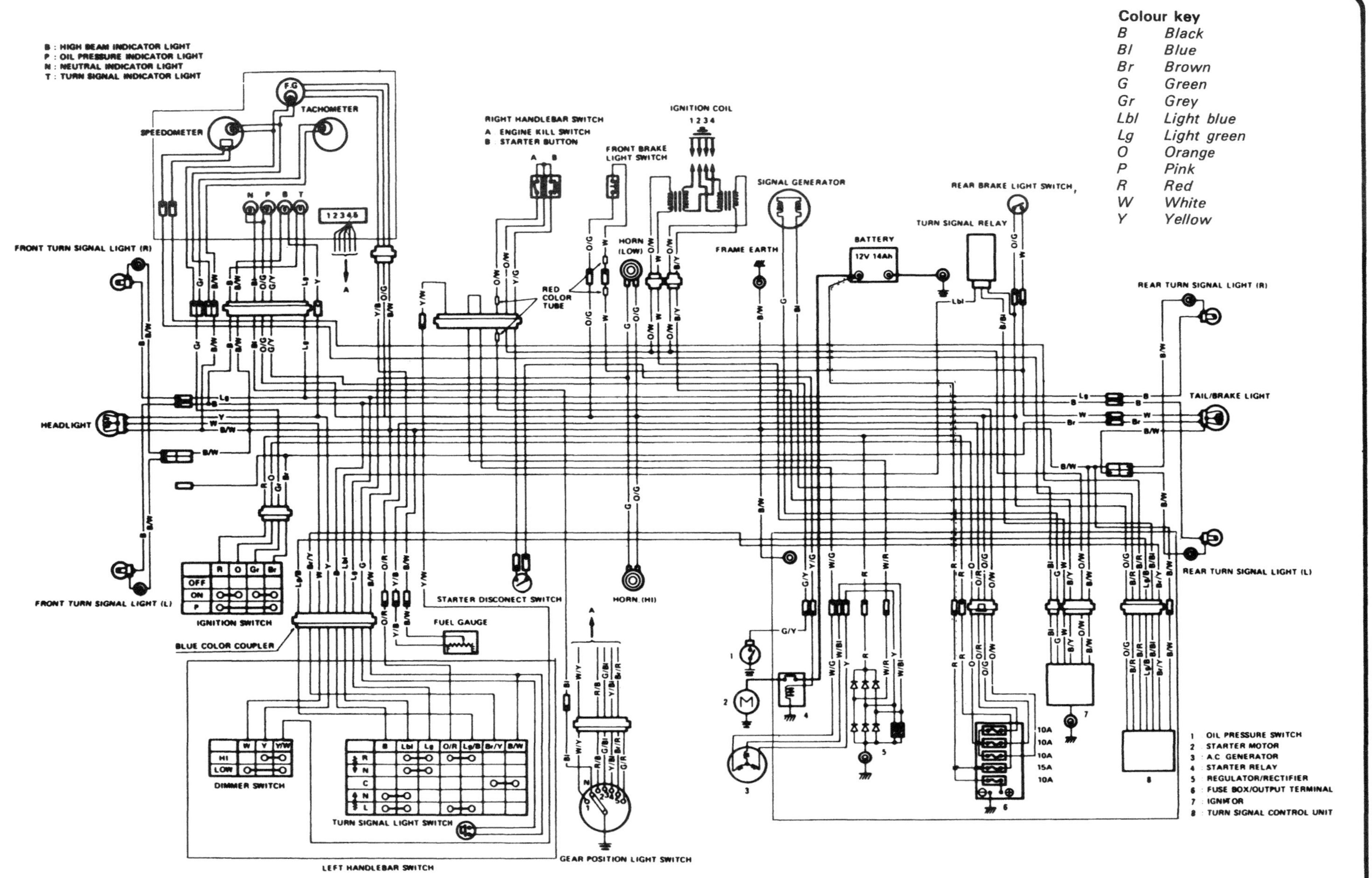

Wiring diagram – GS1100 LT US model

**Colour key**

| | |
|---|---|
| B | Black |
| Bl | Blue |
| Br | Brown |
| G | Green |
| Gr | Grey |
| Lbl | Light blue |
| Lg | Light green |
| O | Orange |
| P | Pink |
| R | Red |
| W | White |
| Y | Yellow |

B : HIGH BEAM INDICATOR LIGHT
P : OIL PRESSURE INDICATOR LIGHT
N : NEUTRAL INDICATOR LIGHT
T : TURN SIGNAL INDICATOR LIGHT

1 OIL PRESSURE SWITCH
2 STARTER MOTOR
3 AC GENERATOR
4 STARTER RELAY
5 REGULATOR/RECTIFIER
6 FUSE BOX/OUTPUT TERMINAL
7 IGNITOR
8 TURN SIGNAL CONTROL UNIT

REAR TURN SIGNAL LIGHT (R)
TAIL/BRAKE LIGHT
REAR TURN SIGNAL LIGHT (L)
REAR BRAKE LIGHT SWITCH
TURN SIGNAL RELAY
BATTERY 12V 14Ah
SIGNAL GENERATOR
FRAME EARTH
IGNITION COIL
FRONT BRAKE LIGHT SWITCH
RIGHT HANDLEBAR SWITCH
A  ENGINE KILL SWITCH
B  STARTER BUTTON
RED COLOR TUBE
HORN (LOW)
HORN (HI)
GEAR POSITION LIGHT SWITCH
STARTER DISCONNECT SWITCH
FUEL GAUGE
TURN SIGNAL LIGHT SWITCH
LEFT HANDLEBAR SWITCH
IGNITION SWITCH
BLUE COLOR COUPLER
DIMMER SWITCH
TACHOMETER
SPEEDOMETER
FRONT TURN SIGNAL LIGHT (R)
HEADLIGHT
FRONT TURN SIGNAL LIGHT (L)

10A 10A 10A 10A 15A 10A

191

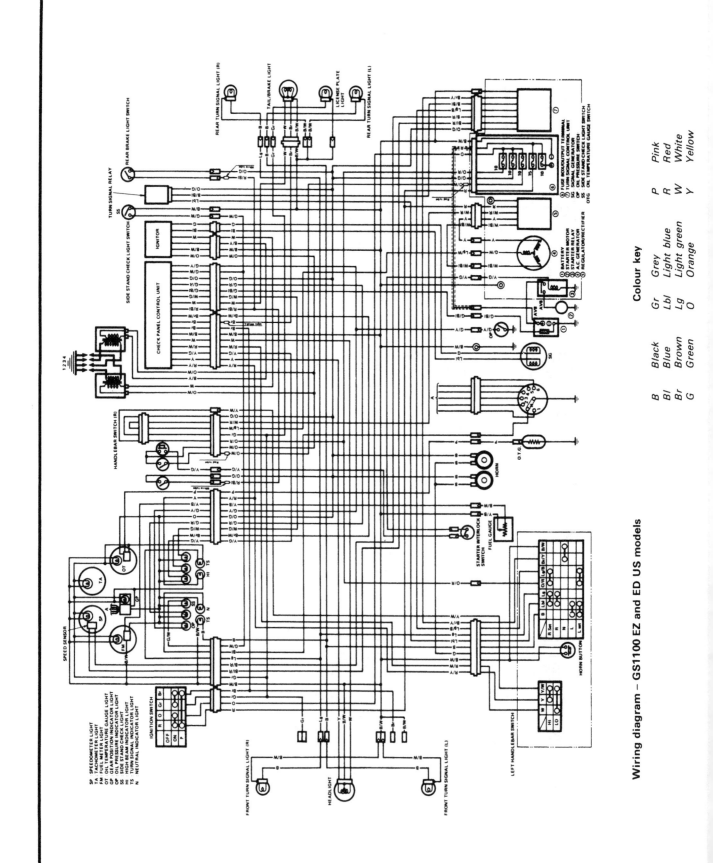

**Wiring diagram – GS1100 EZ and ED US models**

**Colour key**

| | | | | | | |
|---|---|---|---|---|---|---|
| B | Black | Gr | Grey | P | Pink |
| Bl | Blue | Lbl | Light blue | R | Red |
| Br | Brown | Lg | Light green | W | White |
| G | Green | O | Orange | Y | Yellow |

SP SPEEDOMETER LIGHT
TA TACHOMETER LIGHT
FM FUEL METER LIGHT
OT OIL TEMPERATURE GAUGE LIGHT
GP GEAR POSITION INDICATOR LIGHT
OP OIL PRESSURE INDICATOR LIGHT
SS SIDE STAND CHECK LIGHT
HI HIGH BEAM INDICATOR LIGHT
TS TURN SIGNAL INDICATOR LIGHT
N NEUTRAL INDICATOR LIGHT

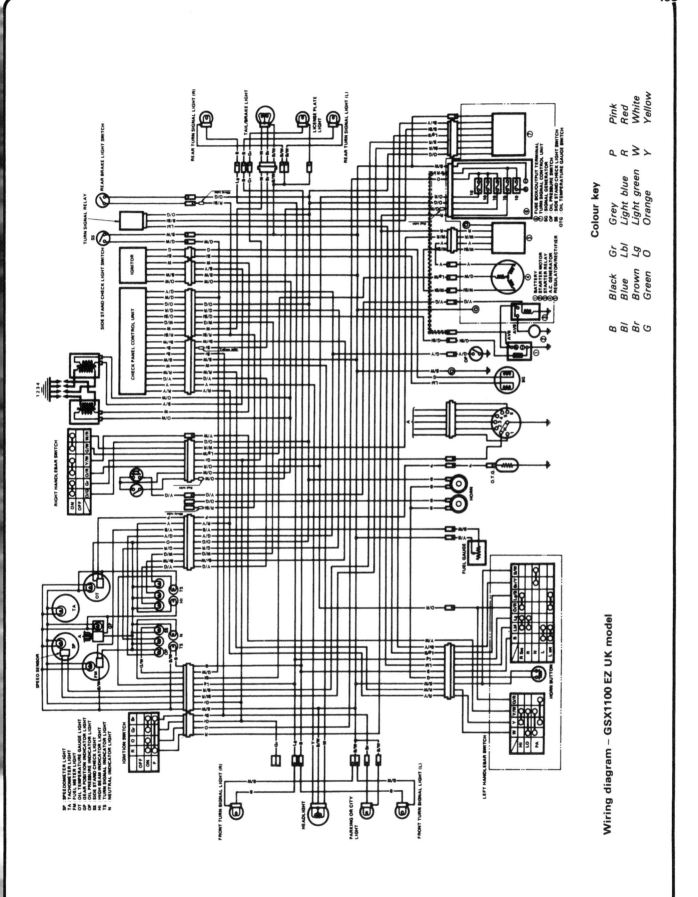

Wiring diagram – GSX1100 EZ UK model

**Colour key**

| B | Black | Gr | Grey | P | Pink |
|---|---|---|---|---|---|
| Bl | Blue | Lbl | Light blue | R | Red |
| Br | Brown | Lg | Light green | W | White |
| G | Green | O | Orange | Y | Yellow |

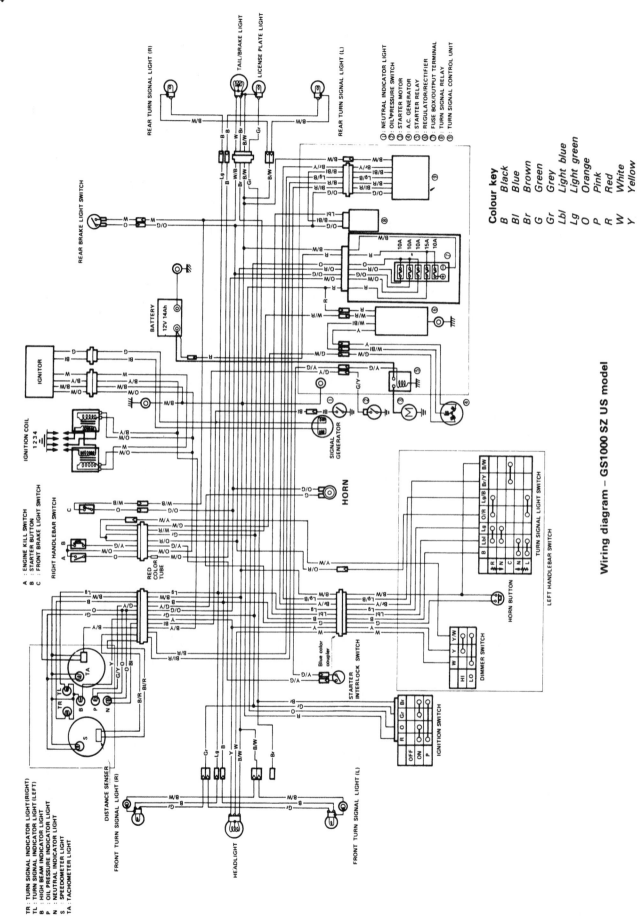

Wiring diagram – GS1000 SZ US model

**Colour key**

| | |
|---|---|
| B | Black |
| Bl | Blue |
| Br | Brown |
| G | Green |
| Gr | Grey |
| Lbl | Light blue |
| Lg | Light green |
| O | Orange |
| P | Pink |
| R | Red |
| W | White |
| Y | Yellow |

① NEUTRAL INDICATOR LIGHT
② OIL PRESSURE SWITCH
③ STARTER MOTOR
④ A.C. GENERATOR
⑤ STARTER RELAY
⑥ REGULATOR/RECTIFIER
⑦ FUSE BOX/OUTPUT TERMINAL
⑧ TURN SIGNAL RELAY
⑨ TURN SIGNAL CONTROL UNIT

TR : TURN SIGNAL INDICATOR LIGHT(RIGHT)
TL : TURN SIGNAL INDICATOR LIGHT (LEFT)
B : HIGH BEAM INDICATOR LIGHT
P : OIL PRESSURE INDICATOR LIGHT
N : NEUTRAL INDICATOR LIGHT
S : SPEEDOMETER LIGHT
TA : TACHOMETER LIGHT

A : ENGINE KILL SWITCH
B : STARTER BUTTON
C : FRONT BRAKE LIGHT SWITCH

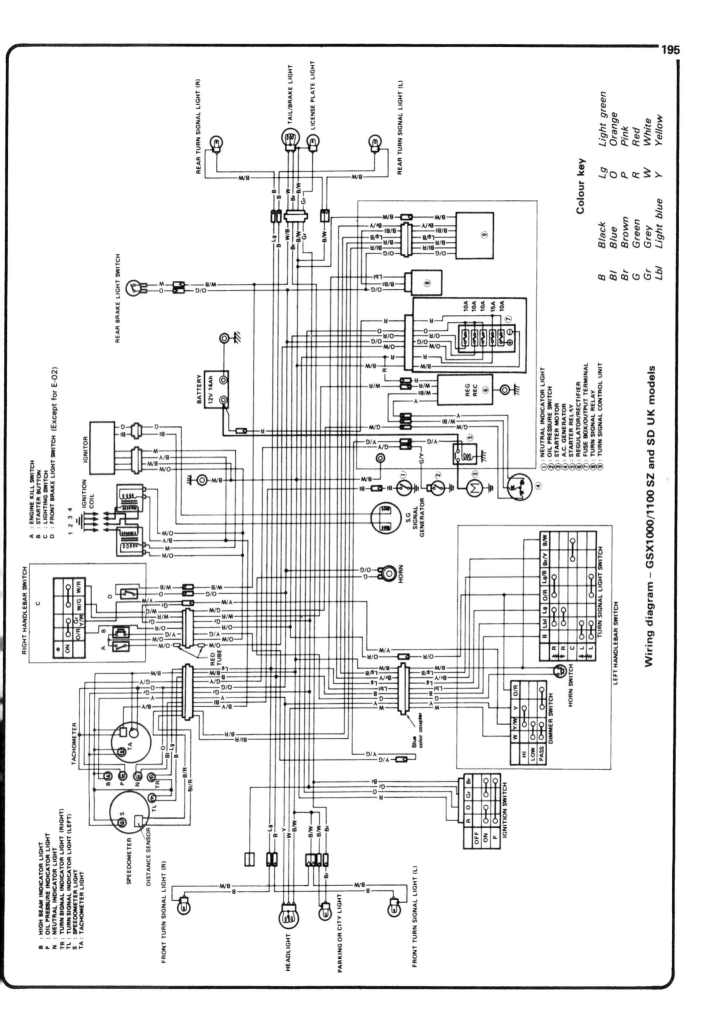

Wiring diagram – GSX1000/1100 SZ and SD UK models

**Colour key**

| | | | |
|---|---|---|---|
| B | Black | Lg | Light green |
| Bl | Blue | O | Orange |
| Br | Brown | P | Pink |
| G | Green | R | Red |
| Gr | Grey | W | White |
| Lbl | Light blue | Y | Yellow |

① : NEUTRAL INDICATOR LIGHT
② : OIL PRESSURE SWITCH
③ : STARTER MOTOR
④ : A.C. GENERATOR
⑤ : STARTER RELAY
⑥ : REGULATOR/RECTIFIER
⑦ : FUSE BOX/OUTPUT TERMINAL
⑧ : TURN SIGNAL RELAY
⑨ : TURN SIGNAL CONTROL UNIT

B : HIGH BEAM INDICATOR LIGHT
P : OIL PRESSURE INDICATOR LIGHT
N : NEUTRAL INDICATOR LIGHT
TR : TURN SIGNAL INDICATOR LIGHT (RIGHT)
TL : TURN SIGNAL INDICATOR LIGHT (LEFT)
S : SPEEDOMETER LIGHT
TA : TACHOMETER LIGHT

A : ENGINE KILL SWITCH
B : STARTER BUTTON
C : LIGHTING SWITCH
D : FRONT BRAKE LIGHT SWITCH (Except for E-02)

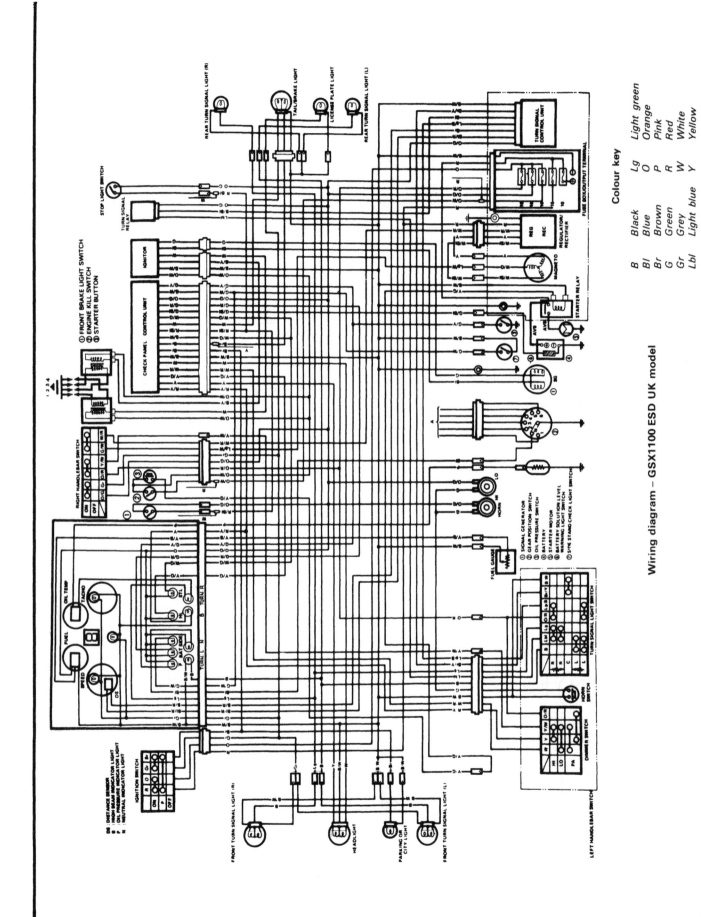

**Wiring diagram – GSX1100 ESD UK model**

Colour key

| | | | |
|---|---|---|---|
| B | Black | Lg | Light green |
| Bl | Blue | O | Orange |
| Br | Brown | P | Pink |
| G | Green | R | Red |
| Gr | Grey | W | White |
| Lbl | Light blue | Y | Yellow |

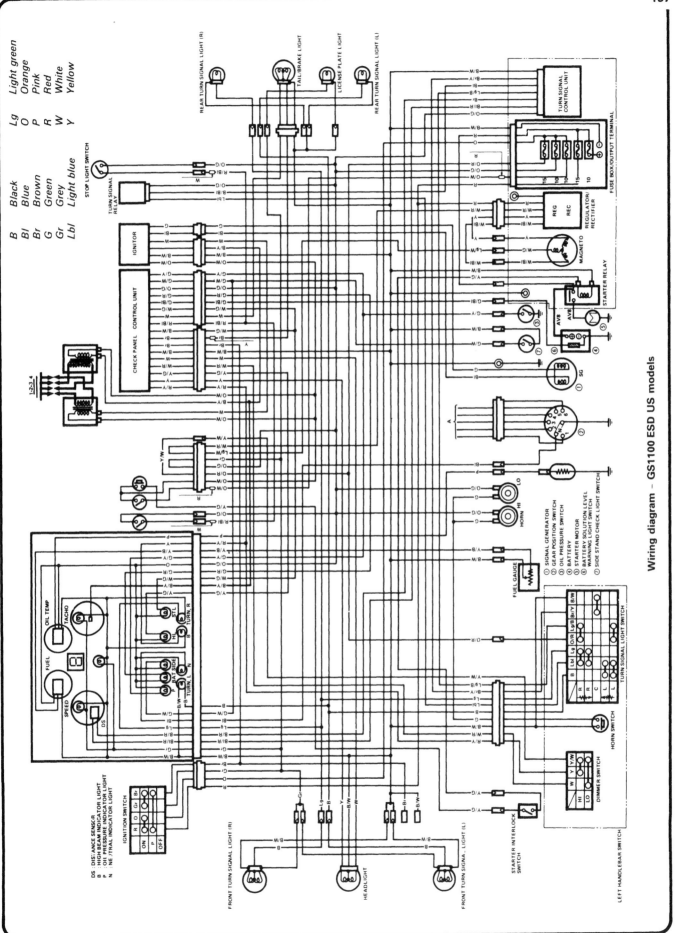

Wiring diagram – GS1100 ESD US models

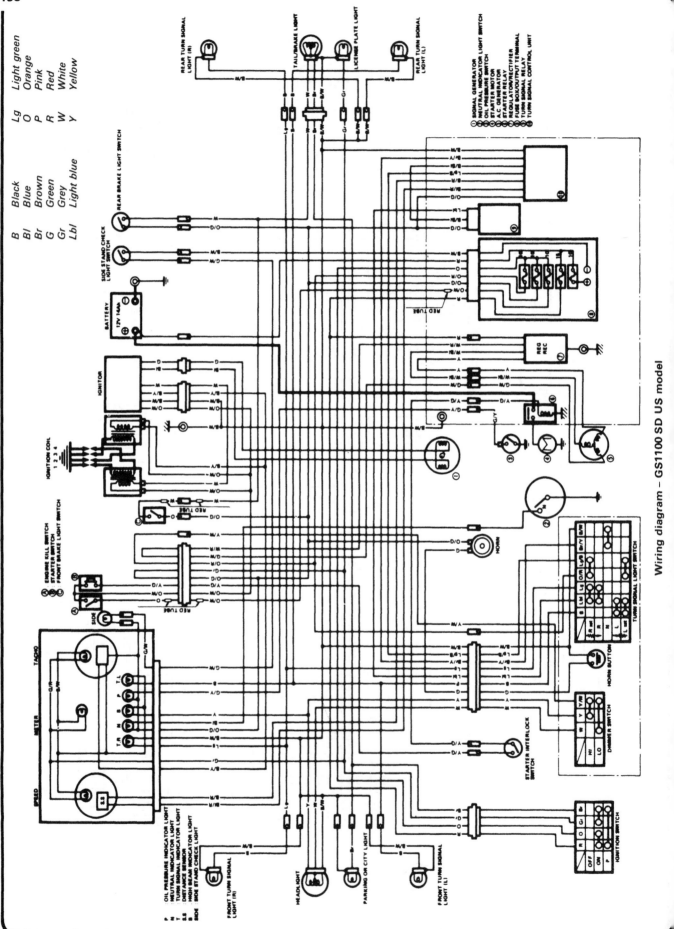

Wiring diagram – GS1100 SD US model

The GS1150 EF US model

The GS1150 ESF US model

# Chapter 7 The 1984 on
# 1135cc-engined GSX1100 and GS1150 models

## Contents

## Specifications

*Information is only given where different from that shown in previous chapters. Unless stated, information applies to all models. For E models refer back to the ED information, and for ES or EF models refer back to the ESD information.*

### Model dimensions and weights

| | |
|---|---|
| Overall length ............................................................................. | 2240 mm (88.2 in) |
| Overall width: | |
| E models...................................................................................... | 815 mm (32.1 in) |
| ES and EF models ........................................................................ | 740 mm (29.1 in) |
| Overall height: | |
| E models...................................................................................... | 1140 mm (44.9 in) |
| ES and EF models ........................................................................ | 1280 mm (50.4 in) |
| Wheelbase................................................................................... | 1550 mm (61.0 in) |
| Dry weight: | |
| E models...................................................................................... | 232 kg (511 lbs) |
| ES models ................................................................................... | 237 kg (522 lbs) |
| EF models ................................................................................... | 238 kg (525 lbs) |

### Specifications relating to Routine maintenance

### Engine

| | |
|---|---|
| Valve clearances ......................................................................... | 0.08 – 0.13 mm (0.003 – 0.005 in) when cold |
| Standard compression ................................................................. | 11 – 15 kg cm$^2$ (156 – 213 psi) |
| Service limit................................................................................. | 9 kg cm (128 psi) |
| Max difference between cylinders.................................................. | 2 kg cm$^2$ (28.4 psi) |
| Spark plug type ........................................................................... | NGK D9EA or ND X27ES-U |
| Oil pressure ................................................................................ | 0.2 – 0.4 kg cm$^2$ (2.82 – 5.64 psi) @ 3000 rpm |

## Cycle parts

| Tyre pressures – psi (kg cm²) | Front | Rear |
|---|---|---|
| Solo | 32 (2.25) | 36 (2.50) |
| Pillion | 36 (2.50) | 42 (2.90) |
| Fork air pressure – US models | 4.3 psi (0.3 kg cm²) | |
| Fork oil capacity: | | |
| UK E models | 268 cc (9.43 Imp fl oz) | |
| UK EF models | 280 cc (9.85 Imp fl oz) | |
| US models | 286 cc (9.66 US fl oz) | |
| Fork oil level: | | |
| UK models | 190 mm (7.48 in) | |
| US models | 180 mm (7.09 in) | |

## Specifications relating to Chapter 1

## Engine

| Bore | 74.0 mm (2.913 in) |
|---|---|
| Capacity | 1135 cc (67.5 cu in) |
| Compression ratio | 9.7 : 1 |

## Camshaft and rockers

| Cam height (inlet) | 35.241 – 35.281 mm (1.387 – 1.389 in) |
|---|---|
| Service limit | 34.950 mm (1.376 in) |
| Cam height (exhaust) | 34.933 – 34.973 mm (1.375 – 1.377 in) |
| Service limit | 34.640 mm (1.362 in) |

## Valves and guides

| Inlet valve head diameter | 28.0 mm (1.10 in) |
|---|---|
| Valve lift, inlet | 8.0 mm (0.31 in) |
| Valve lift, exhaust | 7.5 mm (0.29 in) |
| Valve clearance (cold) – inlet and exhaust | 0.08 – 0.13 mm (0.003 – 0.005 in) |
| Valve stem end length (In/Ex) – service limit | 3.8 mm (0.15 in) |
| Valve spring pressure: | |
| Inner | 4.9 – 5.8 kg (10.8 – 12.8 lb) @ 28.5 mm (1.12 in) |
| Outer | 7.2 – 8.1 kg (15.8 – 17.8 lb) @ 32.0 mm (1.26 in) |

## Cylinder bores

| Compression pressure | 11–15 kg cm² (156–213 psi) |
|---|---|
| Service limit | 9 kg cm² (128 psi) |
| Bore size (Std) | 74.000 – 74.015 mm (2.913 – 2.914 in) |
| Service limit | 74.080 mm (2.916 in) |

## Pistons

| Piston diameter (Std) | 73.945 – 73.960 mm (2.911 – 2.912 in) |
|---|---|
| Service limit | 73.880 mm (2.909 in) |
| Ring end gap (free): | |
| Top | About 10.0 mm (0.39 in) |
| Service limit | 8.0 mm (0.31 in) |
| Ring end gap (installed): | |
| Top and 2nd ring | 0.10 – 0.25 mm (0.004 – 0.010 in) |
| Service limit | 0.70 mm (0.028 in) |
| Ring groove width: | |
| Top | 1.01 – 1.03 mm (0.040 – 0.041 in) |
| Oil | 2.01 – 2.03 mm (0.079 – 0.080 in) |
| Gudgeon pin bore ID | 20.000 – 20.006 mm (0.787 – 0.788 in) |
| Service limit | 20.030 mm (0.789 in) |
| Gudgeon pin OD | 19.996 – 20.000 mm (0.786 – 0.787 in) |
| Service limit | 19.980 mm (0.785 in) |

## Crankshaft and connecting rods

| Small end ID | 20.006 – 20.014 mm (0.788 – 0.789 in) |
|---|---|
| Service limit | 20.040 mm (0.790 in) |

## Clutch

| Friction plate thickness – US models | See EZ model specifications |
|---|---|

## Transmission

| Primary reduction | 1.780 (89/50T) |
|---|---|
| Final reduction | 2.800 (42/15T) |
| Final drive chain: | |
| Length | 102 links |
| 20 links length (max) | 383.2 mm (15.09 in) |
| Free play | 15 – 25 mm (0.59 – 0.98 in) |

## Torque settings

| Component | kgf m | lbf ft |
| --- | --- | --- |
| Cylinder head cover bolt | 1.3 – 1.5 | 9.5 – 11.0 |
| Upper front engine mounting bolt | 7.0 – 8.8 | 50.4 – 63.4 |
| Lower front engine mounting bolts | 5.0 – 6.0 | 36.0 – 43.2 |
| Upper rear engine mounting bolt | 7.0 – 8.8 | 50.4 – 63.4 |
| Lower rear engine mounting bolt | 5.5 – 6.6 | 39.6 – 47.5 |
| Cam chain tensioner locknut | 0.8 – 1.0 | 5.8 – 7.2 |
| Ignition pickup rotor retaining bolt | 2.5 – 3.5 | 18.0 – 25.2 |
| Starter clutch Allen bolt – US models | 2.3 – 2.8 | 16.5 – 20.0 |

## Specifications relating to Chapter 2

### Fuel tank

| | |
| --- | --- |
| Overall capacity | 20.0 litre (5.2/4.4 US/Imp gal) |
| Reserve capacity | 4.9 litre (10.3/8.6 US/Imp pint) |

### Carburettors

| | |
| --- | --- |
| Type | BS36SS |
| Size | 36 mm (1.43 in) |
| Identity number | 00A30 |
| Idle speed | 1100 ± 50 rpm |
| Fuel level | 3.0 ± 0.5 mm (0.12 ± 0.02 in) |
| Float height | 21.4 ± 1.0 mm (0.84 ± 0.04 in) |
| Main jet: | |
| Cylinders 1 and 4 | 120 |
| Cylinders 2 and 3 | 122.5 |
| Main air jet | 1.2 mm |
| Jet needle | 5D59 |
| Clip position | 2nd groove from the top |
| Needle jet | X-5 |
| Throttle valve | 130 |
| Pilot jet | 47.5 |
| By pass | 0.9, 0.8, 0.9 mm |
| Pilot outlet | 0.9 mm |
| Float valve seat | 2.0 mm |
| Starter jet | 32.5 |
| Pilot screw (pre-set) | 2 turns out |
| Pilot air jet | 125 |
| Cut-away | Not available |
| Throttle cable free play | 2 – 3 mm (0.08 – 0.12 in) |
| Choke cable free play | 0.5 – 1.0 mm (0.02 – 0.04 in) |

### Torque settings

| | |
| --- | --- |
| Oil cooler union bolts | 2.5 – 3.0 kgf m (18.0 – 21.6 lbf ft) |

## Specifications relating to Chapter 3

### Ignition timing

| | |
| --- | --- |
| Below 1500 ± 100 rpm | 10° BTDC |
| Above 3500 ± 100 rpm | 32° BTDC |

### Spark plug

NGK D9EA or ND X27ES-U

### Ignition coil resistances

| | |
| --- | --- |
| Primary windings | 2 – 5 ohms |
| Secondary windings | 30 – 40 K ohms |

### Ignition source coil resistance

100 – 300 ohms

## Specifications relating to Chapter 4

### Front forks

| | |
| --- | --- |
| Stroke | 150 mm (5.91 in) |
| Oil level: | |
| UK models | 190 mm (7.48 in) |
| US models | 180 mm (7.09 in) |
| Oil grade | 15W fork oil |
| Oil quantity (per leg): | |
| UK E models | 268 cc (9.43 Imp fl oz) |

| | |
|---|---|
| UK EF models | 280 cc (9.85 Imp fl oz) |
| US models | 286 cc (9.66 US fl oz) |
| Fork spring free length – service limit | 500 mm (19.6 in) |
| Fork air pressure – US models only | 0.3 kg cm² (4.26 psi) |

## Rear suspension

| | |
|---|---|
| Type | Full Floater, monoshock |
| Travel | 115 mm (4.53 in) |
| Suspension unit | Oil damped, coil spring with adjustable damping and preload |

## Torque settings

| Component | kgf m | lbf ft |
|---|---|---|
| Anti-dive unit bolt | 0.6 – 0.8 | 4.5 – 6.0 |
| Front fork air valve union nut | 1.0 – 1.3 | 7.0 – 9.5 |
| Handlebar clamp bolt | 5.0 – 6.0 | 36.0 – 43.4 |
| Handlebar holder nut | 2.0 – 3.0 | 14.5 – 21.5 |
| Rear suspension unit mounting bolts | 4.8 – 7.2 | 34.6 – 51.8 |
| All suspension linkage pivot bolts | 8.4 – 10.0 | 60.5 – 72.0 |
| Swinging arm pivot shaft | 5.0 – 8.0 | 36.0 – 57.6 |

## Specifications relating to Chapter 5

## Tyre sizes

| | |
|---|---|
| Front | 110/90V16 |
| Rear | 130/90V17 |

| Tyre pressures, psi kg cm²) | Front | Rear |
|---|---|---|
| Solo | 32 (2.25) | 36 (2.50) |
| Dual | 36 (2.50) | 42 (2.90) |

## Front brake

| | |
|---|---|
| Master cylinder piston OD | 15.827 – 15.854 mm (0.623 – 0.624 in) |
| Caliper bore ID | 38.180 – 38.256 mm (1.503 – 1.506 in) |
| Caliper piston OD | 38.098 – 38.148 mm (1.500 – 1.502 in) |

## Rear brake

| | |
|---|---|
| Master cylinder bore ID | 15.870 – 15.913 mm (0.625 – 0.626 in) |
| Master cylinder piston OD | 15.827 – 15.854 mm (0.623 – 0.624 in) |

## Torque settings

| | |
|---|---|
| Brake caliper housing bolts | 1.8 – 2.3 kgf m (13.0 – 16.6 lbf ft) |
| Brake pedal pinch bolt | 1.5 – 2.5 kgf m (11.0 – 18.0 lbf ft) |
| Caliper bleed valve | 0.6 – 0.9 kgf m (4.3 – 6.5 lbf ft) |

## Specifications relating to Chapter 6

**Starter motor brush length** .......... 9.0 mm (0.35 in) minimum

**Starter relay resistance** .......... 3 – 7 ohms

## 1   Introduction

This chapter covers the UK GSX1100 and US 1150 models introduced in 1984. Although seemingly of different engine capacity both share the 1135cc engine unit derived from the 1075cc engine fitted to previous models.

Apart from the obvious increase in bore size, changes to the engine unit are few. An oil cooler has been added and the ignition system is now advanced electronically, rather than by the mechanical advance/retard unit fitted previously. As far as the cycle components are concerned, the conventional twin-shock rear suspension has been superseded by Suzuki's Full Floater monoshock system. Tyres are now of the tubeless type. A monitoring system is incorporated in the instrument panel to warn the rider of headlamp, tail lamp or brake lamp bulb failure, low battery electrolyte level, low oil pressure, and whether the side stand has been retracted.

When working on one of the later machines, refer first to this chapter to check for any modifications to the component concerned. If no mention is made then the task will be as given in the previous chapters. Always refer back to the previous model, thus for an ES model, refer back to the information given for the ES model in chapters 1 to 6.

Throughout this manual the models are identified by their model

suffix letters, rather than their dates of import. To assist in identification, each model is listed below, together with its initial frame number and approximate dates of import for UK models or model year for US models. Note that the latter may not necessarily coincide with the machine's actual date of registration or sale.

| Model | Initial frame no. | Dates of import |
|---|---|---|
| UK GSX1100 EFE | GV71B-100001 | Mar 84 to Mar 86 |
| UK GSX1100 EF | GV71B-101973 | Jul 85 to Mar 86 |
| UK GSX1100 EFF | GV71B-101973 | Mar 86 to Feb 87 |
| UK GSX1100 EG | GV71B-103359 | Mar 85 to Oct 88 |
| UK GSX1100 EFG | GV71B-103359 | Sep 86 to Oct 88 |
| US GS1150 ESE | JS1GV71A E2100001 | 1984 |
| US GS1150 EF/ESF | JS1GV71A F2100001 | 1985 |
| US GS1150 EG | JS1GV71A G2100001 | 1986 |

## 2   Routine maintenance: service schedule – UK model

Owners of GSX1100 EF, EFF, EG and EFG models should note that the manufacturer recommends a different service schedule from that shown at the beginning of Routine maintenance for the earlier models. Instead, refer to the service schedule given for US models.

---

### 3  Engine/gearbox unit: removal and refitting

**Cylinder head**

1   When refitting the head gasket note that it must be positioned the correct way up; this is denoted by the word HEAD on its top surface. Refit the cylinder head and install the nuts, finger-tight at this stage. The eight domed nuts and copper washers are fitted to the inner studs, and the four plain nuts and steel washers are fitted to the outer studs. Tighten all nuts evenly in the sequence shown. After tightening, refit the single bolt to the front of the head at the joint between the head and block and tighten it to the specified torque setting.

**Cylinder head cover**

2   Prior to fitting the cover check the condition of the main sealing gasket. If torn or cracked at any point, it should be renewed. Similarly check the condition of the sealing washers fitted to the eight shorter retaining bolts and renew if necessary. Note that the two longer bolts must be positioned at the front and rear of the cam chain tunnel. Tighten all bolts evenly to the specified torque setting.

**Engine removal and refitting**

3   The engine/gearbox unit can be removed from the frame as described in Chapter 1 noting the following.

4   After draining the engine oil position the container underneath the oil cooler hose/engine union bolts and remove both bolts and sealing washers. When all the oil has drained from the hoses clean both connections and place them inside a clean polythene bag along with the union bolts and sealing washers and secure the bag with an elastic band. This will keep them clean and prevent the ingress of dirt whilst the engine is out of the frame.

5   It will be necessary to remove both ignition HT coils and the side stand switch to provide enough clearance for the engine to be removed from the frame.

6   Note that the engine mounting bolts are removed and refitted as described in Chapter 1, although the spacer fitted to the upper rear mounting bolt is no longer fitted.

7   On refitting follow the instructions given in Chapter 1, Section 42. In addition, refit the side stand switch and ignition HT coils ensuring that the HT leads and wiring connectors are correctly positioned as shown in the accompanying illustration. Position a sealing washer on each side of the oil cooler hose connections and refit the union bolts. Tighten both to the specified torque setting.

8   Refill the crankcases with oil and make final adjustments as described in Chapter 1, Sections 43 and 44.

---

### 4  Clutch: modifications

1   The clutch assembly is similar to the earlier type shown in Fig. 1.22, the only differences being the addition of a thrust washer between the clutch centre (5) and tab washer (7) and a shim positioned between the sleeve (3) and thrust washer (8) to adjust the thrust clearance. The clutch can be removed and refitted as described in Chapter 1, noting that before refitting it will be necessary to check the clutch drum thrust clearance as described below.

2   Place the oil pump drive gear and its spacer, the clutch drum and its spacer, shim and large thrust washer on a flat surface as shown in Fig. 1.9.

3   Using feeler gauges measure the clearance between the large thrust washer and clutch drum as shown in the illustration. This clearance should be 0.03 – 0.08 mm (0.0012 – 0.0032 in). If not, replace the shim with one of the required thickness until the correct clearance is obtained. Shims of various thicknesses can be purchased from an official Suzuki dealer.

4   Note that failure to check the thrust clearance could lead to excessive clutch noise. This check is particularly important if either the oil pump drive gear, its spacer, the clutch drum/primary driven gear, its spacer, the shim itself or the thrust washer are renewed at anytime.

---

### 5  Oil cooler: maintenance, removal and refitting

1   The oil cooler matrix is mounted on the frame front downtubes to gain the best possible airflow. The oil feed and return hoses are connected to the engine on each side of the oil filter housing. Access to the unit is restricted by the fairing side sections on ES models, and side and lower sections on EF models; remove these components first.

2   To maintain peak efficiency, the matrix should be kept clear of any debris, preferably by using an air jet directed from behind the matrix. Avoid using sharp instruments to dislodge any foreign matter as the vanes are easily damaged. If leakage of the matrix does occur, it must be renewed; repairs are unlikely to be successful.

3   Before removing the oil cooler, place a suitably-sized container underneath it to catch the small amount of oil which will be released when the union bolts are removed. Remove the two union bolts together with their sealing washers, which retain the hoses to the underside of the oil cooler. Remove the three bolts which secure it to the mounting bracket and remove the oil cooler from the machine.

4   If the oil feed or return hoses are damaged or show signs of leakage they should be renewed immediately. Leakage at the hose unions will

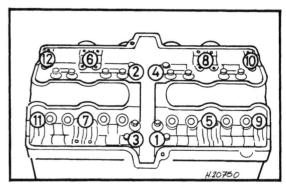

Fig. 7.1 Cylinder head nut tightening sequence (Sec 3)

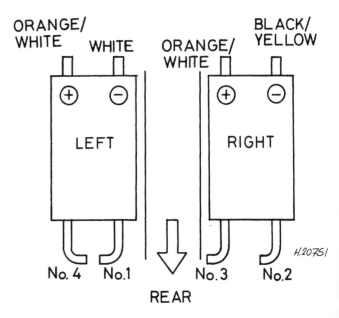

Fig. 7.2 Ignition HT coil lead and wire positions (Sec 3)

probably be due to a damaged sealing washer; these should be renewed if at all suspect. To remove the hoses, release both the top and bottom union bolts, noting that some provision must first be made to catch the escaping oil. Remove the bolts from the hose retaining clamp to release the hoses.

5 The oil cooler and hoses are refitted by a reversal of the removal procedure. Position a sealing washer on each side of the hose connections and refit the union bolts. Tighten all bolts to their specified torque settings.

## 6 Ignition system: modifications

1 The ignition system fitted to later models has no provision for adjustment and, with the exception of the spark plugs, is maintenance free. The automatic timing unit (ATU) fitted to earlier models has been deleted, and ignition advance is now controlled by a modified amplifier unit.

2 When working on the ignition system proceed as described in Chapter 3, taking note of the different specifications at the start of this Chapter. When checking the ignition amplifier unit (Chapter 3, Section 9), connect the positive (red) meter probe to the black wire terminal and not the blue as stated.

3 Due to the fact that the automatic timing unit is no longer fitted, the procedure for removal and refitting of the pickup assembly differs from that given in Chapter 1; refer to the following Section for details.

## 7 Ignition pickup assembly: removal and refitting

1 The ignition pickup assembly is located behind the small circular cover on the right-hand end of the crankshaft.

2 To remove the pickup assembly slacken the four retaining screws and remove the cover. Release the rotor mounting Allen bolt whilst holding the crankshaft, via the moulded hexagon on the rotor end, with a suitably-sized spanner and remove the signal generator rotor.

3 Trace the pickup wiring back along the crankcase and free it from its guide clips. Release the three backplate retaining screws and remove the pickup assembly from the machine. The pickup wiring can be freed by disconnecting its terminal block at the ignition amplifier.

4 Refit the pickup assembly by reversing the removal procedure. When refitting the rotor, ensure the slot in the back of the rotor locates correctly with the pin in the crankshaft end. Tighten the rotor mounting bolt to the specified torque setting, reconnect the wiring and secure it with the guide clips.

## 8 Front forks: removal and refitting

1 On models fitted with full or half fairings it is recommended that the fairing is removed to improve access to the top and bottom yokes (see Section 14).

2 Release the handlebar pad retaining screw and remove the handlebar pad. Slacken the handlebar retaining nuts situated on the underside of the top yoke and remove the mounting Allen bolts, having first prised out their head inserts. Lift the handlebars off the fork stanchions, positioning them clear of the forks. Support the handlebars to avoid straining the hydraulic hose and cables and also to keep the master cylinder reservoir in an upright position. The front forks can now be removed as described in Section 2 of Chapter 4. Note that it is not necessary to disconnect the anti-dive units on these models.

3 The forks are refitted by a reversal of the removal sequence. Refit the fork legs in the yokes and lightly tighten the pinch bolts. Slide the handlebars over the fork stanchions and install the mounting bolts and nuts, securing them finger-tight at this stage. Position each stanchion so that its upper edge is in line with the top face of the handlebar casting and then tighten all bolts to their specified torque settings. Refit the mudguard, wheel and fairing (where applicable) and thoroughly check

the operation of the forks before taking the machine on the road. Check the suspension settings as described in Sections 9 and 10, and set the fork air pressure on US models.

## 9 Front forks: modifications

### Dismantling and reassembly

1 The front forks are similar in construction to those fitted to the earlier EZ, ED, ESD and Katana models, except for the anti-dive unit, which is no longer interconnected with the hydraulic front brake. Refer to the accompanying illustration for details of the later type anti-dive unit, and Fig. 4.3 for the main fork components; note that the floating piston has been replaced with a spacer. The fork legs can be dismantled as described in Section 5 of Chapter 4.

### Anti-dive unit

2 The anti-dive unit is of the Posi-damp type, which is not linked to the hydraulic front brake as on previous models. Instead it relies on increased fork pressure to activate a valve in the unit, which produces a constricting effect, thereby giving increased damping. Note that no replacement parts are available, except for the two O-rings between the unit and fork lower leg. Therefore if it is faulty or fluid leakage occurs, it must be renewed as an assembly.

3 The Posi-damp unit has four damping settings, designed to vary the duration of the damping effect. These are shown in the following table, together with comparable spring preload and rear suspension settings. When making adjustment, depress the top of the unit and turn the adjuster wheel until the required damping position is shown at the front of the unit; make sure that it locates in the detent provided. Always set each anti-dive unit adjuster to the same position, otherwise uneven damping and poor roadholding will result.

| | Front fork | | Rear suspension unit | |
|---|---|---|---|---|
| Solo riding: | Preload | Anti-dive | Preload | Damping |
| Standard | 1 (2) | 3 (3) | 3 (2) | 3 (2) |
| Softer | 1 (1) | 2 (2) | 3 (1) | 2 (1) |
| Stiffer | 2 (3) | 4 (3) | 4 (3) | 4 (4) |
| Dual riding | 2 (3) | 4 (3) | 4 (3) | 4 (4) |

*Figures in parenthesis relate to US models*

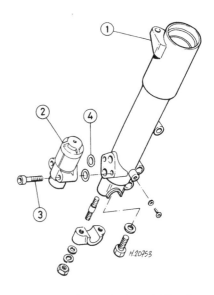

**Fig. 7.3 Front fork anti-dive unit (Sec 9)**

1 Fork lower leg      3 Allen bolt – 2 off
2 Anti-dive unit      4 O-ring – 2 off

## 10   Swinging arm, rear suspension unit and linkage: removal and refitting

1   Place the machine on its centre stand and remove the seat and both sidepanels. Remove the rear wheel as described in Section 15. Slacken the bolt which secures the brake caliper to the torque arm and remove the caliper. Position the caliper clear of the swinging arm and tie it to the frame to avoid straining the hydraulic hose.

2   Remove the four bolts which secure the remote damping and preload adjusters to the frame and position them clear of the swinging arm. *On no account attempt to separate the adjusters from the suspension unit. If the hose connections are unscrewed, pressure will be lost and the assembly will have to be renewed.* Free the suspension linkage from its mounting bracket by unscrewing the single pivot bolt and nut situated directly above the rear master cylinder. Remove the swinging arm pivot shaft nut and withdraw the pivot shaft. If stuck in place, the pivot shaft can be tapped out using a hammer and a suitable drift. As the shaft comes free, support the swinging arm, then lower it clear of the frame, together with the suspension unit remote adjusters.

3   Separate the rear suspension unit from the swinging arm and linkage by releasing both its upper and lower mounting bolts. Likewise remove the bolts which secure the bellcrank to the cushion rod and the cushion rod to the swinging arm to separate the linkage components and swinging arm.

4   Examine and lubricate (where applicable) the various bearings and bushes in the assembly. (Refer to the following Sections for details).

5   Refitting is a direct reversal of the removal sequence. Reassemble the suspension linkage and unit in the swinging arm and tighten all the mounting bolts to the specified torque settings. Refer to the accompanying illustration as a guide to the correct washer and dust cover positions. Check that the pivot shaft is straight, clean and free from corrosion, then smear a small quantity of general purpose grease over its shank. Insert the swinging arm and suspension assembly into the frame and refit the pivot shaft and nut, followed by the suspension linkage to mounting bracket pivot bolt. Tighten both to their specified torque settings. Remount the remote damping and preload adjusters to the frame, ensuring that the hose and operating cable are correctly routed.

6   Refit the rear brake caliper and wheel as described in Section 15 and secure both retaining nuts with new split pins. Reset the suspension preload and damping (see table in Section 9) and thoroughly check the operation of the rear suspension and brakes before taking the machine on the road.

## 11   Swinging arm bearings: examination and renewal

1   Lift off the dust covers and washers and remove the bearing inner sleeves. Remove all traces of old grease and inspect the sleeves and needle roller bearings for signs of scoring or corrosion. If damage is discovered both the bearings and sleeves must be renewed as follows. It should be noted that the bearings will probably be destroyed during removal, so do not attempt to remove them unless renewal is definitely required.

2   Lever out the stepped bushes using a large flat-bladed screwdriver, examine these bushes for signs of damage and renew if necessary. The swinging arm bearings can then be drifted out of position using a hammer and a suitably-sized socket as a drift.

3   Ensure the swinging arm lugs are clean and undamaged, and apply a smear of grease to the outer races of the new bearings. The bearings should be fitted with their marked face outermost (bearings will be punch marked on outer face). Use a drawbolt arrangement as shown in the accompanying illustration to draw them into position. Refit the stepped bushes, grease the bearings and inner sleeves, then slide the inner sleeves into place. Examine the dust covers, renewing them if necessary, and fit a washer inside each cap. Apply a small amount of grease to the inside of the dust seals and fit them to the swinging arm lugs.

## 12   Suspension linkage bearings: examination and renewal

1   The bellcrank is fitted with a pair of needle roller bearings in its wider boss and a single needle roller bearing in its rear boss. The cushion rod is fitted with a needle roller bearing at each end. All bearings are fitted with inner sleeves and are sealed at each end by dust covers.

2   Lift off the dust covers and washers and push out the inner sleeves. Wipe off all traces of old grease and inspect the components for signs of corrosion or damage. If wear or damage is present in any of the bearings or inner sleeves both components must be renewed. Note that the bearings will probably be damaged on removal, so do not disturb them unless renewal is definitely required.

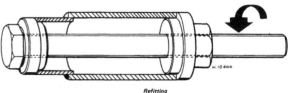

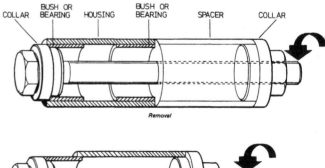

**Fig. 7.4 Swinging arm (Sec 10)**

| | |
|---|---|
| 1   Cap – 2 off | 6   Headed spacer – 4 off |
| 2   Pivot shaft | 7   Needle roller bearing – 2 |
| 3   Dust cover – 4 off |    off |
| 4   Washer – 4 off | 8   Nut |
| 5   Inner sleeve – 4 off | 9   Torque arm |

**Fig. 7.5 Drawbolt tool for removing and refitting bearings (Sec 11)**

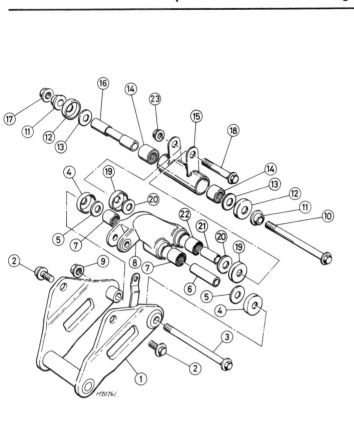

**Fig. 7.6 Suspension linkage**

| | |
|---|---|
| 1   Mounting bracket | 13  Washer – 2 off |
| 2   Bolt – 2 off | 14  Needle roller bearing – 2 |
| 3   Pivot bolt | off |
| 4   Dust cover – 2 off | 15  Cushion rod |
| 5   Washer – 2 off | 16  Inner sleeve |
| 6   Inner sleeve | 17  Nut |
| 7   Needle roller bearing – 2 | 18  Pivot bolt |
| off | 19  Dust cover – 2 off |
| 8   Bellcrank | 20  Washer – 2 off |
| 9   Nut | 21  Inner sleeve |
| 10  Pivot shaft | 22  Needle roller bearing |
| 11  Headed spacer – 2 off | 23  Nut |
| 12  Dust cover – 2 off | |

3   The needle roller bearings can be removed using a bearing puller with slide-hammer attachment, a drawbolt tool, or if care is exercised they can be drifted from position. If using the latter method, ensure that the bearing leaves its housing squarely and that the housing surface is not damaged. Ensure the bosses are clean and free from scoring before refitting the new bearings. Where two bearings are fitted, they should be fitted with their punch-marked ends facing outwards. Use a drawbolt tool to draw them into place. Grease the bearings and insert the inner sleeves. Apply a small amount of grease to the washers and dust cover lips before fitting them over the ends of the bellcrank or cushion rod assemblies.

## 13   Rear suspension unit: examination, renovation and adjustment

1   The rear suspension unit is of the oil-damped, coil spring type, located centrally in the rear suspension linkage. The unit features a remote hydraulically-operated spring preload adjuster and a cable-operated damping adjuster, mounted on the left-hand side of the machine. *On no account should the hydraulic preload adjuster be separated from the suspension unit; they are joined by a pressurised hose.* It is not possible to repair the suspension unit or preload adjuster; if a fault develops the complete assembly must be renewed. The damping

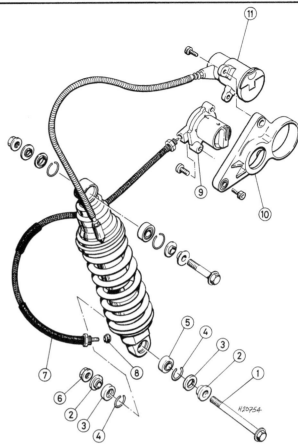

**Fig. 7.7 Rear suspension unit**

| | |
|---|---|
| 1   Mounting bolt – 2 off | 7   Damping adjuster cable |
| 2   Headed spacer – 2 off | 8   Drive gear |
| 3   Oil seal – 2 off | 9   Remote damping adjuster |
| 4   Circlip – 2 off | 10  Mounting bracket |
| 5   Spherical bearing – 2 off | 11  Remote spring preload |
| 6   Nut – 2 off | adjuster |

adjuster, however, can be dismantled and checked as described below, after removing the suspension unit from the machine as described in Section 10.

2   Remove the dust cover from the bottom of the suspension unit to reveal the damping adjusting ring. Turn the remote damping adjuster knob whilst observing the adjusting ring. If working correctly, the ring should move freely in each direction when the remote adjuster is operated, if not, the mechanism should be dismantled for further examination.

3   Detach the cable ends from the suspension unit and damping adjuster. Hold the outer cable and twist one end of the inner cable whilst watching the opposite end. If the cable is broken or fails to operate smoothly it must be renewed. If all is well the fault must lie in either the remote adjuster or suspension unit gears. The small adjusting gear fitted in the suspension unit can simply be tipped out of position. Examine the gear for signs of wear or damage and renew if necessary. Apply general purpose grease to the gear prior to refitting. To refit the gear the damping adjusting ring on the suspension unit must be positioned so that the mark located between the III and IIII positions aligns with the mark on the suspension unit itself. This will ensure that the gears mesh correctly. If the fault still exists attention should be turned to the remote adjuster unit. Although the adjuster can be dismantled for lubrication purposes, note that no spare parts are available with which it can be repaired. Dismantling the unit is a straightforward process, although when refitting the adjuster unit and cable care must be taken to ensure that the components are correctly set up as follows.

4   Check that the suspension adjusting gear is still in the position described above and refit the dust cover and operating cable to the suspension unit. To reassemble the remote damping adjuster, first refit the driven gear housing to the control cable. Slide the spring onto the adjuster ring gear shaft and refit the adjuster knob, ensuring that the

hole in the shaft aligns with that on the side of the knob; secure the knob with the retaining screw. Align the mark on the side of the adjuster knob with that on the drive gear housing and refit the assembly to the driven gear housing. Before proceeding any further check that the position indicated by the adjuster knob is the same as that of the suspension unit adjusting ring.

5   If worn, the spherical bearings fitted in the suspension unit mounting eyes can be renewed. To remove the bearings lift off the headed spacer on each side of the mounting, then prise out the dust seals. Remove the wire circlips using a small electrical screwdriver to displace them. Draw out the spherical bearing using a bearing extractor, a drift or a drawbolt arrangement. Fit the new bearing using a drawbolt arrangement. Retain the bearing using new wire circlips, ensuring that they seat fully and press the new dust seals into position. Do not apply any grease to the bearings; only a smear of grease should be applied to the dust seal lips before refitting the headed spacers.

### Adjustment

6   Preload and damping adjustment can be made to suit the rider's weight and riding style. To preserve good roadholding, it is important that the front and rear suspension settings are correctly matched; see table in Section 9. Ensure that the machine is positioned upright, on its centre stand, during adjustment.

7   To adjust the preload, lift the adjuster handle and turn it clockwise to increase preload or anticlockwise to decrease it. The adjuster has five positions which are marked on the side of the unit, position 1 being the softest, and five being the hardest.

8   Damping adjustment can be made by turning the adjuster knob so that its index mark aligns with either of the four positions, and is heard to click into place. Position 1 gives the softest damping and position 4, the hardest.

---

### 14   Fairing: removal and refitting

1   The US GS1150 ES models have a half (sports) fairing fitted, whereas the UK GSX1100 EF models are fitted with a full fairing.

2   Refer to the accompanying illustration for details of removal and refitting, paying particular attention to the spacer and washer positions at all mounting points. Note that damage can result from overtightening the fairing panel fasteners.

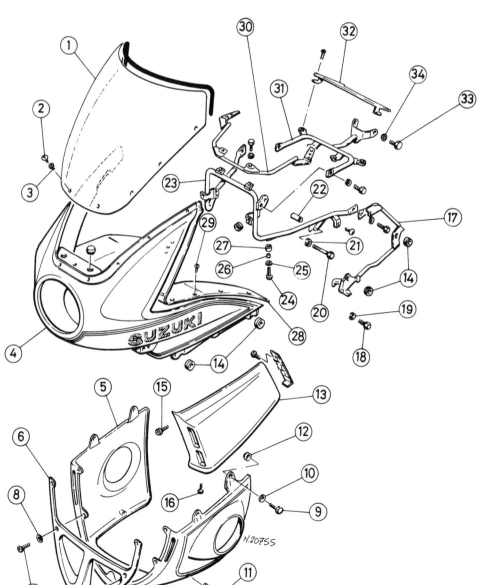

**Fig. 7.8 Fairing – ES and EF models**

| | |
|---|---|
| 1 | Windscreen |
| 2 | Screw – 8 off |
| 3 | Spacer – 8 off |
| 4 | Fairing upper section |
| 5 | Fairing lower section * |
| 6 | Lower section front cover * |
| 7 | Screw – 5 off* |
| 8 | Washer – 5 off* |
| 9 | Bolt – 4 off* |
| 10 | Washer – 6 off |
| 11 | Screw – 2 off |
| 12 | Washer – 6 off |
| 13 | Fairing side section – 2 off |
| 14 | Grommet – 8 off |
| 15 | Screw – 4 off |
| 16 | Screw – 2 off |
| 17 | Side section mounting bracket |
| 18 | Bolt – 2 off |
| 19 | Spring washer – 2 off |
| 20 | Bolt – 2 off |
| 21 | Spring washer – 2 off |
| 22 | Spacer – 2 off |
| 23 | Upper section mounting bracket |
| 24 | Screw – 2 off |
| 25 | Washer – 2 off |
| 26 | Spacer – 2 off |
| 27 | Grommet – 2 off |
| 28 | Fairing upper section inner panel |
| 29 | Screw – 18 off |
| 30 | Instrument bracket |
| 31 | Mounting bracket to headstock stay |
| 32 | Cover |
| 33 | Bolt – 2 off |
| 34 | Spring washer – 2 off |
| * | GSX1100 EF models only |

## 15 Wheels and brakes: modifications

### Front wheel

1 Refer to Chapter 5, Section 4, noting that it will be necessary to remove both brake calipers to allow the wheel to pass through the fork legs.

### Rear wheel

2 Prior to removing the rear wheel it will be necessary to remove the wheel spindle covers fitted to the swinging arm ends; each cover is retained by two screws. Proceed as described in Chapter 5, Section 6, noting that the chain adjuster bracket bolts should be slackened and the wheel moved forward so that the chain can be disengaged from the sprocket.

### Brake calipers

3 The front and rear calipers are similar to the rear caliper unit fitted to the earlier models. Refer therefore to the information given for the rear brake in Chapter 5, Section 15. Note the Specifications at the beginning of this Chapter.

## 16 Tubeless tyres: removal and refitting

1 It is strongly recommended that should a repair to a tubeless tyre be necessary, the wheel is removed from the machine and taken to a tyre fitting specialist or an authorized dealer. This is because the force required to break the seal between the wheel rim and tyre bead is considerable and is considered to be beyond the capabilities of an individual working with normal tyre removing tools. Any abortive attempt to break the rim to bead seal may also cause damage to the wheel rim, resulting in expensive wheel renewal. If, however, a suitable bead releasing tool is available, and experience has already been gained in its use, tyre removal and refitting can be accomplished as follows.

2 Remove the wheel from the machine. Deflate the tyre by removing the valve core and when it is fully deflated, push the bead of the tyre away from the wheel rim on both sides so that the bead enters the well of the rim. As noted, this operation will almost certainly require the use of a bead releasing tool.

3 Insert a tyre lever close to the valve and lever the edge of the tyre over the outside of the wheel rim. Very little force should be necessary; if resistance is encountered it is probably due to the fact that the tyre beads have not entered the well of the wheel rim all the way round the tyre. Should the initial problem persist, lubrication of the tyre bead and the inside edge and lip of the rim will facilitate removal. Use a recommended lubricant, a diluted solution of washing-up liquid or french chalk. Lubrication is usually recommended as an aid to tyre fitting but its use is equally desirable during removal. The risk of lever damage to wheel rims can be minimised by the use of proprietary plastic rim protectors placed over the rim flange at the point where the tyre levers are inserted. Suitable rim protectors can be fabricated very easily from short lengths (4 – 6 inches) of thick-walled nylon petrol pipe which have been split down one side using a sharp knife. The use of rim protectors should be adopted whenever levers are used and, therefore, when the risk of damage is likely.

4 Once the tyre has been edged over the wheel rim, it is easy to work around the wheel rim so that the tyre is completely free on one side.

5 Working from the other side of the wheel, ease the other edge of the tyre over the outside of the wheel rim, which is furthest away. Continue to work around the rim until the tyre is freed completely from the rim.

6 Refer to the following sections for details of puncture repair, tyre renewal and valves.

7 Tyre refitting is virtually a reversal of the removal procedure. If the tyre has a balance mark (usually a spot of coloured paint), this must be positioned alongside the tyre valve. Similarly, any arrow indicating direction of rotation must face the right way.

8 Starting at the point furthest from the valve, push the tyre bead over the edge of the wheel rim until it is located in the well. Continue to work around the tyre in this fashion until the whole of one side of the tyre is on the rim. It may be necessary to use a tyre lever during the final stages. Here again, the use of a lubricant will aid fitting. It is strongly recommended that when fitting the tyre only a recommended lubricant is used because such lubricants also have sealing properties. Do not be over generous in the application of lubricant or tyre creep may occur.

9 Fitting the second bead is similar to fitting the first. Start by pushing the bead over the rim and into the well at a point diametrically opposite the tyre valve. Continue working around the tyre, each side of the starting point, ensuring that the bead opposite the working area is always in the well. Apply lubricant as necessary. Avoid using tyre levers unless absolutely essential, to help reduce damage to the soft wheel rim. Use of the levers should be required only when the final portion of bead is to be pushed over the rim.

10 Lubricate the tyre beads again prior to inflating the tyre, and check that the wheel rim is evenly positioned in relation to the tyre beads. Inflation of the tyre may well prove impossible without the use of a high pressure air hose. The tyre will retain air completely only when the beads are pressed firmly against the rim edges at all points and it may be found when using a foot pump that air escapes at the same rate as it is pumped in. This problem may also be encountered when using an air hose on new tyres which have been compressed in storage and by virtue of their profile hold the beads away from the rim edges. To overcome this difficulty, a tourniquet may be placed around the circumference of the tyre, over the central area of the tread. The compression of the tread in this area will cause the beads to be pushed outwards in the desired direction. The type of tourniquet most widely used consists of a length of hose closed at both ends, with a suitable clamp fitted to enable both ends to be connected. An ordinary tyre valve is fitted at one end of the tube so that after the hose has been secured around the tyre it may be inflated, giving a constricting effect. Another possible method of seating beads to obtain initial inflation is to press the tyre into the angle between a wall and the floor. With the airline attached to the valve additional pressure is then applied to the tyre by the hand and shin, as shown in the accompanying illustration. The application of pressure at four points around the tyre's circumference whilst simultaneously applying the airline will often effect an initial seal between the tyre beads and wheel rim, thus allowing inflation to occur.

11 Having successfully accomplished inflation, increase the pressure to 40 psi and check that the tyre is evenly disposed on the wheel rim. This may be judged by checking that the thin positioning line found on each tyre wall is equidistant from the wheel rim around the total circumference of the tyre. If this is not the case, deflate the tyre, apply additional lubrication and reinflate. Minor adjustments to the tyre position may be made by bouncing the wheel on the ground.

12 Always run the tyres at the recommended pressures and never under- or over-inflate. The correct pressures are given in the Specifications at the start of this chapter. Note that if non-standard tyres are fitted check with the tyre manufacturer or supplier for recommended pressures. Finally refit the valve dust cap.

## 17 Tubeless tyres: puncture repair and tyre renewal

1 If a puncture occurs, the tyre should be removed for inspection of damage before any attempt is made at remedial action. The temporary repair of a punctured tyre by inserting a plug from the outside should not be attempted. The manufacturers strongly recommend that no such

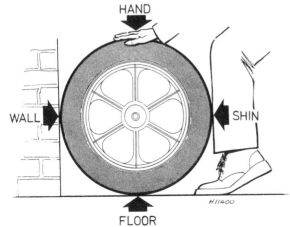

**Fig. 7.9 Method of seating the beads on tubeless tyres (Sec 16)**

repair is carried out on a motorcycle tyre. Not only does the tyre have a thin carcass, which does not give sufficient support to the plug, but the consequences of a sudden deflation are often sufficiently serious that the risk of such an occurrence should be avoided at all costs.

2    The tyre should be inspected both inside and out for damage to the carcass. Unfortunately the inner lining of the tyre – which takes the place of the inner tube – may easily obscure any damage and some experience is required in making a correct assessment of the tyre condition.

3    There are two main types of repair which are considered safe for adoption in repairing tubeless motorcycle tyres. The first type of repair consists of inserting a mushroom-headed plug into the hole from the inside of the tyre. The hole is prepared for insertion of the plug by reaming and the application of an adhesive. The second repair is carried out by buffing the inner lining in the damaged area and applying a cold or vulcanised patch. Because both inspection and repair, if they are to be carried out safely, require experience in this type of work, it is recommended that the tyre be placed in the hands of a repairer with the necessary skills, rather than repaired in the home workshop.

4    In the event of an emergency, the only recommended 'get-you-home' repair is to fit a standard inner tube of the correct size. If this course of action is adopted, care should be taken to ensure that the cause of the puncture has been removed before the inner tube is fitted. It may be found that the valve hole in the rim is considerably larger than the diameter of the inner tube valve stem. To prevent the ingress of road dirt, and to help support the valve, a spacer should be fitted over the valve.

5    In the event of the unavailability of tubeless tyres, ordinary tubed tyres fitted with inner tubes of the correct size may be fitted. Refer to the manufacturer or a tyre fitting specialist to ensure that only a tyre and tube of equivalent type and suitability is fitted, and also to advise on the fitting of a valve nut to the rim hole.

6    **Note:** *The manufacturer recommends that if the tyre has been repaired, speed should not exceed 50 mph (80 kph) for the first 24 hours after the repair, and thereafter should not exceed 80 mph (130 kph).*

## 18    Tubeless tyres: valves – examination and renewal

1    It will be appreciated from the preceding Sections that the adoption of tubeless tyres has made it necessary to modify the valve arrangement, as there is no longer an inner tube which can carry the valve core. The problem has been overcome by fitting a separate tyre valve which passes through a close-fitting hole in the rim, and which is secured by a nut and locknut. The valve is fitted from the rim well, and it follows that the valve can be removed and refitted only when the tyre has been removed from the rim. Leakage of air from around the valve body is likely to occur only if the sealing seat fails or if the nut and locknut become loose.

2    The valve core is of the same type as that used with tubed tyres, and screws into the valve body. The core can be removed with a small slotted tool which is normally incorporated in plunger type pressure gauges. Some valve dust caps incorporate a key for removing valve cores. Although tubeless tyre valves seldom give trouble, it is possible for a leak to develop if a small particle of grit lodges on the sealing face. Occasionally, an elusive slow puncture can be traced to a leaking valve core, and this should be checked before a genuine puncture is suspected.

3    The valve dust caps are a significant part of the tyre valve assembly. Not only do they prevent the ingress of road dirt in the valve, but also act as a secondary seal which will reduce the risk of sudden deflation if a valve core should fail.

## 19    Regulator/rectifier unit: testing

The regulator/rectifier unit can be tested as described in Section 7 of Chapter 6, but refer to the accompanying table for test values.

## 20    Headlamp: beam adjustment

1    On models fitted with a full or half fairing, adjustment is controlled

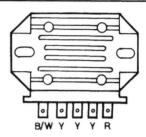

UNIT: Ω

| | | (+) PROBE OF TESTER TO: | | | | |
|---|---|---|---|---|---|---|
| | | R | Y | Y | Y | B/W |
| PROBE OF TESTER TO: (−) | R | | ∞ | ∞ | ∞ | ∞ |
| | Y | 5-10 | | ∞ | ∞ | ∞ |
| | Y | 5-10 | ∞ | | ∞ | ∞ |
| | Y | 5-10 | ∞ | ∞ | | ∞ |
| | B/W | 50-200 | 5-10 | 5-10 | 5-10 | |

*H.20756*

**Fig. 7.10 Regulator/rectifier unit test table (Sec 19)**

by two knobs situated at the bottom of the headlamp unit which can be accessed from inside the fairing upper section. The right-hand knob controls the horizontal adjustment and the left-hand knob controls the vertical adjustment.

2    On unfaired models, vertical adjustment can be made by slackening the two bolts which secure it to the mounting brackets and moving the headlamp unit to the desired position. Horizontal adjustment is controlled by way of a screw in the edge of the rim.

## 21    Tail/brake lamp and licence plate lamp: bulb renewal

1    To gain access to the tail/brake lamp bulbs it will be necessary to remove both parts of the seat. Turn the bulb holders anticlockwise and remove them from the tail lamp unit. The bulbs are a bayonet fit in their holders. They are of the twin filament type and have offset pins to ensure they are fitted correctly.

2    Remove the two screws from the underside of the reflector cover, and lift off the cover to gain access to the licence plate lamp. The lens can be withdrawn after removing the two screws. Push the bulb inwards and twist it anticlockwise to release it from the bulbholder. When refitting, check that the lens gasket is intact and correctly positioned.

## 22    Instrument panel: general

The instrument panel fitted to these models is very similar to that fitted to the ESD model (Fig. 6.13) apart from the speedometer and tachometer heads being available as separate items. The accompanying figure shows the relative positions of the panel components.

## 23    Monitor system: operation and testing

1    The various warning lamps mounted in the instrument panel are controlled by a monitor unit situated behind the right-hand side panel. All lamps should come on when the ignition is switched on and all but

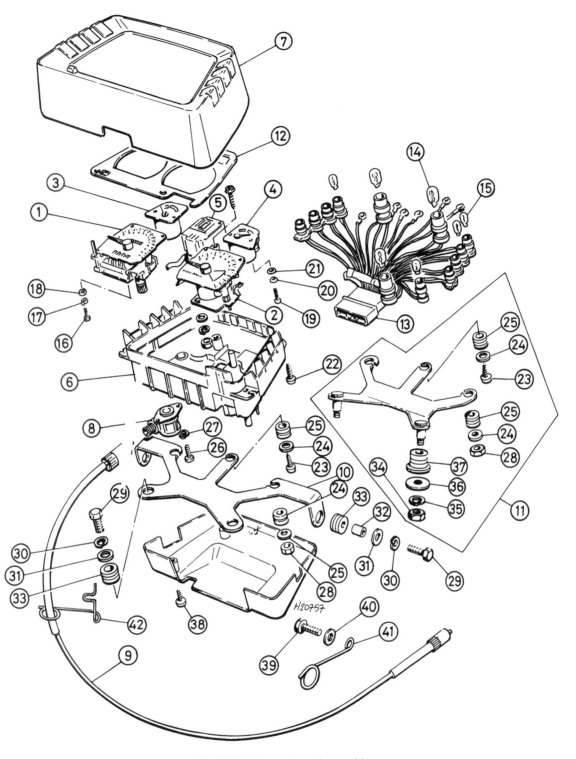

**Fig. 7.11 Instrument panel assembly**

| | | | |
|---|---|---|---|
| 1 Speedometer | 12 Instrument panel | 23 Screw | 34 Nut – 2 off |
| 2 Tachometer | 13 Wiring | 24 Washer – 3 off | 35 Spring washer – 2 off |
| 3 Fuel gauge | 14 Bulb | 25 Grommet – 3 off | 36 Washer – 2 off |
| 4 Oil temperature gauge | 15 Bulb | 26 Screw – 2 off | 37 Grommet – 2 off |
| 5 Gear position indicator | 16 Screw – 2 off | 27 Spring washer – 2 off | 38 Screw – 4 off |
| 6 Panel housing | 17 Spring washer – 2 off | 28 Nut – 2 off | 39 Bolt |
| 7 Top cover | 18 Washer – 2 off | 29 Bolt – 4 off | 40 Cable guide – E model |
| 8 Speedometer gearbox | 19 Screw – 6 off | 30 Spring washer – 4 off | 41 Cable guide – ES and EF |
| 9 Speedometer cable | 20 Spring washer – 6 off | 31 Washer – 4 off | models |
| 10 Mounting bracket – E model | 21 Washer – 6 off | 32 Spacer – 4 off | |
| 11 Mounting bracket – ES and EF models | 22 Screw – 4 off | 33 Grommet – 4 off | |

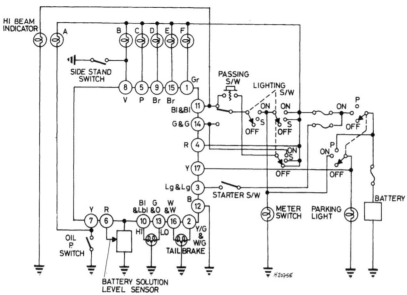

Bl  Blue
Br  Brown
G   Green
Lbl Light blue
Lg  Light green
O   Orange
P   Pink
R   Red
V   Violet
W   White
Y   Yellow

A  Oil pressure indicator
B  Side stand indicator
C  Tail lamp circuit indicator
D  Brake lamp circuit indicator
E  Battery electrolyte level indicator
F  Headlamp circuit indicator

**Fig. 7.12 Monitor system circuit diagram (Sec 23)**

Unit: X kΩ±40% kΩ

| ⊖＼⊕ | 1 | 2 | 3 | 4 | 5 | 6 | 7 | 8 | 9 | 10 | 11 | 12 | 13 | 14 | 15 | 16 | 17 |
|---|---|---|---|---|---|---|---|---|---|---|---|---|---|---|---|---|---|
| 1 | | ∞ | ∞ | ∞ | ∞ | ∞ | ∞ | ∞ | ∞ | ∞ | ∞ | ∞ | ∞ | ∞ | ∞ | ∞ | ∞ |
| 2 | 9 | | 0 | ∞ | 8.5 | 20 | 7.5 | ∞ | 8.5 | 7.5 | 7.5 | 3.5 | 0.8 | 0.8 | 8.5 | 0.8 | ∞ |
| 3 | 9 | 0 | | ∞ | 8.5 | 20 | 7.5 | ∞ | 8.5 | 7.5 | 7.5 | 3.5 | 0.8 | 0.8 | 8.5 | 0.8 | ∞ |
| 4 | 22 | 3.5 | 3.5 | | 20 | 30 | 16 | ∞ | 20 | 18 | 18 | 7.5 | 3.5 | 3.5 | 20 | 3.5 | ∞ |
| 5 | ∞ | ∞ | ∞ | ∞ | | ∞ | ∞ | ∞ | ∞ | ∞ | ∞ | ∞ | ∞ | ∞ | ∞ | ∞ | ∞ |
| 6 | 28 | 14 | 14 | ∞ | 28 | | 28 | ∞ | 25 | 25 | 25 | 14 | 14 | 14 | 26 | 14 | ∞ |
| 7 | 28 | 13 | 14 | ∞ | 28 | 30 | | ∞ | 25 | 25 | 25 | 16 | 13 | 13 | 26 | 13 | ∞ |
| 8 | 50 | 20 | 20 | ∞ | 48 | 45 | 3 | | 50 | 40 | 40 | 25 | 20 | 20 | 50 | 20 | ∞ |
| 9 | ∞ | ∞ | ∞ | ∞ | ∞ | ∞ | ∞ | ∞ | | ∞ | ∞ | ∞ | ∞ | ∞ | ∞ | ∞ | ∞ |
| 10 | 12 | 10 | 10 | ∞ | 12 | 25 | 17 | ∞ | 12 | | 0 | 5 | 10 | 10 | 13 | 10 | ∞ |
| 11 | 12 | 10 | 10 | ∞ | 12 | 25 | 17 | ∞ | 12 | 0 | | 5 | 10 | 10 | 13 | 10 | ∞ |
| 12 | 3.5 | 2.5 | 2.8 | ∞ | 3.8 | 16 | 9 | ∞ | 3.6 | 2.8 | 2.8 | | 2.6 | 2.5 | 3.5 | 2.6 | ∞ |
| 13 | 8.5 | 0.8 | 0.8 | ∞ | 8.5 | 20 | 7.5 | ∞ | 8.5 | 7.5 | 7.5 | 3.2 | | 0 | 8.5 | 0.8 | ∞ |
| 14 | 8.5 | 0.8 | 0.8 | ∞ | 8.5 | 20 | 7.5 | ∞ | 8.5 | 7.5 | 7.5 | 3.2 | 0 | | 8.5 | 0.8 | ∞ |
| 15 | 500 | ∞ | ∞ | ∞ | ∞ | ∞ | ∞ | ∞ | ∞ | ∞ | ∞ | ∞ | ∞ | ∞ | | ∞ | ∞ |
| 16 | 8.5 | 0.8 | 0.8 | ∞ | 8.5 | 20 | 7.5 | ∞ | 8.5 | 7.5 | 7.5 | 3.2 | 0.8 | 0.8 | 8.5 | | ∞ |
| 17 | 20 | 3.5 | 3.8 | ∞ | 20 | 30 | 16 | ∞ | 20 | 17 | 17 | 7.5 | 3.6 | 3.6 | 20 | 2.8 | |

**Fig. 7.13 Monitor unit test table (Sec 23)**

the side stand lamp should go out as soon as the engine is started. The side stand lamp remains on until the stand is retracted.

2    The head, tail and brake lamp warning lights will come on if there is an open circuit (eg faulty bulb) in the relevant circuit. The battery lamp will illuminate if the electrolyte falls below the minimum level and the side stand lamp should come on every time the stand is down. The engine oil pressure switch controls all the warning lamps except the side stand lamp and if the oil pressure falls to an unsafe level all lamps except the side stand lamp will illuminate. If the system is suspected of being faulty it can be checked as follows.

3    If one or more of the lamps fail to come on when the ignition is switched on, dismantle the instrument panel and check the bulb, renewing it if necessary. If the bulbs are intact go on to check the relevant circuit wiring and switches as described in Chapter 6, Sections 2 and 3.

4    If the wiring, switches and bulbs are in order check the voltage between the electrolyte level sensor and the negative terminal of the battery. This should be over 3.0 volts; if not, the level sensor must be renewed.

5    If the preceding tests have failed to isolate the fault, the monitor control unit should be removed from the machine and tested. If the values obtained differ greatly from those in the accompanying table the unit should be renewed, although it is advisable to have your findings confirmed by a Suzuki dealer.

## 24  Fuel gauge circuit: testing

The fuel gauge circuit should be tested as described in Section 21 of Chapter 6, but refer to the accompanying figure and test values given below.

| Float position | Resistance | Fuel quantity |
|---|---|---|
| Full | 5–10 ohms | Approx 19.2 lit |
| Half | 30–35 ohms | Approx 10.0 lit |
| Empty | 40–60 ohms | Approx 4.7 lit |

## 25  Oil temperature circuit: testing

1    The oil temperature circuit consists of a crankcase mounted sender unit, the relevant wiring and the gauge in the instrument panel. If the circuit becomes faulty a simple test to help track down the faulty component is to switch the ignition on, disconnect the pink lead from the temperature sender unit and ground it to earth. If the needle fails to move this indicates that the wiring or gauge is at fault, whereas if the

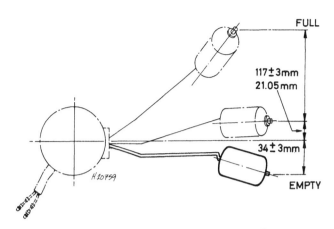

**Fig. 7.14 Fuel gauge sender unit float positions (Sec 24)**

needle swings over to the H position this indicates that the sender unit is at fault.

2    To test fully the sender unit remove it from the machine; it can be checked by measuring its resistance at various temperatures. The task must be carried out in a well-ventilated area and great care must be taken to avoid the risk of personal injury. The equipment necessary is a heatproof container, a small gas-powered camping stove or similar, a thermometer capable of reading up to 150°C (302°F) and an ohmmeter or a multimeter set to the appropriate resistance scale.

3    Fill the container with oil and suspend the sender unit on some wire so that the probe end is immersed in it. Connect one of the meter leads to the unit body and the other to its terminal. Suspend the thermometer so that the bulb is close to the sender probe.

4    Start to heat the oil, and make a note of the resistance reading at each temperature shown in the table below. If the unit does not give readings which approximate quite closely to those shown, it must be renewed.

| Oil temperature | Standard resistance |
|---|---|
| 50°C (122°F) | Approx 155 ohms |
| 100°C (212°F) | Approx 38 ohms |
| 130°C (266°F) | Approx 19 ohms |
| 150°C (302°F) | Approx 13 ohms |

5    If the above test indicates the gauge to be faulty it will first be necessary to check the relevant wiring as described in Sections 2 and 3 of Chapter 6. If the wiring is found to be correct the gauge is proven faulty and should be renewed.

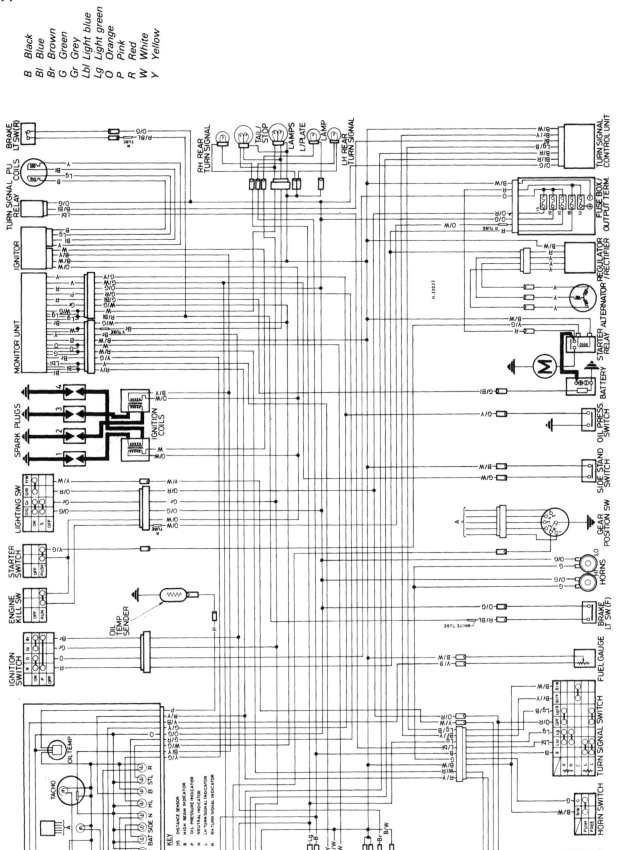

Wiring diagram - UK models

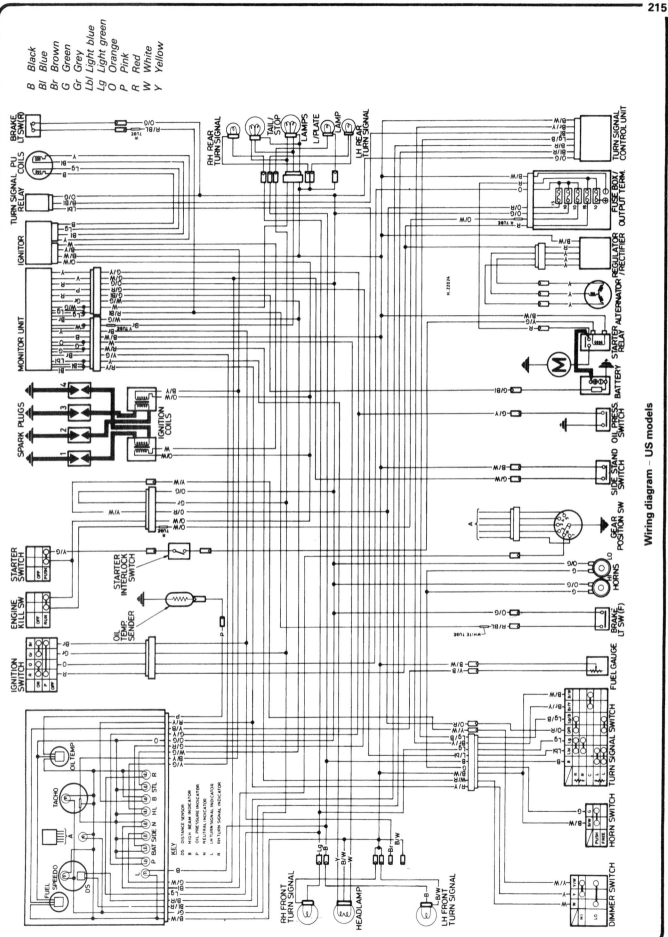

**Wiring diagram – US models**

# Conversion factors

**Length (distance)**

| | | | | | | | |
|---|---|---|---|---|---|---|---|
| Inches (in) | X | 25.4 | = Millimetres (mm) | X | 0.0394 | = Inches (in) |
| Feet (ft) | X | 0.305 | = Metres (m) | X | 3.281 | = Feet (ft) |
| Miles | X | 1.609 | = Kilometres (km) | X | 0.621 | = Miles |

**Volume (capacity)**

| | | | | | | | |
|---|---|---|---|---|---|---|---|
| Cubic inches (cu in; in³) | X | 16.387 | = Cubic centimetres (cc; cm³) | X | 0.061 | = Cubic inches (cu in; in³) |
| Imperial pints (Imp pt) | X | 0.568 | = Litres (l) | X | 1.76 | = Imperial pints (Imp pt) |
| Imperial quarts (Imp qt) | X | 1.137 | = Litres (l) | X | 0.88 | = Imperial quarts (Imp qt) |
| Imperial quarts (Imp qt) | X | 1.201 | = US quarts (US qt) | X | 0.833 | = Imperial quarts (Imp qt) |
| US quarts (US qt) | X | 0.946 | = Litres (l) | X | 1.057 | = US quarts (US qt) |
| Imperial gallons (Imp gal) | X | 4.546 | = Litres (l) | X | 0.22 | = Imperial gallons (Imp gal) |
| Imperial gallons (Imp gal) | X | 1.201 | = US gallons (US gal) | X | 0.833 | = Imperial gallons (Imp gal) |
| US gallons (US gal) | X | 3.785 | = Litres (l) | X | 0.264 | = US gallons (US gal) |

**Mass (weight)**

| | | | | | | | |
|---|---|---|---|---|---|---|---|
| Ounces (oz) | X | 28.35 | = Grams (g) | X | 0.035 | = Ounces (oz) |
| Pounds (lb) | X | 0.454 | = Kilograms (kg) | X | 2.205 | = Pounds (lb) |

**Force**

| | | | | | | | |
|---|---|---|---|---|---|---|---|
| Ounces-force (ozf; oz) | X | 0.278 | = Newtons (N) | X | 3.6 | = Ounces-force (ozf; oz) |
| Pounds-force (lbf; lb) | X | 4.448 | = Newtons (N) | X | 0.225 | = Pounds-force (lbf; lb) |
| Newtons (N) | X | 0.1 | = Kilograms-force (kgf; kg) | X | 9.81 | = Newtons (N) |

**Pressure**

| | | | | | | | |
|---|---|---|---|---|---|---|---|
| Pounds-force per square inch (psi; lbf/in²; lb/in²) | X | 0.070 | = Kilograms-force per square centimetre (kgf/cm²; kg/cm²) | X | 14.223 | = Pounds-force per square inch (psi; lbf/in²; lb/in²) |
| Pounds-force per square inch (psi; lbf/in²; lb/in²) | X | 0.068 | = Atmospheres (atm) | X | 14.696 | = Pounds-force per square inch (psi; lbf/in²; lb/in²) |
| Pounds-force per square inch (psi; lbf/in²; lb/in²) | X | 0.069 | = Bars | X | 14.5 | = Pounds-force per square inch (psi; lbf/in²; lb/in²) |
| Pounds-force per square inch (psi; lbf/in²; lb/in²) | X | 6.895 | = Kilopascals (kPa) | X | 0.145 | = Pounds-force per square inch (psi; lbf/in²; lb/in²) |
| Kilopascals (kPa) | X | 0.01 | = Kilograms-force per square centimetre (kgf/cm²; kg/cm²) | X | 98.1 | = Kilopascals (kPa) |
| Millibar (mbar) | X | 100 | = Pascals (Pa) | X | 0.01 | = Millibar (mbar) |
| Millibar (mbar) | X | 0.0145 | = Pounds-force per square inch (psi; lbf/in²; lb/in²) | X | 68.947 | = Millibar (mbar) |
| Millibar (mbar) | X | 0.75 | = Millimetres of mercury (mmHg) | X | 1.333 | = Millibar (mbar) |
| Millibar (mbar) | X | 0.401 | = Inches of water (inH₂O) | X | 2.491 | = Millibar (mbar) |
| Millimetres of mercury (mmHg) | X | 0.535 | = Inches of water (inH₂O) | X | 1.868 | = Millimetres of mercury (mmHg) |
| Inches of water (inH₂O) | X | 0.036 | = Pounds-force per square inch (psi; lbf/in²; lb/in²) | X | 27.68 | = Inches of water (inH₂O) |

**Torque (moment of force)**

| | | | | | | | |
|---|---|---|---|---|---|---|---|
| Pounds-force inches (lbf in; lb in) | X | 1.152 | = Kilograms-force centimetre (kgf cm; kg cm) | X | 0.868 | = Pounds-force inches (lbf in; lb in) |
| Pounds-force inches (lbf in; lb in) | X | 0.113 | = Newton metres (Nm) | X | 8.85 | = Pounds-force inches (lbf in; lb in) |
| Pounds-force inches (lbf in; lb in) | X | 0.083 | = Pounds-force feet (lbf ft; lb ft) | X | 12 | = Pounds-force inches (lbf in; lb in) |
| Pounds-force feet (lbf ft; lb ft) | X | 0.138 | = Kilograms-force metres (kgf m; kg m) | X | 7.233 | = Pounds-force feet (lbf ft; lb ft) |
| Pounds-force feet (lbf ft; lb ft) | X | 1.356 | = Newton metres (Nm) | X | 0.738 | = Pounds-force feet (lbf ft; lb ft) |
| Newton metres (Nm) | X | 0.102 | = Kilograms-force metres (kgf m; kg m) | X | 9.804 | = Newton metres (Nm) |

**Power**

| | | | | | | | |
|---|---|---|---|---|---|---|---|
| Horsepower (hp) | X | 745.7 | = Watts (W) | X | 0.0013 | = Horsepower (hp) |

**Velocity (speed)**

| | | | | | | | |
|---|---|---|---|---|---|---|---|
| Miles per hour (miles/hr; mph) | X | 1.609 | = Kilometres per hour (km/hr; kph) | X | 0.621 | = Miles per hour (miles/hr; mph) |

**Fuel consumption***

| | | | | | | | |
|---|---|---|---|---|---|---|---|
| Miles per gallon, Imperial (mpg) | X | 0.354 | = Kilometres per litre (km/l) | X | 2.825 | = Miles per gallon, Imperial (mpg) |
| Miles per gallon, US (mpg) | X | 0.425 | = Kilometres per litre (km/l) | X | 2.352 | = Miles per gallon, US (mpg) |

**Temperature**

Degrees Fahrenheit = (°C x 1.8) + 32          Degrees Celsius (Degrees Centigrade; °C) = (°F - 32) x 0.56

*It is common practice to convert from miles per gallon (mpg) to litres/100 kilometres (l/100km), where mpg (Imperial) x l/100 km = 282 and mpg (US) x l/100 km = 235

# English/American terminology

Because this book has been written in England, British English component names, phrases and spellings have been used throughout. American English usage is quite often different and whereas normally no confusion should occur, a list of equivalent terminology is given below.

| English | American | English | American |
|---|---|---|---|
| Air filter | Air cleaner | Number plate | License plate |
| Alignment (headlamp) | Aim | Output or layshaft | Countershaft |
| Allen screw/key | Socket screw/wrench | Panniers | Side cases |
| Anticlockwise | Counterclockwise | Paraffin | Kerosene |
| Bottom/top gear | Low/high gear | Petrol | Gasoline |
| Bottom/top yoke | Bottom/top triple clamp | Petrol/fuel tank | Gas tank |
| Bush | Bushing | Pinking | Pinging |
| Carburettor | Carburetor | Rear suspension unit | Rear shock absorber |
| Catch | Latch | Rocker cover | Valve cover |
| Circlip | Snap ring | Selector | Shifter |
| Clutch drum | Clutch housing | Self-locking pliers | Vise-grips |
| Dip switch | Dimmer switch | Side or parking lamp | Parking or auxiliary light |
| Disulphide | Disulfide | Side or prop stand | Kick stand |
| Dynamo | DC generator | Silencer | Muffler |
| Earth | Ground | Spanner | Wrench |
| End float | End play | Split pin | Cotter pin |
| Engineer's blue | Machinist's dye | Stanchion | Tube |
| Exhaust pipe | Header | Sulphuric | Sulfuric |
| Fault diagnosis | Trouble shooting | Sump | Oil pan |
| Float chamber | Float bowl | Swinging arm | Swingarm |
| Footrest | Footpeg | Tab washer | Lock washer |
| Fuel/petrol tap | Petcock | Top box | Trunk |
| Gaiter | Boot | Torch | Flashlight |
| Gearbox | Transmission | Two/four stroke | Two/four cycle |
| Gearchange | Shift | Tyre | Tire |
| Gudgeon pin | Wrist/piston pin | Valve collar | Valve retainer |
| Indicator | Turn signal | Valve collets | Valve cotters |
| Inlet | Intake | Vice | Vise |
| Input shaft or mainshaft | Mainshaft | Wheel spindle | Axle |
| Kickstart | Kickstarter | White spirit | Stoddard solvent |
| Lower leg | Slider | Windscreen | Windshield |
| Mudguard | Fender | | |

# Index